Studies in Infrastructure and Control

Series Editors

Dipankar Deb, Department of Electrical Engineering, Institute of Infrastructure Technology Research and Management, Ahmedabad, India

Akshya Swain, Department of Electrical, Computer and Software Engineering, University of Auckland, Auckland, New Zealand

Alexandra Grancharova, Department of Industrial Automation, University of Chemical Technology and Metallurgy, Sofia, Bulgaria

Editorial Board

Faizal Hafiz, Université Côte d'Azur, Sophia-Antipolis, France

Chathura Wanigasekara, German Aerospace Centre (DLR), Institute for the Protection of Maritime Infrastructures, Bremerhaven, Germany

Dhafer Al-Makhles, Prince Sultan University, Riyadh, Saudi Arabia

Badri Narayana Subudhi, Indian Institute of Technology Jammu, Jammu, Jammu and Kashmir, India

Gayadhar Panda, Department of Electrical Engineering, National Institute of Technology Meghalaya, Shillong, Meghalaya, India

Tor Arne Johansen, Norwegian University of Science and Technology, Trondheim, Norway

Sorin Olaru, CentraleSupélec—CNRS, Université Paris-Saclay, Gif sur Yvette, France

Juš Kocijan, Jožef Stefan Institute, University of Nova Gorica, Ljubljana, Slovenia

Ionela Prodan, Conception and Integration of Systems, University Grenoble Alpes, Valence, France

Damir Vrančić, Jožef Stefan Institute, Ljubljana, Slovenia

Ilhami Colak, Department of Electrical and Electronics Engineering, Nisantasi University, İstanbul, Türkiye

Gang Tao, Department of Electrical and Computer Engineering, University of Virginia, Charlottesville, VA, USA

Ralph Kennel, Fakultät für Elektrotechnik und Informationstechnik, Technische Universität München, München, Germany

S. M. Muyeen, Department of Electrical Engineering, Qatar University, Berlin, Berlin, Germany

Francisco M. Gonzalez-Longatt, Department of Electrical Power Engineering, University of South-Eastern Norway, Notodden, Norway

The book series aims to publish top-quality state-of-the-art textbooks, research monographs, edited volumes and selected conference proceedings related to infrastructure, innovation, control, and related fields. Additionally, established and emerging applications related to applied areas like smart cities, internet of things, machine learning, artificial intelligence, etc., are developed and utilized in an effort to demonstrate recent innovations in infrastructure and the possible implications of control theory therein. The study also includes areas like transportation infrastructure, building infrastructure management and seismic vibration control, and also spans a gamut of areas from renewable energy infrastructure like solar parks, wind farms, biomass power plants and related technologies, to the associated policies and related innovations and control methodologies involved.

Gowrisankar Arulprakash · Kishore Bingi ·
Cristina Serpa
Editors

Mathematical Modelling of Complex Patterns Through Fractals and Dynamical Systems

Springer

Editors
Gowrisankar Arulprakash
Department of Mathematics
School of Advanced Sciences
Vellore Institute of Technology
Vellore, Tamil Nadu, India

Kishore Bingi
Department of Electrical and Electronics Engineering
Universiti Teknologi PETRONAS
Seri Iskandar, Malaysia

Cristina Serpa
Centro de Matemática
Aplicações Fundamentais e Investigação Operacional
Instituto Superior de Engenharia de Lisboa
Lisbon, Portugal

ISSN 2730-6453 ISSN 2730-6461 (electronic)
Studies in Infrastructure and Control
ISBN 978-981-97-2342-3 ISBN 978-981-97-2343-0 (eBook)
https://doi.org/10.1007/978-981-97-2343-0

This Springer imprint is published by the registered company Springer Nature Singapore Pte Ltd.
The registered company address is: 152 Beach Road, #21-01/04 Gateway East, Singapore 189721, Singapore

Preface

Modelling complex patterns through fractals and dynamical systems involves using mathematical tools to describe intricate and often self-replicating structures. Fractals are mathematical objects with self-similar patterns at different scales, while dynamical systems study how variables evolve over time. Combining these concepts allows for the representation of complex and dynamic patterns found in nature, such as the Mandelbrot set in fractals or the Lorenz attractor in dynamical systems. The interaction between these elements can lead to rich mathematical models that capture the complexity observed in various phenomena.

Fractals have evolved since their definition in the 1970s into a key component of nonlinear dynamics and chaos theory, in addition to the striking beauty inherent in their complexity. Fractals are self-similar patterns that repeat at different scales, and they are often used to describe complex nonlinear systems that exhibit similar patterns of behaviour at different levels of magnification. By studying the fractal geometry of chaotic dynamics associated with nonlinear systems, researchers can gain insights into how small perturbations can lead to large-scale changes in the system's behaviour. Fractals also help to shed light on the idea of "strange attractors", which are a key concept in dynamical systems. Strange attractors are complex, non-repeating patterns that emerge in linear as well as in nonlinear systems, and they play a crucial role in determining the system's long-term behaviour. Fractal geometry is often used to describe the structure of strange attractors, as they typically exhibit self-similar patterns at different levels of magnification. Overall, the implications of fractals in the study of nonlinear systems are significant. By providing a powerful visual language for describing complex and nonlinear chaotic systems, fractals have helped researchers to gain a deeper understanding of the underlying patterns and structures that govern these systems' behaviour. This, in turn, has led to new insights into a wide range of scientific fields, from physics and biology to economics and social science. The book on *Mathematical Modelling of Complex Patterns Through Fractals and Dynamical Systems* aims to bring together cutting-edge research works proposing the features of fractals and dynamical systems in both traditional and scientific disciplines as well as in applications. This book contains ten chapters that are concisely described as follows:

In chapter 1, the authors delve into the realm of initial value problems in fractal delay equations, offering a comprehensive exploration of their mathematical properties and solution methodologies. Fractal delay equations play a crucial role in modelling systems with fractal time and intricate delay dynamics, making them indispensable tools in various scientific and engineering disciplines. The chapters in this book are organized to provide a systematic understanding of initial value problems in fractal delay equations. We begin by establishing the theoretical foundations, discussing the fundamental concepts of fractal calculus, and elucidating the mathematical properties of fractal delay equations.

In chapter 2, the concept of fractal analysis of biodiversity: the Living Planet Index (LPI) is described. The LPI is a global index that measures the state of the world's biodiversity. Analysing the LPI solely by statistical trends provides, however, limited insight. Fractal regression analysis is a recently developed tool that has been successfully applied in multidisciplinary scientific contexts for time series analysis. This method is based on the construction of a fractal function tailored to the dataset from which directional coefficients, representing trends, and fractal coefficients, representing oscillations, may be computed. These coefficients allow us to classify the world's regions according to the progression of the LPI, helping us to identify and mathematically characterize the region of Latin America and the Caribbean and the category of freshwater as worst-case scenarios concerning the evolution of biodiversity.

In chapter 3, the authors endeavour to create the Rational Cubic Fractal Interpolation Function (RCFIF) by employing an iterated function system that integrates cubic polynomial numerators and linear polynomial denominators within a rational function framework. The objective is to identify appropriate shape parameters and scaling factors, aligning the RCFIF with a predefined piecewise plane. This chapter delivers an in-depth exploration of the theoretical foundations, supplemented by numerical examples to enhance understanding and visualization.

In chapter 4, a novel approach to constructing models of coherent upper conditional previsions for both bounded and unbounded random variables in a metric space is presented. The models are built upon the concept of different dimensional outer measures, and the relationships between these measures are thoroughly examined. By utilizing a dimensional measure, it becomes possible to determine the dimensional outer measure of the conditioning event, even in cases where the metric space is not second countable. This measure can indicate whether the conditioning event has a zero-, positive-, finite-, or infinite-dimensional outer measure in its respective dimension.

In chapter 5, the author discusses the magnetohydrodynamic heat and mass transfer flow of a Casson nanofluid over linear and non-linear stretching surfaces embedded in a porous medium. The study takes into account the effects of Brownian motion, thermophoresis, thermal radiation, and chemical reaction. Proper transformations are employed to convert the non-linear partial differential systems into ordinary differential equations. The Runge–Kutta method, in combination with the shooting technique, is utilized for finding the numerical solutions to the momentum, temperature, and concentration equations, subject to the boundary conditions. The impact of different relevant parameters on the velocity, temperature, and concentration profiles of the fluid is analysed and the findings are presented in graphical form through plots. The study highlights the intricate dynamics of nanofluids, where the Casson, magnetic, buoyancy ratio, and mixed convection parameters collectively influence behaviour. Notably, contrasting effects emerge in the realm of mixed convection, adding depth to our understanding of these influential parameters. It is noted that as the Brownian motion parameter increases, the velocity and temperature profiles of the fluid increase, whereas the opposite behaviour is observed in the concentration profile. Moreover, a detailed investigation is conducted on the skin-friction coefficient, the Nusselt number and the Sherwood number and the results are presented in tabular format.

In chapter 6, a bulk arrival retrial queueing model that offers multi-phases of heterogeneous service is investigated. Moreover, as soon as the orbit becomes empty, the server goes on a Bernoulli working vacation (BWV), where it works at a lower speed. At certain times, consumers are also permitted to balk and renege. Further, the busy server may experience a failure at any time. Here the supplementary variable approach has been incorporated in order to derive the probability generating function (PGF) of the number of clients in the system as well as in the orbit. The impact of particular parameters on the system's overall performance has been demonstrated with the aid of a few numerical examples. Finally, this research is accelerated in order to bring about the best possible (optimal) cost for the system by adopting a range of optimization approaches, namely, particle swarm optimization (PSO), artificial bee colony (ABC), grey wolf optimizer (GWO), and differential evolution (DE).

In chapter 7, Sierpiński and Koch Snowflake are the two most studied topics in fractal geometry. Sierpinski Rhombus (SR_n) is formed by a pattern of n sequences of a graph that results in a planar fractal. Koch Snowflake (KS_n) is also formed by some patterns similar to the Sierpiński Rhombus. In this study, topological indices are used to study fractal structures. This paper is the first to derive a closely related formula of M-polynomials and entropy measures for the fractal structures SR_n and KS_n. The calculated topological value is usually correlated with the physical properties of the structure. The scope of this work is to relate the derivation of topological indices with the fractal dimension for the graph sequences SR_n and KS_n.

In chapter 8, the convection dynamics of superparamagnetic iron oxide nanoparticles (MNPs) in blood fluid is examined. The system under study consists of a liquid with a layer of blood and MNPs with an iron oxide core travelling through a blood artery with cholesterol deposits on the inner lining. It is heated and subjected to an external magnetic field as it passes through the blood vessel. The partial differential equations of momentum and energy conservation were utilized to create the nonlinear three-dimensional governing equations for the system under consideration. Phase portrait, time series, and Hartmann number calculations are used to examine the impact of the magnetic field on the chaotic convection of the MNPs in blood fluid. It has been found that the convection of MNPs stabilizes in a specific direction when magnetized by an external magnetic field. This points to the expanding use of magnetic resonance imaging (MRI) and magnetic drug delivery (MDD) as contrast agents in non-invasive imaging techniques.

The fractal dimension of linear fractal interpolation with various settings is investigated in the chapter 9. Additionally, the Riemann–Liouville fractional integral of a linear fractal interpolation function with variable scaling factors is studied. Also, the fractal approximation of the Rössler attractor based on different types of parameters and how the parameters affect the fractal dimension for the time series of the Rössler attractor are investigated.

In chapter 10, the dynamic behaviour of ocean waves described by the geophysical-Burgers' equation is presented. Phase plane analysis is applied to study nonlinear waves described by the geophysical-Burgers' equation. A travelling wave transformation is considered to convert the geophysical-Burgers' equation to a dynamical system. All equilibrium points of the corresponding dynamical system are obtained and analysed based on the corresponding eigenvalues. Phase portrait for the dynamical system is plotted. Solutions of kink and anti-kink waves corresponding to heteroclinic orbits and periodic waves corresponding to periodic orbits are obtained. Effect of various parameters on these wave solutions is shown.

Vellore, India — Gowrisankar Arulprakash
Seri Iskandar, Malaysia — Kishore Bingi
Lisbon, Portugal — Cristina Serpa

Contents

About the Editors

Gowrisankar Arulprakash received his Ph.D. degree (in Mathematics) from the Gandhigram Rural Institute (Deemed to be University), Gandhigram, Dindigul, India, in 2017. He was employed as an institution postdoctoral fellow at the Indian Institution of Technology Guwahati (IITG), in Guwahati, Assam, India. He is currently employed as an Assistant Professor in the Department of Mathematics, School of Advanced Sciences, Vellore Institute of Technology, Vellore, Tamil Nadu, India with six years of experience. His broad areas of research include Fractal Analysis, Fractal Interpolation Functions, Fractional Calculus, Nonlinear Dynamics and Climate Change. He has contributed to more than 40 research articles published in reputable international journals. He was successful in getting three books published by the following publishers: CRC Press, 2019, Springer: Complexity, 2021 and SpringerBriefs in Complexity, 2024. He was one of the editors for two edited books in CRC Press and Taylor & Francis Group. He served as a guest editor for two special issues in "The European Physical Journal Special Topics". He is acting as a reviewer for more than 10 international journals and has received the title of "Distinguished EPJ Referees-2021". Recently, he has received a TARE fellowship under SERB, Government of India, for the period 2024–2026.

Kishore Bingi received the B.Tech. Degree in Electrical and Electronics Engineering from Acharya Nagarjuna University, India, in 2012. He received the M.Tech. Degree in Instrumentation and Control Systems from the National Institute of Technology Calicut, India, in 2014, and a Ph.D. in Electrical and Electronic Engineering from Universiti Teknologi PETRONAS, Malaysia, in 2019. From 2014 to 2015, he worked as an Assistant Systems Engineer at TATA Consultancy Services Limited, India. From 2019 to 2020, he worked as Research Scientist and Post-doctoral Researcher at the Universiti Teknologi PETRONAS, Malaysia. From 2020 to 2022, he served as an Assistant Professor at the Process Control Laboratory, School of Electrical Engineering, Vellore Institute of Technology, India. Since 2022 he has been working as a faculty member at the Department of Electrical and Electronic Engineering at Universiti Teknologi PETRONAS, Malaysia. His research area is developing fractional-order neural networks, including fractional-order systems and controllers, chaos prediction and forecasting, and advanced hybrid optimization techniques. He is an IEEE and IET Member and a registered Chartered Engineer (CEng) from Engineering Council UK. He serves as an Editorial Board Member for the *International Journal of Applied Mathematics and Computer Science* and Academic Editor for *Mathematical Problems in Engineering* and the *Journal of Control Science and Engineering*.

Cristina Serpa has a degree in Applied Mathematics—Fundamental Applications, a Master's and a Ph.D. in Mathematics, in Mathematical Analysis, from the Faculty of Sciences of the University of Lisbon. She is an invited professor in the Mathematics Department of the Instituto Superior de Engenharia de Lisboa of the Instituto Politécnico de Lisboa. Her teaching duties are usually focused on calculus curricular units for engineering. She is an integrated member of CMAFcIO—Center for Mathematics, Fundamental Applications and Operational Research, of the University of Lisbon and collaborates with the Portuguese Mathematical Society. He researches Fractals at the fundamental theoretical level and in the context of their applications. He created the Fractal Regression method, developing corresponding software to obtain fractal functions from real data.

Chapter 1
Fractal Delay Equations

Alireza Khalili Golmankhaneh, Inés Tejado, Hamdullah Sevli, and Juan Napoles

Abstract This chapter delves into the realm of initial value problems in fractal delay equations, offering a comprehensive exploration of their mathematical properties and solution methodologies. Fractal delay equations play a crucial role in modeling systems with fractal time and intricate delay dynamics, making them indispensable tools in various scientific and engineering disciplines. The chapters in this book are organized to provide a systematic understanding of initial value problems in fractal delay equations. We begin by establishing the theoretical foundations, discussing the fundamental concepts of fractal calculus, and elucidating the mathematical properties of fractal delay equations.

1 Introduction

Fractals are fascinating mathematical objects with irregular and zigzag shapes, often characterized by a fractal dimension that exceeds their topological dimension [1, 2]. Their intricate structures make them challenging to analyze using conventional calculus methods. To address this, various mathematical approaches have been for-

A. K. Golmankhaneh (✉)
Department of Physics, Urmia Branch, Islamic Azad University, 63896 Urmia, Iran
e-mail: alirezakhalili2005@gmail.com

I. Tejado
Escuela de Ingenierías Industriales, Universidad de Extremadura, 06006 Badajoz, Spain
e-mail: itejbal@unex.es

H. Sevli
Department of Computer Engineering, Faculty of Engineering,
Van Yuzuncu Yil University, Campus, 65080 Van, Turkey

J. Napoles
Departamento de Matematicas, Universidad Nacional de Nordeste,
Avenida de la Libertad 5450, 3400 Corrientes, Argentina
e-mail: jnapoles@exa.unne.edu.ar

G. Arulprakash et al. (eds.), *Mathematical Modelling of Complex Patterns Through Fractals and Dynamical Systems*, Studies in Infrastructure and Control,
https://doi.org/10.1007/978-981-97-2343-0_1

mulated, including measure theory [3], probability theory [4], harmonic analysis [5], fractional space [6], and fractional calculus [7].

Fractal calculus, or F^{α}-calculus, is a generalization of ordinary calculus designed to handle functions with fractal support and fractal spaces [8, 9]. It has been applied to explore phenomena like sub- and super-diffusion on fractal mediums and its effects on statistical mechanics [10], solving fractal differential equations using Laplace and Fourier transforms [9], and studying fractal random variables and stable distributions [11, 12]. Nonlocal fractal calculus has been used for the analysis of electrical circuits with noise [13, 14], and it has also played a crucial role in understanding the Devil's staircase function [15].

In this context, we aim to introduce fractal delay differential equations (FDDEs), which model phenomena with time delays between cause and effect. FDDEs extend the concept of delay differential equations (DDEs) to handle systems involving fractal spaces and supports [16–18]. Discrete constant delay differential equations have been classified into different types, such as retarded, neutral, and advanced [19]. FDDEs are a powerful tool for modeling real-life problems with complex temporal dependencies [20]. They involve infinite-dimensional systems, unlike ordinary differential equations (ODEs), which describe finite-dimensional systems [21, 22].

The main objective of this chapter is to introduce FDDEs and explore their solutions using various mathematical techniques. We investigate the method of steps and fractal Laplace transform as viable methods for solving FDDEs and understanding their behavior in different scenarios.

The structure of this chapter is organized as follows: We begin with a review of fractal calculus on Cantor sets in Sect. 2 [9, 23, 24]. In Sect. 3, we introduce fractal functional differential equations and solve fractal retarded, neutral, and renewal functional differential equations using the method of steps [25, 26]. Section 4 explores the use of fractal Laplace transform for solving fractal functional differential equations [9].

2 Background Information

In this section, we review some fundamental definitions of fractal calculus on fractal sets [9, 27]. These concepts provide the mathematical tools necessary for understanding and analyzing fractal delay differential equations.

Definition 1 The flag function of a set F and a closed set I is denoted by $\rho(F, I)$ and is defined as follows:

$$\rho(F, I) = \begin{cases} 1, \text{ if } F \cap I \neq \emptyset; \\ 0, \text{ otherwise}. \end{cases} \tag{1}$$

Definition 2 For a fractal set F, a subdivision $P_{[a,b]}$ of $[a, b]$, and a given $\delta > 0$, the coarse-grained mass of $F \cap [a, b]$ is defined as

$$\gamma_\delta^\alpha(F,a,b) = \inf_{|P|\le\delta} \sum_{i=0}^{n-1} \Gamma(\alpha+1)(t_{i+1}-t_i)^\alpha \rho(F,[t_i,t_{i+1}]), \tag{2}$$

where $|P|$ represents the maximum size of the subdivision, and $0 < \alpha \le 1$.

Definition 3 The mass function of a fractal set F is given by

$$\gamma^\alpha(F,a,b) = \lim_{\delta\to 0} \gamma_\delta^\alpha(F,a,b). \tag{3}$$

Definition 4 The γ-dimension of $F \cap [a,b]$ is defined as

$$\begin{aligned}\dim_\gamma(F\cap[a,b]) &= \inf \alpha : \gamma^\alpha(F,a,b) = 0\\ &= \sup \alpha : \gamma^\alpha(F,a,b) = \infty.\end{aligned}$$

Definition 5 The integral staircase function of order α for a fractal set F is denoted as $S_F^\alpha(t)$ and defined as

$$S_F^\alpha(t) = \begin{cases} \gamma^\alpha(F,a_0,t), & if\ t \ge a_0;\\ -\gamma^\alpha(F,t,a_0), & otherwise, \end{cases} \tag{4}$$

where a_0 is an arbitrary and fixed real number.

Definition 6 For a function y on an α-perfect fractal set F, the fractal derivative or F^α-derivative at t is denoted by $D_F^\alpha y(t)$ and defined as

$$D_F^\alpha y(t) = \begin{cases} F\_lim_{x\to t} \frac{y(x)-y(t)}{S_F^\alpha(x)-S_F^\alpha(t)}, & if\ t \in F;\\ 0, & otherwise. \end{cases} \tag{5}$$

if the fractal limit (denoted by F_lim) exists [8].

Definition 7 Let F be such that S_F^α is finite on the interval $I = [a,b]$, and let f be a bounded function of F (denoted by $f \in B(F)$). Then, the fractal integral or F^α-integral is defined as

$$\begin{aligned}\int_a^b y(t) d_F^\alpha t &= \sup_{P_{[a,b]}} \sum_{i=0}^{n-1} \inf_{t\in F\cap I} y(t)(S_F^\alpha(t_{i+1}) - S_F^\alpha(t_i))\\ &= \inf_{P_{[a,b]}} \sum_{i=0}^{n-1} \sup_{t\in F\cap I} y(t)(S_F^\alpha(t_{i+1}) - S_F^\alpha(t_i)),\end{aligned} \tag{6}$$

where $t \in F$, and we take the infimum and supremum over all subdivisions $P_{[a,b]}$ of $[a,b]$.

These definitions lay the groundwork for understanding fractal calculus on fractal sets, which is essential for our subsequent discussions on fractal delay differential equations.

3 Fractal Functional Differential Equations

In this section, we introduce fractal functional differential equations (FDDEs) and explore their various types. FDDEs are used to model phenomena that depend on both the present and the past, where time is represented in a fractal manner. These equations find applications in various fields such as physics, biology, and finance.

Non-linear Fractal Delay Differential Equation

We start with the general form of a non-linear fractal delay differential equation:

$$D_F^\alpha y(t) + D_F^\alpha y(t-\tau) = f(S_F^\alpha(t), y(t), y(S_F^\alpha(t) - S_F^\alpha(\tau))), \tag{7}$$

where D_F^α is the fractional derivative of order α and S_F^α is the fractional integral of order α. The function f represents a non-linear relationship, and $\tau > 0$ is the delay or lag. The equation has an initial or history function $\psi : [-\tau, 0] \to \mathbb{R}$. The function $f : \mathbb{R} \to \mathbb{R}$ belongs to the class of F-continuous functions ($f \in C(F)$). The equation can be classified into different types based on the dependencies of the fractional derivative $D_F^\alpha y(t)$ on the past values of $y(t)$ and its derivatives. We have three types:

In Table 1, we categorize different types of fractal functional differential equations as follows:

1. Retarded fractal functional differential equation: In this type, the fractal derivative of $y(t)$ depends solely on its past value.
2. Neutral fractal functional differential equation: In this type, the fractal derivative of $y(t)$ depends on both its past values and its fractal derivative.
3. Advanced fractal functional differential equation: In this type, the fractal derivative of $y(t)$ depends on its past values alone.

These categories describe the different ways in which the fractal derivative of $y(t)$ can be related to its past behavior in the context of functional differential equations. In the retarded FDDE, the fractal derivative of $y(t)$ depends on its past value. In the neutral FDDE, the fractal derivative of $y(t)$ depends on the past values of both $y(t)$ and its fractal derivative. In the advanced FDDE, the fractal derivative of $y(t)$ on the past depends on the past values of $y(t)$.

Table 1 Different kinds of fractal functional differential equations

Classifications of fractal non-linear functional differential equations	
Retarded	$D_F^\alpha y(t) = f(S_F^\alpha(t), y(t), y(S_F^\alpha(t) - S_F^\alpha(\tau)))$
Neutral	$D_F^\alpha y(t) = f(S_F^\alpha(t), y(t), y(S_F^\alpha(t) - S_F^\alpha(\tau)), D_F^\alpha y(S_F^\alpha(t) - S_F^\alpha(\tau)))$
Advanced	$D_F^\alpha y(S_F^\alpha(t) - S_F^\alpha(\tau)) = f(S_F^\alpha(t), y(t), y(S_F^\alpha(t) - S_F^\alpha(\tau)))$

Table 2 Different kinds of fractal linear delay differential equations, where b_0, b_1 are constants

Classifications of fractal linear delay differential equations	
Retarded	$D_F^\alpha y(t) = b_0 y(t) + b_1 y(t-\tau) + f(t)$
Neutral	$D_F^\alpha y(t) + D_F^\alpha y(t-\tau) = b_0 y(t) + b_1 y(t-\tau) + f(t)$
Advanced	$D_F^\alpha y(t-\tau) = b_0 y(t) + b_1 y(t-\tau) + f(t)$

Fractal Linear Delay Differential Equations

Next, we explore the general form of fractal linear delay differential equations (FLDDEs) [28]:

$$D_F^\alpha y(t) + D_F^\alpha y(t-\tau) = Ay(t) + By(t-\tau) + f(t), \tag{8}$$

where A and B are constants, and $f : \mathbb{R} \to \mathbb{R}$ is an F^α-continuous function with $\mathrm{Sch} f = F$. Here, $\mathrm{Sch} f$ denotes the Schwartz space of tempered distributions. To solve Eq. (8), we can use the method of steps, which involves dividing the time interval into smaller sub-intervals and finding solutions for each interval.

In Table 2, we present the three types of linear fractal delay differential equations:

1. The retarded fractal linear delay differential equation, where the fractal derivative of $y(t)$ is linearly dependent on its past value.
2. The neutral fractal linear delay differential equation, where the fractal derivative of $y(t)$ is linearly dependent on both its past values and its fractal derivative.
3. The advanced fractal linear delay differential equation, where the fractal derivative of $y(t)$ on the past is linearly dependent on its past values.

These three types of equations describe different ways in which the fractal derivative of $y(t)$ can be related to its past behavior. Similar to FDDEs, FLDDEs can also be classified into retarded, neutral, and advanced types based on the dependencies of the fractional derivative $D_F^\alpha y(t)$ on the past values of $y(t)$ and its derivatives.

Solutions of Fractal Linear Ordinary Differential Equations

A special case of fractal linear delay differential equations is when $\tau = 0$, and the equation reduces to a fractal linear ordinary differential equation (FLODE) [28]:

$$D_F^\alpha y(t) = Ay(t) + f(t), \tag{9}$$

where A is a constant and $f : \mathbb{R} \to \mathbb{R}$ is an F^α-continuous function.

The solution of Eq. (9) with initial condition $y(0) = c$ is given by

$$y(t) = ce^{AS_F^\alpha(t)} + \int_0^t e^{AS_F^\alpha(t) - AS_F^\alpha(s)} f(s) d_F^\alpha s. \tag{10}$$

This equation provides a solution for FLODEs in terms of fractal derivatives and integrals.

Examples of Fractal Functional Differential Equations

Here are some examples of fractal functional differential equations:

Example 1 Fractal Retarded Functional Differential Equation: Consider a fractal retarded functional differential equation of the following form:

$$D_F^\alpha y(t) = ky(S_F^\alpha(t) - S_F^\alpha(\tau)), \tag{11}$$

where $k \in \mathbb{R}$ is a constant, and $y(t) = \phi(t) = a$ for $t \in [-\tau, 0]$ represents a history function. Let ϕ be the solution on the interval $[0, \tau]$. In this equation, the fractal derivative of $y(t)$ at any given time t is determined by the value of $y(t)$ at a previous time $S_F^\alpha(t) - S_F^\alpha(\tau)$. The constant k influences the rate of change of $y(t)$, and the initial condition ϕ specifies the function $y(t)$ on the interval $[-\tau, 0]$ [28]. The solution ϕ on the interval $[0, \tau]$ is governed by the differential equation, and it describes how the function $y(t)$ evolves based on its own past values within the fractal calculus framework. If we set $\tau = 1, k = a = 1$, the solution to the fractal retarded functional differential Eq. (11) becomes

$$y(t) = \begin{cases} 1, & -1 \le t \le 0 \\ S_F^\alpha(t) + 1, & 0 \le t \le 1 \\ 1 + S_F^\alpha(t) + \frac{1}{2}(S_F^\alpha(t) - \Gamma(\alpha+1))^2, & 1 \le t \le 2. \end{cases} \tag{12}$$

In Fig. 1, we can observe the behavior of the solutions over the respective intervals for different values of the fractional order α. The exact solution $y(t)$ is given by Eq. (12).

By noting that $S_F^\alpha(t) \le t^\alpha$, we can express it differently as follows:

$$y(t) = \begin{cases} 1, & -1 \le t \le 0 \\ t^\alpha + 1, & 0 \le t \le 1 \\ 1 + t^\alpha + \frac{1}{2}(t^\alpha - \Gamma(\alpha+1))^2, & 1 \le t \le 2. \end{cases} \tag{13}$$

This observation is significant when analyzing and solving functional differential equations involving fractal derivatives. It implies that the influence of past values of $y(t)$ on its present behavior, as characterized by the fractional derivative of order α, is restricted by the growth rate given by the function t^α.

Example 2 Fractal Neutral Functional Differential Equation: Consider a fractal delay differential equation as

$$D_F^\alpha y(t) = y(t) + y(t-1) + S_F^\alpha(t), \tag{14}$$

with the history function

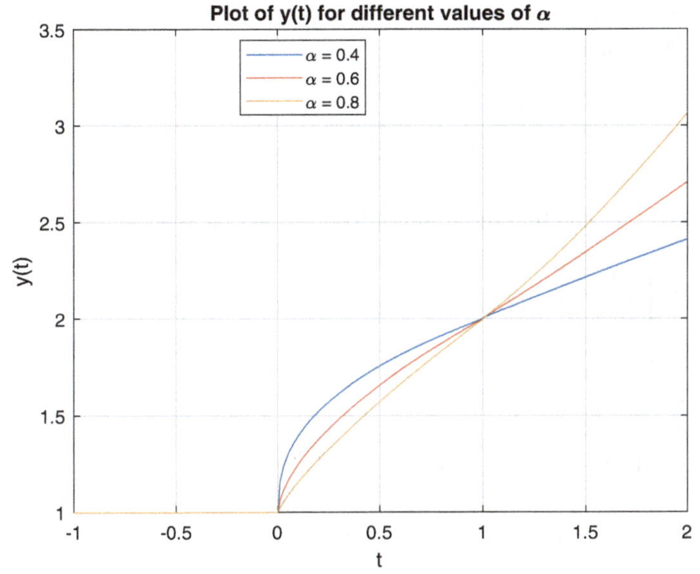

Fig. 1 Graph of Eq. (11), illustrating its behavior

$$y(t) = \phi(t) = S_F^\alpha(t), \quad -1 \le t \le 0. \tag{15}$$

To find the solution of Eq. (14), we use the expression from Eq. (10) [28]:

$$\begin{aligned} y(t) &= e^{S_F^\alpha(t)}\phi(0) + \int_0^t e^{S_F^\alpha(t)-S_F^\alpha(s)}[y(S_F^\alpha(s)-1)+f(s)]d_F^\alpha s \\ &= \int_0^t e^{S_F^\alpha(t)-S_F^\alpha(s)}[y(S_F^\alpha(s)-1)+S_F^\alpha(s)]d_F^\alpha s. \end{aligned} \tag{16}$$

Since $s-1 \in [-1, 0]$, we have $y(S_F^\alpha(s)-1) = \phi(S_F^\alpha(s)-1) = S_F^\alpha(s)-1$. This allows us to simplify the expression further:

$$y(t) = e^{S_F^\alpha(t)} \int_0^t e^{S_F^\alpha(s)} \left[2S_F^\alpha(s)-1\right] d_F^\alpha s. \tag{17}$$

Applying the fractal integral by parts, we obtain

$$\begin{aligned} y(t) &= e^{S_F^\alpha(t)} \left\{ (2S_F^\alpha(s)-1)e^{-S_F^\alpha(s)}\Big|_0^t - \int_0^t 2e^{-S_F^\alpha(t)} d_F^\alpha s \right\} \\ &= e^{S_F^\alpha(t)} - 2S_F^\alpha(t) - 1 \qquad (18) \\ &\propto e^{t^\alpha} - 2t^\alpha - 1. \qquad (19) \end{aligned}$$

Let's consider the interval $t \in [1, 2]$, which implies that $s - 1 \in [0, 1]$. For this interval, we need to calculate the expression for $y(S_F^\alpha(s) - 1)$:

$$\begin{aligned} y(S_F^\alpha(s) - 1) &= y_1(S_F^\alpha(s) - 1) \\ &= e^{S_F^\alpha(t)-1} - 2S_F^\alpha(t) + 1. \end{aligned} \tag{20}$$

Using Eq. (10), we can find the solution for this interval as follows:

$$\begin{aligned} y(t) &= e^{S_F^\alpha(t)} \left\{ \int_0^1 e^{-S_F^\alpha(s)}[2S_F^\alpha(s) - 1]d_F^\alpha s + \int_1^t e^{-S_F^\alpha(s)}[e^{S_F^\alpha(s)-1} - 2S_F^\alpha(s) + 1]d_F^\alpha s \right\} \\ &= e^{S_F^\alpha(t)} \left\{ S_F^\alpha(t)e^{-S_F^\alpha(t)} + S_F^\alpha(t)e^{-1} + -5e^{-1} + 1 \right\}. \end{aligned} \tag{21}$$

Therefore, we have the solution for $t \in [1, 2]$ as [28]:

$$\begin{aligned} y_2(t) &= e^{S_F^\alpha(t)} \left\{ S_F^\alpha(t)e^{-S_F^\alpha(t)} + S_F^\alpha(t)e^{-1} + -5e^{-1} + 1 \right\}, \quad t \in [1, 2] \\ &\propto e^{t^\alpha} \left\{ t^\alpha e^{-t^\alpha} + t^\alpha e^{-1} + -5e^{-1} + 1 \right\}. \end{aligned} \tag{22}$$

So we can write

$$y(t) = \begin{cases} e^{t^\alpha} - 2t^\alpha - 1, & \text{for } 0 \le t \le 1 \\ e^{t^\alpha} \left\{ t^\alpha e^{-t^\alpha} + t^\alpha e^{-1} - 5e^{-1} + 1 \right\}, & \text{for } 1 < t \le 2 \end{cases}. \tag{23}$$

In Fig. 2, we present the solutions of Eq. 14 for various values of α, illustrating how the order of the fractal differential equation influences the resulting solution.

Example 3 Fractal Neutral Functional Differential Equation Consider a fractal neutral functional differential equation of the form:

$$D_F^\alpha y(t) = y(S_F^\alpha(t) - 1) + 2D_F^\alpha y(S_F^\alpha(t) - 1),$$

where y is F-continuous and has a history function $y(t) = 1$ for $t \in [0, 1]$.

If the equation holds for $t > 1$, then it follows that $D_F^\alpha y(t) = 1$ for $1 \le t \le 2$, and

$$y(t) = S_F^\alpha(t) \propto t^\alpha, \quad 1 \le t \le 2.$$

Similarly, if the equation holds for $2 < t \le 3$, then $D_F^\alpha y(t) = S_F^\alpha(t) + 1$ for $2 < t \le 3$, and

$$y(t) = \frac{1}{2}(S_F^\alpha(t))^2 + S_F^\alpha(t) - 2 \propto \frac{1}{2}t^{2\alpha} + t^\alpha, \quad 2 < t \le 3.$$

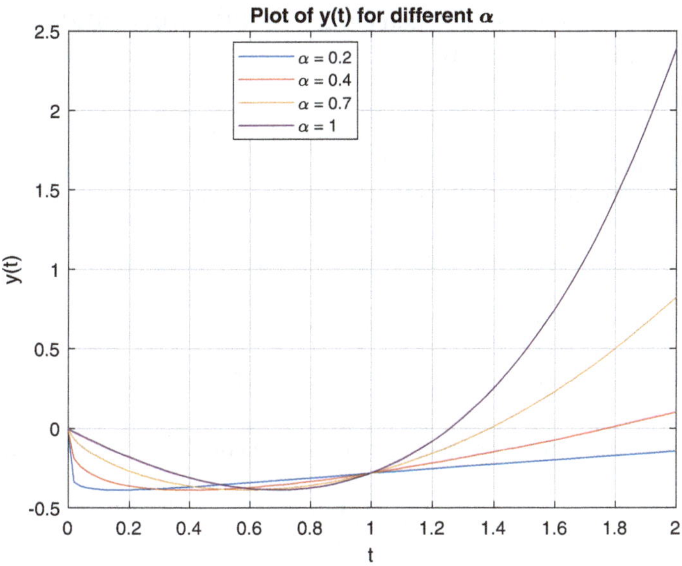

Fig. 2 The graph depicts the behavior of Eq. (23), showcasing its characteristics and variations

Also we can write

$$y(t) = \begin{cases} S_F^\alpha(t) \propto t^\alpha, & \text{for } 1 \leq t \leq 2 \\ \frac{1}{2}(S_F^\alpha(t))^2 + S_F^\alpha(t) - 2 \propto \frac{1}{2}t^{2\alpha} + t^\alpha - 2, & \text{for } 2 < t \leq 3 \end{cases}. \tag{24}$$

Example 4 Fractal Functional Differential Equation with Integral Term

$$D_F^\alpha y(t) = f(S_F^\alpha(t), y(t), \int_{-\infty}^{0} y(t+\tau)e^{\nu\tau} d_F^\alpha \tau), \tag{25}$$

where $\nu \in \mathbb{R}$ and $x(t) = \int_{-\infty}^{0} y(t+\tau)e^{\nu\tau} d_F^\alpha \tau$. The equation leads to a system of coupled fractal ordinary differential equations.

These examples demonstrate the diversity and richness of solutions that can arise in fractal functional differential equations. In summary, fractal functional differential equations are powerful tools for modeling complex dynamic systems that depend on both the present and the past, with time represented in a fractal manner. They offer various types of solutions and have applications in different fields of science and engineering.

4 Laplace Transform for Solving Fractal Functional Differential Equations

In this section, we explore the use of the Laplace transform to solve fractal functional differential equations (FDEs) [28, 29].

Consider a fractal delay differential equation given by

$$D_F^\alpha y(t) = Ay(t) + By(S_F^\alpha(t) - S_F^\alpha(\tau)). \tag{26}$$

The characteristic equation of Eq. (26) is given by

$$h^\alpha(\omega) \equiv S_F^\alpha(\omega) - A - Be^{-S_F^\alpha(\omega)S_F^\alpha(\tau)}. \tag{27}$$

Now, let's introduce the fundamental solution $Y(t)$ of Eq. (26) with the initial condition:

$$Y(t) = \begin{cases} 0, \ t < 0; \\ 1, \ t = 0. \end{cases} \tag{28}$$

The Laplace transform $\mathcal{L}$ of $Y(t)$ is such that:

$$\mathcal{L}(Y)(\omega) = \frac{1}{h^\alpha(\omega)}. \tag{29}$$

Additionally, for any $c > b$, where b is the exponent associated with the bound $|Y(t)| \leq ae^{bS_F^\alpha(t)}$ for $t \geq 0$, the solution of Eq. (26) can be expressed as

$$Y(t) = F\_lim_{T\to\infty} \frac{1}{2\pi i} \int_{c-iT}^{c+iT} e^{S_F^\alpha(\omega)S_F^\alpha(t)} h^\alpha(\omega) d_F^\alpha \omega. \tag{30}$$

Next, we apply the Laplace transform technique to solve two examples of fractal functional differential equations.

Example 5 Fractal Renewal Functional Differential Equation Consider the fractal renewal functional differential equation:

$$y(t) = f(t) + \int_0^t y(t-\tau)\phi(\tau) d_F^\alpha \tau. \tag{31}$$

By taking the fractal Laplace transform on both sides of Eq. (31), we get

$$\mathcal{L}(y) = \mathcal{L}(f) + \mathcal{L}(y)\mathcal{L}(\phi), \tag{32}$$

where $\mathcal{L}$ indicates the fractal Laplace transform and uses fractal convolution [28]. Solving for $\mathcal{L}(y)$, we obtain

$$\mathcal{L}(y) = \frac{\mathcal{L}(f)}{1 - \mathcal{L}(\phi)}. \tag{33}$$

By applying the inverse fractal Laplace transform $\mathcal{L}^{-1}$, we find the solution $y(t)$:

$$y(t) = \frac{1}{2\pi i} \int_c \frac{e^{S_F^\alpha(\omega) S_F^\alpha(t)} \left[\int_0^\infty e^{-S_F^\alpha(\omega) S_F^\alpha(t_1)} f(t_1) d_F^\alpha t_1\right] d_F^\alpha \omega}{1 - \int_0^\infty e^{-S_F^\alpha(\omega) S_F^\alpha(t_1)} \phi(t_1) d_F^\alpha t_1}, \tag{34}$$

where $\int_c$ is a contour integral, and $i = \sqrt{-1}$.

Example 6 Fractal Functional Differential Equation with Delay Consider the fractal functional differential equation with delay:

$$D_F^\alpha y(t) = 2y(S_F^\alpha(t) - 1). \tag{35}$$

To solve Eq. (35), we apply the fractal Laplace transform and obtain [28]:

$$\int_1^\infty D_F^\alpha y(t) e^{-S_F^\alpha(\omega) S_F^\alpha(t)} d_F^\alpha \omega = 2 \int_1^\infty y(S_F^\alpha(t) - 1) e^{-S_F^\alpha(\omega) S_F^\alpha(t)} d_F^\alpha \omega. \tag{36}$$

By performing fractal integration by parts and a change of variable, we arrive at

$$\int_1^\infty y(t) e^{-S_F^\alpha(\omega) S_F^\alpha(t)} d_F^\alpha t = \frac{y(1) e^{-S_F^\alpha(\omega)} + 2e^{-S_F^\alpha(\omega)} \int_0^1 y(u) e^{-S_F^\alpha(\omega) S_F^\alpha(u)} d_F^\alpha u}{S_F^\alpha(\omega) - 2e^{-S_F^\alpha(\omega)}}. \tag{37}$$

Finally, applying the inverse Laplace transform $\mathcal{L}^{-1}$ to both sides, we obtain the solution $y(t)$ [28]:

$$y(t) = \int_c \frac{y(1) e^{-S_F^\alpha(\omega)} + 2e^{-S_F^\alpha(\omega)} \int_0^1 y(u) e^{-S_F^\alpha(\omega) S_F^\alpha(u)} d_F^\alpha u}{S_F^\alpha(\omega) - 2e^{-S_F^\alpha(\omega)}} e^{S_F^\alpha(\omega) S_F^\alpha(t)} d_F^\alpha \omega. \tag{38}$$

These examples illustrate the application of the Laplace transform to solve fractal functional differential equations. The Laplace transform offers an effective method to obtain solutions for such complex equations in the fractional domain.

Appendix

In this appendix, we provide definitions and properties related to the fractal Laplace transform and convolution, which are essential tools for solving fractal functional differential equations.

Definition 8 The fractal Laplace transform of an F-continuous or piecewise F-continuous function f is defined as [9]:

$$F(\omega) = \mathcal{L}(f)(\omega) = \int_0^\infty \exp(-S_F^\alpha(\omega)S_F^\alpha(t))f(t)d_F^\alpha t, \quad 0 \leq t \leq \infty, \tag{39}$$

where $|f(t)| \leq a e^{bS_F^\alpha(t)}$ for some positive constants a and b.

Definition 9 The inverse fractal Laplace transform is defined as [9]:

$$f(t) = \mathcal{L}^{-1}(F(\omega)) = F\_lim_{T\to\infty} \frac{1}{2\pi i}\int_{c-iT}^{c+iT} e^{S_F^\alpha(\omega)S_F^\alpha(t)}F(\omega)d_F^\alpha \omega, \tag{40}$$

where $c \in \mathbb{R}$.

Definition 10 The fractal convolution of functions f and g is defined as [9]:

$$(f * g)(t) = \int_0^t f(t-s)g(s)d_F^\alpha s. \tag{41}$$

Lemma 1 *The fractal Laplace transform convolution of functions is given by [9]:*

$$\mathcal{L}(f * g) = \mathcal{L}(f)\mathcal{L}(g). \tag{42}$$

Definition 11 A point t is a change point of a function f if f is not constant over any open interval containing t. The set of change points of f is denoted by Sch$f$ [9].

These definitions and results provide the mathematical foundation for utilizing fractal Laplace transforms and convolutions in solving fractal functional differential equations. They play a crucial role in transforming and analyzing functions in the fractional domain, enabling the solution of complex problems involving fractal calculus.

References

1. Barnsley MF (2014) Fractals everywhere. Academic Press
2. Mandelbrot BB (1982) The fractal geometry of nature. WH freeman, New York
3. Freiberg U, Zähle M (2002) Harmonic calculus on fractals-a measure geometric approach I. Potential Anal 16(3):265–277
4. Barlow MT, Perkins EA (1988) Brownian motion on the Sierpinski gasket. Probab Theory Relat Field 79(4):543–623
5. Kigami J (2001) Analysis on fractals. Cambridge University Press
6. Stillinger FH (1977) Axiomatic basis for spaces with noninteger dimension. J Math Phys 18(6):1224–1234
7. Tarasov VE (2010) Fractional dynamics. Springer, Berlin, Heidelberg
8. Parvate A, Gangal AD (2009) Calculus on fractal subsets of real line-I: formulation. Fractals 17(01):53–81

9. Golmankhaneh AK (2022) Fractal calculus and its applications. World Scientific
10. Golmankhaneh AK, Balankin AS (2018) Sub-and super-diffusion on Cantor sets: beyond the paradox. Phys Lett A 382(14):960–967
11. Golmankhaneh AK, Fernandez A (2019) Random variables and stable distributions on fractal Cantor sets. Fractal Fract 3(2):31
12. Golmankhaneh AK, Tunç C (2020) Stochastic differential equations on fractal sets. Stochastics 92(8):1244–1260
13. Banchuin R (2022) Noise analysis of electrical circuits on fractal set. COMPEL Int J Comput Math Electr Electron Eng 41(5):1464–1490
14. Banchuin R (2022) Nonlocal fractal calculus based analyses of electrical circuits on fractal set. COMPEL Int J Comput Math Electr Electron Eng 41(1):528–549
15. Wibowo S, Indrati CR et al (2021) The relationship between a fractal F^{α}–absolutely continuous function and a fractal bounded p–variation function. In: International conference on science and engineering (ICSE-UIN-SUKA 2021). Atlantis Press, pp 35–38
16. Hale JK (1971) Functional differential equations. Springer
17. Kolmanovskii V, Myshkis A (2012) Applied theory of functional differential equations, vol 85. Springer Science & Business Media
18. Hale JK, Lunel SMV (2013) Introduction to functional differential equations, vol 99. Springer Science & Business Media
19. Kuang Y (1993) Delay differential equations: with applications in population dynamics. Academic Press
20. Rihan FA, Abdelrahman D, Al-Maskari F, Ibrahim F, Abdeen MA (2014) Delay differential model for tumour-immune response with chemoimmunotherapy and optimal control. Comput Math Methods Med. https://doi.org/10.1155/2014/982978
21. Driver RD (2012) Ordinary and delay differential equations, vol 20. Springer Science & Business Media
22. Erneux T (2009) Applied delay differential equations, vol 3. Springer Science & Business Media
23. Rahman G, Nisar KS, Golamankaneh AK (2021) The nonlocal fractal integral reverse Minkowski's and other related inequalities on fractal sets. Math Probl Eng
24. Valério D, Ortigueira MD, Lopes AM (2022) How many fractional derivatives are there? Mathematics 10(5):737
25. Xu C, Farman M, Akgül A, Nisar KS, Ahmad A (2022) Modeling and analysis fractal order cancer model with effects of chemotherapy. Chaos Solitons Fractals 161:112325
26. Khan D, Ali G, Khan A, Khan I, Chu Y-M, Nisar KS (2020) A new idea of fractal-fractional derivative with power law kernel for free convection heat transfer in a channel flow between two static upright parallel plates. Comput Mater Contin 65(2):1237–1251
27. Parvate A, Gangal A (2011) Calculus on fractal subsets of real line-II: conjugacy with ordinary calculus. Fractals 19(03):271–290
28. Golmankhaneh AK, Tejado I, Sevli H, Valdés JEN (2023) On initial value problems of fractal delay equations. Appl Math Comput 449:127980
29. Kerr G, González-Parra G (2022) Accuracy of the Laplace transform method for linear neutral delay differential equations. Math Comput Simulation 197:308–326

Chapter 2
A Fractal Analysis of Biodiversity: The Living Planet Index

Cristina Serpa and Jorge Buescu

Abstract The Living Planet Index (LPI) is a global index which measures the state of the world's biodiversity. Analyzing the LPI solely by statistical trends provides, however, limited insight. Fractal Regression Analysis is a recently developed tool that has been successfully applied in multidisciplinary scientific contexts for time series analysis. This method is based on the construction of a fractal function tailored to the dataset from which directional coefficients, representing trends, and fractal coefficients, representing oscillations, may be computed. These coefficients allow us to classify the world's regions according to the progression of the LPI, helping us to identify and mathematically characterize the region of Latin America and Caribbean and the category of freshwater as worst-case scenarios with respect to the evolution of biodiversity.

1 Introduction

The Living Planet Index (LPI) is a global index that provides a comprehensive measure of the state of the world's biodiversity. It was developed in 1997 by the World Wildlife Fund (WWF) in collaboration with the Zoological Society of London (ZSL) as a tool to quantitatively track changes in the abundance of wildlife populations over time and to provide a comprehensive measure of the state of the world's biodiversity. It uses time-series data to calculate average rates of change in a large number

C. Serpa
Instituto Superior de Engenharia de Lisboa and CMAFcIO - Centro de Matemática, Aplicações Fundamentais e Investigação Operacional, Campo Grande, 16, Lisboa 1749-016, Portugal

C. Serpa (✉) · J. Buescu
Faculdade de Ciências da Universidade de Lisboa and CMAFcIO - Centro de Matemática, Aplicações Fundamentais e Investigação Operacional, Campo Grande, 16, Lisboa 1749-016, Portugal
e-mail: mcserpa@fc.ul.pt

J. Buescu
e-mail: jsbuescu@ciencias.ulisboa.pt

G. Arulprakash et al. (eds.), *Mathematical Modelling of Complex Patterns Through Fractals and Dynamical Systems*, Studies in Infrastructure and Control,
https://doi.org/10.1007/978-981-97-2343-0_2

of populations of terrestrial, freshwater, and marine vertebrate species. The dataset used in the LPI contains about 3000 population time series for over 1100 species.

The first LPI report was published in 1998 and provided data on trends in the abundance of vertebrate species from 1970 to 1995. Since then, the LPI has been updated every 2 years to provide the latest information on global biodiversity trends. The most recent report, released in September 2020 [31], revealed that the world's wildlife populations have declined by an average of 68% since 1970, showing that the current rate of biodiversity loss is unprecedented in human history and constitutes a most serious and urgent concern.

Since the LPI is based on data from thousands of species and covers a wide range of ecosystems and regions, it provides a tool allowing the identification of trends and patterns in biodiversity loss that would not be visible by considering only a small number of species or ecosystems. We next describe some key studies on biodiversity issues using the LPI as a key tool.

Loh et al., in [14], describe the development of the LPI and outline two methods of effectively calculating it. The results show a decline in terrestrial species of 25% on average between 1970 and 2000, indicating a significant decline in global biodiversity over that time period. Collen et al. [5] describe the methodology used to calculate the LPI, including the use of standardized population indices and statistical techniques to account for variations in sampling effort and detect significant changes in population trends. The authors also provide examples of how the LPI has been used to track changes in biodiversity over time, and to inform conservation efforts at a global scale. Newbold et al. [16] present an analysis of the global effects of land use on local terrestrial biodiversity, based on data from the Living Planet Index (LPI) and other quantitative indicators. In [26], Jenkins et al. present an analysis of global patterns of terrestrial vertebrate diversity based on data from several quantitative indicators, among which is the LPI. The authors analyzed trends in population sizes of more than 25,000 species of mammals, birds, reptiles, and amphibians, showing that terrestrial vertebrate diversity is highly unevenly distributed across the globe, with hotspots of high diversity in tropical regions such as the Amazon Basin, Southeast Asia, and the Congo Basin and identifying areas of high conservation priority. In [4], Butchart et al. present an analysis of trends in global biodiversity based on an extensive set of quantitative indicators. The authors analyzed data on population sizes of vertebrate species from the LPI, as well as data on species richness, ecosystem area, and genetic diversity, finding that all of these indicators showed significant declines over time, making it "highly unlikely" that the United Nations Convention on Biological Diversity (CBD) 2010 target had been met. Later, Tittensor et al. [27] analyze data on trends in population sizes of vertebrate species from the LPI, as well as data on habitat loss, protected areas, and other indicators of biodiversity, finding that progress towards the 2010 target CBD had indeed been poor overall, with most indicators showing negative trends. The LPI showed a decline of 28% between 1970 and 2008, indicating a significant decline in global biodiversity over that time period.

In this paper, we shall study the time series for the LPI with the dataset taken from the database *Our world in data* [30], freely accessible online at https://ourworldindata.org/grapher/global-living-planet-index. This database lists the time series of the LPI for several world regions and biosphere groups, together with their upper and lower confidence intervals.

The classical mathematical methods used to analyze real-world data are sometimes insufficient to characterize and describe them in a meaningful way. In this respect, the study of fractals brings a fresh perspective which is frequently quite useful. A fractal is an object, function, or structure whose key feature is self-similarity. This means that some global configuration is replicated over and over, at ever smaller scales, as part of the whole structure.

Fractal structures have been recognized to arise naturally in many natural and real-world systems (see [2] for specific examples). The challenge that has been faced by scientists is to characterize mathematically real-world approximate fractal structures, displaying self-similarity through a finite range of scales, in order to better understand them. A common approach in the literature is to use a quantitative measure that, in some reasonable sense, allows us to gauge the fractal properties. Fractal dimensions (see e.g., [12]) are the most usual tools for this purpose, the fractal nature in the phenomena typically being signaled by non-integer values of the dimension. The box dimension (see e.g., [7]), which is relatively easy to compute in practice, is the most usual indicator for these purposes. The Hausdorff dimension (see e.g., [18]) is the fundamental theoretical measure representing the fractal features; it may in many cases of interest be estimated using the box dimension.

Computation of the Hausdorff dimension is generally challenging to obtain in practical cases. However, an alternative approach using machine learning has been proposed (Fernandez-Martinez [8]). On the other hand, the estimation of the box dimension is comparatively easier (see Freiberg [9] for a numerical procedure). In the literature, we can find estimations for applied science studies as well (e.g., Alonso [1, 28], Palanivel [17]).

Although having a quantitative measure is an important step towards identifying potential fractal structures, it is also important to identify the inherent fractal shape. We consider a family of functions with fractal structure (to be defined below) to model real-world data. These functions are mathematically well-defined and display intrinsic self-similarity properties (see [20–22]).

They are characterized by a set of coefficients, which we estimate using the affine fractal regression method, theoretically formulated in [19]. This tool has been recently developed and has been successfully used in some multidisciplinary scientific contexts and applications, such as nanomaterials [15, 25], physics [11] and corporate studies [24]. This novel method allows for a broad range of applications; in this paper, we use it, for the first time, to approach the extremely important issue of the global evolution of biodiversity as measured by the LPI.

2 Methods

Self-similarity is a fundamental feature of fractal structures. We illustrate this fact with the construction of a classical example, the Koch snowflake (see Fig. 1). The con-

Fig. 1 The construction of the Koch snowflake

struction of this geometrical fractal is performed iteratively. Starting with a straight-line segment, the first level of the construction corresponds to juxtaposing a triangle over it. The second fractal level replicates the construction of the first level in each of the four newly generated line segments. This construction then proceeds iteratively: at each level, the configuration replicates basic geometrical construction in each of the line segments generated in the previous level.

The affine fractal regression method, described in [19], is based on the system of functional iterative equations

$$\varphi\left(\frac{x+j}{p}\right) = a_j \varphi(x) + b_j x + c_j, \; x \in [0, 1), \; 0 \leq j \leq p-1 \tag{1}$$

where $a_j, b_j, c_j \in \mathbb{R}$, $0 < |a_j| < 1$, $\forall\, 0 \leq j \leq p-1$, and $p > 1$ is an integer. As shown in [20], the solutions φ of such a problem have fractal structure. The standard domain here is $[0, 1)$. This is the affine case of a more general iterative functional system

$$\varphi\left(f_j(x)\right) = F_j(x, \varphi(x)), \; x \in X_j, \; j = 0, 1, \ldots p-1, \tag{2}$$

where X and Y are non-empty sets, $X_j \subset X$, $f_j : X_j \to X$, $F_j : X_j \times Y_j \to Y$ are given functions, and $\varphi : \cup_{j=0}^{p-1} X_j = X \to Y$ is the unknown function. This system is studied and its solution constructed, in very general conditions, in [21]. The development of these problems has its origins in Barnsley's foundational paper on fractal interpolation [3]. Related theoretical studies may be found in the literature (see e.g., [10, 23]).

The explicit solution φ of (1) is constructed in [19]). Given $x \in [0, 1)$, we consider the p-expansion

$$x = \sum_{n=1}^{+\infty} \frac{\xi_n}{p^n},$$

where $0 \leq \xi_i < p$ is an integer (note that this expansion is not necessarily unique). In the case where infinitely many $\xi_i \neq 0$, $\varphi(x)$ is given by the series

$$\varphi\left(\sum_{n=1}^{+\infty}\frac{\xi_n}{p^n}\right)=\sum_{n=1}^{\infty}\left(\prod_{k=1}^{n-1}a_{\xi_k}\right)\left(b_{\xi_n}\sum_{m=1}^{\infty}\frac{\xi_{m+n}}{p^m}+c_{\xi_n}\right),$$

while, if x is represented by a finite p-expansion, $\varphi(x)$ is given by the finite sum

$$\varphi\,(0)=\frac{c_0}{1-a_0},$$

$$\varphi\left(\sum_{n=1}^{m}\frac{\xi_n}{p^n}\right)=\left(\prod_{j=1}^{m}a_{\xi_j}\right)\frac{c_0}{1-a_0} \tag{3}$$

$$+\sum_{k=1}^{m}\left(\prod_{j=1}^{k-1}a_{\xi_j}\right)c_{\xi_k}+\sum_{n=1}^{m-1}\sum_{k=1}^{n}\left(\prod_{j=1}^{k-1}a_{\xi_j}\right)b_{\xi_k}\frac{\xi_{k+m-n}}{p^{m-n}}.$$

Equation (3) will be instrumental in what follows.

In system (1), the self-similarity of the solutions results from the fact that each subinterval of length $1/p$ is mapped by the system onto the whole unit interval, and thus the restriction of the solution to that subinterval, $\varphi\left((x+j)/p\right)$, is mapped onto the solution on the whole unit interval $\varphi\,(x)$ with a scaling coefficient a_j. Since the fractal nature of the solutions is solely determined by the coefficients a_j, we call these the *fractal coefficients*. The parameters b_j are the *directional coefficients*: if all a_j were zero, then the solution would be an affine piecewise linear function, with each b_j representing the slope of the corresponding branch of the function. Finally, the c_j are *offset coefficients*; these ensure the well-posedness of the system, namely existence, uniqueness, and continuity of solutions, but will otherwise play a minor role in what follows.

The fractal coefficients a_j measure the strength of the fractal oscillations present in data. Smoother data yield fractal coefficients close to zero, while stronger fluctuations produce larger a_j (in absolute value). In extremely oscillating cases, these fitting coefficients may be out of the theoretically defined bounds ($|a_j|\leq 1$). This may occur either because we are considering a small number of fractal levels, or possibly because the oscillation in the data does not exhibit a fractal structure which may be properly modeled by a function which is a solution of Eq. (2); in that case the estimated coefficients a_j are generally large and/or outside of the bounds, even in the presence of smooth data. In this last case, we say that the data does not have the feature of self-similarity.

The process of construction of these functions is iterative, very much in the spirit of the construction of the Koch snowflake. We start off by subdividing the domain interval into p equal length subintervals; p is called number of fractal periods. Each of these intervals is in turn subdivided again into p equal-length subintervals, and the process is then iterated, yielding a sequence of smaller and smaller intervals. Each step in this replication procedure corresponds to what is called a *fractal level*.

The rigorous mathematical construction of the solution to the iterative functional Eq. (1) requires an infinite number of replications. However, when dealing with real-

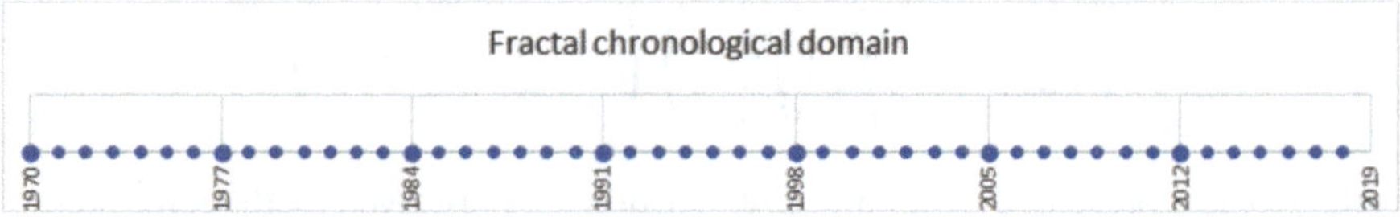

Fig. 2 Fractal chronological domain

world time series, the dataset is obviously finite, so we need to fix a certain maximal fractal level L and then construct the corresponding best approximation. For instance, in Fig. 1, $L = 1$ (initial configuration) corresponds to the "tent" shape and $L = 2$ is the next step in the construction of the Koch snowflake (Figs. 2).

In general, working with a finite time series with fractal period p and (maximal) level L requires p^L data points. In this paper, we will work with $L = 2$, so we will need p^2 data points.

Having the function (3), solution of (1), as a model, we now determine the values of the $3p$ coefficients a_j, b_j and c_j, $0 \le j \le p - 1$, such that the corresponding function best approximates the data. This is performed via the following least-squares method: for each set of possible coefficients, there exists a sum of squared residuals (SSR) between the real data and the approximating fractal function. The goal is to minimize SSR. This is performed by finding the critical points of the error originating by solving the least-squares system

$$\begin{aligned} \frac{\partial SSR}{\partial a_j} &= 0, \ 0 \le j \le p - 1 \\ \frac{\partial SSR}{\partial b_j} &= 0, \ 0 \le j \le p - 1 \\ \frac{\partial SSR}{\partial c_j} &= 0, \ 0 \le j \le p - 1. \end{aligned} \tag{4}$$

Since system (4) does not have a closed-form solution, we use a numerical approach to obtain the best approximation. The software Fractal Real Finder[1] is a tool developed by the first author to perform the search for the coefficients that best fit real-world data to the fractal function model.

3 Results

We work with the dataset obtained from the online database "Our World in Data"[2]. The available data correspond to the upper, lower, and average LPI, and are subdivided into the following global macroregions: Africa; Asia and Pacific; Europe

[1] Fractal Real Finder is copyright registered by Assoft nr. 2576/D/20.

[2] https://ourworldindata.org/grapher/global-living-planet-index, accessed on 2 February 2023.

and Central Asia; Latin America and Caribbean; North America; and Freshwater. Except for Africa, all data correspond to the period 1970–2018. This provides the exact number of data to perform a fractal analysis of 7^2 points. Since in the case of Africa there is a missing year in the dataset, we excluded this macroregion from our study. For each sequence of yearly LPI, we search for the best Fractal Regression Function (FRF) that fits each sequence of data, and obtain the estimates for the corresponding coefficients.

The availability of data for the LPI is a period of $49 = 7^2$ years, which corresponds, in this model, to $p = 7$ fractal periods and $L = 2$ fractal levels (see Fig. 2). The domain of the dataset is the period 1970–2018 and will play the role of the standard domain $[0, 1)$ in Eq. (1).

For the fractal levels up to $L = 2$, the explicit formulation of Eq. (3) leads to

$$\varphi(0) = \frac{c_0}{1 - a_0},$$
$$\varphi\left(\frac{\xi_1}{p}\right) = a_{\xi_1}\frac{c_0}{1 - a_0} + c_{\xi_1},\ \xi_1 \neq 0,$$
$$\varphi\left(\sum_{n=1}^{2}\frac{\xi_n}{p^n}\right) = a_{\xi_1}a_{\xi_2}\frac{c_0}{1 - a_0} + a_{\xi_1}c_{\xi_2} + b_{\xi_1}\frac{\xi_2}{p} + c_{\xi_1},\ \xi_2 \neq 0.$$

We perform a fractal reconstruction, identifying a self-replication up to the second fractal level. This reconstruction will be successful if this type of fractal feature is present in the data. The key coefficients analyzed are the fractal coefficients a_j. These measure the intensity of the fractal oscillations. For a formal and controlled theoretical environment, these should satisfy the bounds $0 < |a_j| < 1$. However, some datasets do not generate convergence to a good fitting function; in other cases, the estimated fractal coefficients lie outside these bounds. Both cases invalidate this approach. As discussed in the previous section, these occur either when the oscillations are too large or the data does not possess a fractal structure adequately modeled by system (1).

From all the sequences studied, two do not originate a good FRF fit: this happens with North America, with both upper CI and average datasets. These two sequences show no fractal structure (see Fig. 3).

The outputs obtained for the other sequences show a good fractal fitting and are given in the tables of the Appendix. For those, we plot the original data in graphs with the estimated corresponding fractal functions (see e.g. Fig. 4). The global scenario (world and freshwater) of species evolution over time is shown in Fig. 5, and is similar to the case of Latin America and the Caribbean.

We now summarize the fractal analysis of the data. The most favorable situation in terms of evolution of biodiversity corresponds to the case where the dynamics of each species interacts with a rich variety of variables, challenging the individual and collective adaptation to survival conditions. We may observe, in such a scenario, oscillations which naturally reflect this adaptive process. This fluctuation situation translates into higher expected fractal coefficients, or even a lack of stable fractal coefficients, indicating the absence of self-similarity characteristics. In the studied examples, we find that in the case of North America (see Fig. 3) no self-similarity is present (for the upper and average cases) and the LPI index has a promising evolution. However, considering the lower CI bound of this index, the plot does not have such favorable features, and a self-similarity is present (see Fig. 7).

Other favorable situations occur when the directional coefficients are non-negative, representing an improvement in the number of living species in the period. These happen over certain time windows, but do not seem possible to maintain stably for a long period. Larger directional coefficients b_j indicate good evolutionary dynamics. In Table 1, we present the cases where the b_j coefficients are positive. The empty spaces correspond to negative evolutions.

The most worrying cases correspond to negative directional coefficients, and the worst case occurs when these are associated with fractal coefficients close to zero, indicating almost complete absence of fluctuations. In Table 2, we include the worst cases for the average LPI when combining these two types of coefficients. We remark that during the first period of LPI monitoring, 1970–1976, no such cases were found (Tables 3, 4, 5, 6, 7, and 8).

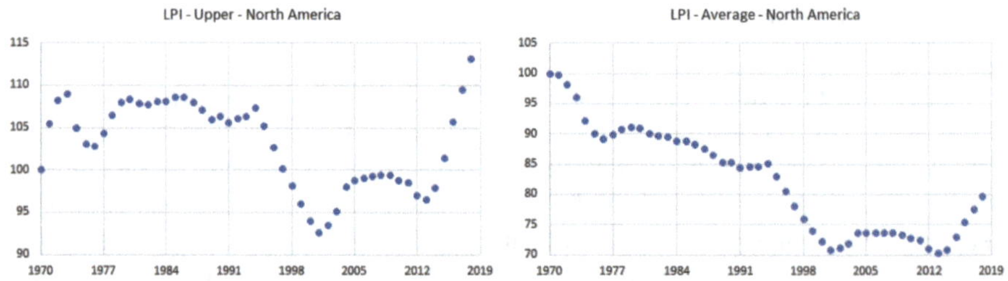

Fig. 3 LPI—Upper and Average—North America

Table 1 Positive directional coefficients for the LPI—Average

	b_j						
	0	1	2	3	4	5	6
	1970–	1977–	1984–	1991–	1998–	2005–	2012–
World			1.044		1.53		
Freshwater					0.845		
Asia and Pacific	16.421		7.451				8.417
Europe and Central Asia	12.289	1.141	22.135				
Latin America and Caribbean							

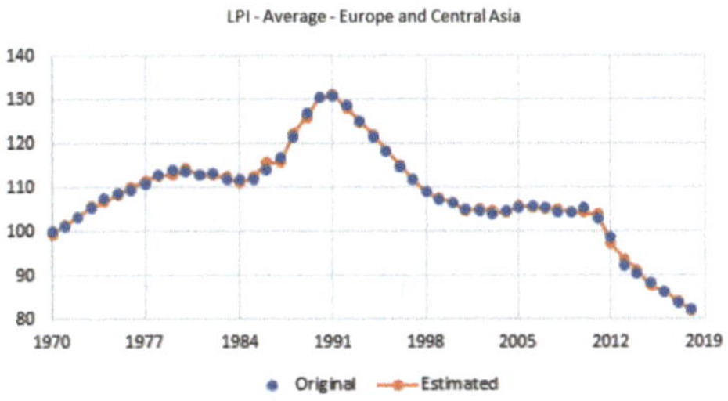

Fig. 4 LPI—Average—Europe and Central Asia

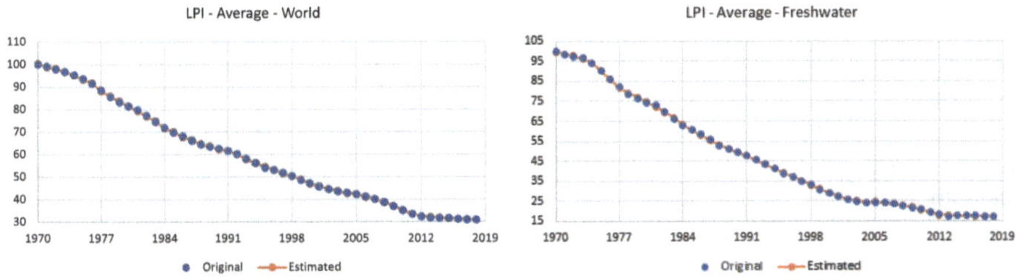

Fig. 5 LPI—Average—World and Freshwater

Fig. 6 LPI—Average—Latin America and the Caribbean

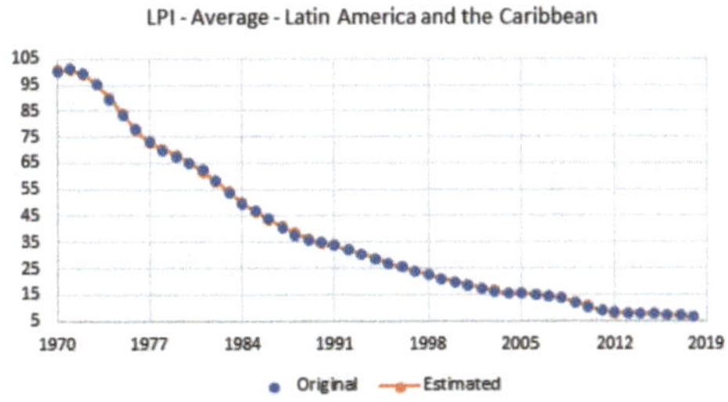

Fig. 7 LPI—Lower—North America

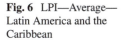

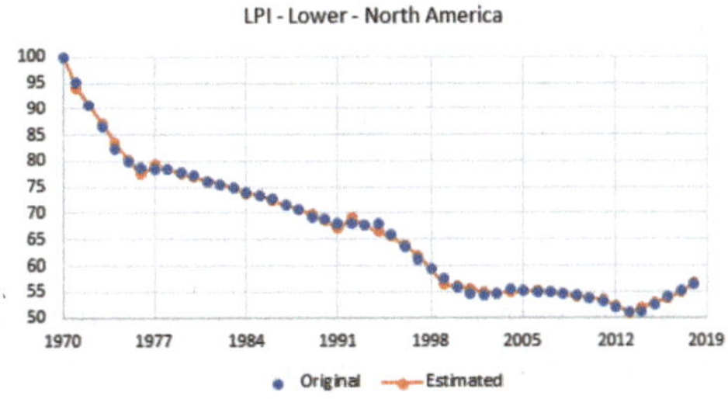

4 Analysis and Conclusions

We now present the main conclusions we may derive from the results summarized in Tables 1 and 2.

Table 2 Negative directional coefficients and weak fractal coefficients $(|a_j| < 0.1)$ for the LPI—Average

		j						
		0	1	2	3	4	5	6
		1970–	1977–	1984–	1991–	1998–	2005–	2012–
World	a_j		–0.011		0.099			0.018
	b_j		–16.323		–4.078			–0.317
Freshwater	a_j		–0.053	0.069	0.004		–0.071	0.003
	b_j		–22.33	–9.876	–14.292		–12.892	–0.413
Asia and Pacific	a_j		0.091					
	b_j		–4.709					
Europe and Central Asia	a_j				0.019	–0.061	0.004	–0.063
	b_j				–22.445	–5.65	–2.195	–17.994
Latin America and Caribbean	a_j		–0.085	0.065	0.01	0.026	–0.069	–0.024
	b_j		–30.466	–11.29	–10.712	–5.527	–15.169	–4.251

Table 3 Fractal Regression coefficients for LPI on World

		j						
		0	1	2	3	4	5	6
		1970–	1977–	1984–	1991–	1998–	2005–	2012–
Upper	a_j	–0.071	–0.111	0.051	0.139	0.056	–0.018	–0.046
	b_j	–7.311	–22.51	–6.828	–2.132	–5.193	–12.66	–5.232
	c_j	108.397	106.573	74.292	55.709	51.746	51.949	43.462
Average	a_j	–0.142	–0.011	0.152	0.099	0.131	–0.135	0.018
	b_j	–21.083	–16.323	1.044	–4.078	1.53	–21.046	–0.317
	c_j	114.403	89.178	56.581	51.849	37.11	55.862	30.459
Lower	a_j	–0.001	0.023	0.093	0.045	0.068	–0.073	0.01
	b_j	–17.359	–14.528	–2.997	–7.561	–2.723	–15.569	–1.131
	c_j	100.018	79.351	55.3	50.029	37.05	43.349	26.029

Recall that, in this fractal analysis, the directional coefficients b_j reflect (but are not equivalent to) statistical trends in the dataset: positive directional coefficients indicate a positive trend in the period under analysis, while negative directional coefficients indicate negative trends. On the other hand, the fractal coefficients a_j reflect, in each data series, the intensity of self-similar oscillations within the time series under analysis. Thus, for example, a linear decrease in biodiversity over time would correspond to negative directional coefficients (reflecting the negative trend) and zero fractal coefficients (reflecting the absence of oscillations in the trend).

Table 4 Fractal Regression coefficients for LPI on Freshwater

		j						
		0	1	2	3	4	5	6
		1970–	1977–	1984–	1991–	1998–	2005–	2012–
Upper	a_j	–0.248	–0.334	0.091	0.065	0.112	–0.054	–0.062
	b_j	–25.876	–43.991	–7.583	–11.679	–1.825	–12.274	–7.176
	c_j	129.305	130.269	69.917	56.903	32.939	41.694	34.052
Average	a_j	–0.314	–0.053	0.069	0.004	0.114	–0.071	0.003
	b_j	–45.923	–22.33	–9.876	–14.292	0.845	–12.892	–0.413
	c_j	130.808	86.781	56.444	47.16	21.545	31.532	17.454
Lower	a_j	–0.078	0.029	0.045	0.009	0.041	–0.034	0.009
	b_j	–35.843	–16.209	–10.726	–10.993	–4.158	–8.133	0.236
	c_j	106.631	66.978	45.635	34.86	19.926	20.178	10.86

Table 5 Fractal Regression coefficients for LPI on Asia and Pacific

		j						
		0	1	2	3	4	5	6
		1970–	1977–	1984–	1991–	1998–	2005–	2012–
Upper	a_j	0.26	0.076	0.145	–0.06	–0.17	0.149	0.123
	b_j	30.19	–7.056	16.093	–22.983	–10.984	–21.393	9.46
	c_j	74.761	108.632	92.8	117.482	109.529	74.615	56.722
Average	a_j	0.152	0.091	0.135	–0.106	–0.225	0.216	0.12
	b_j	16.421	–4.709	7.451	–28.25	–20.246	–8.731	8.417
	c_j	85.04	94.479	79.235	97.105	89.804	41.117	31.874
Lower	a_j	0.013	–0.023	0.009	–0.005	–0.068	0.194	0.036
	b_j	–5.826	–13.68	–7.781	–20.764	–10.269	–3.431	1.466
	c_j	97.378	94.875	79.352	67.39	54.254	26.184	24.723

Thus, we see that the most critical case in a time series from the point of view of biodiversity corresponds to consistently negative directional coefficients coupled with near-zero fractal coefficients, which together indicate a steady, permanent, and practically monotone decline in biodiversity. We will refer to this situation below as the *worst-case scenario*.

From Tables 1 and 2, we see that the global situation of the world with respect to biodiversity is extremely worrying: in the seven fractal periods under consideration, three of them fall into the worst-case scenario of negative directional coefficients coupled with near-zero fractal coefficients. Only in two of the periods (1984–1990 and 1998–2004) were the directional coefficients positive.

Table 6 Fractal Regression coefficients for LPI on Europe and Central Asia

		j						
		0	1	2	3	4	5	6
		1970–	1977–	1984–	1991–	1998–	2005–	2012–
Upper	a_j	0.062	0.067	–0.142	0.022	–0.058	–0.038	–0.047
	b_j	20.664	3.354	33.727	–24.288	–1.859	5.068	–13.439
	c_j	93.776	115.357	140.839	150.679	135.619	132.466	129.889
Average	a_j	0.031	0.069	–0.153	0.019	–0.061	0.004	–0.063
	b_j	12.289	1.141	22.135	–22.445	–5.65	–2.195	–17.994
	c_j	96.4	104.819	126.401	129.123	115.076	105.437	103.45
Lower	a_j	0.003	0.071	–0.164	0.012	–0.063	0.044	–0.087
	b_j	3.023	–0.615	12.694	–20.829	–8.366	–4.793	–21.231
	c_j	98.627	95.066	113.453	111.31	97.998	82.995	84.278

Table 7 Fractal Regression coefficients for LPI on Latin America and Caribbean

		j						
		0	1	2	3	4	5	6
		1970–	1977–	1984–	1991–	1998–	2005–	2012–
Upper	a_j	–0.45	–0.206	0.118	0.034	0.042	–0.137	–0.055
	b_j	–55.934	–45.976	–7.808	–11.648	–5.658	–25.366	–7.958
	c_j	150.212	110.947	52.387	42.91	26.873	37.534	18.904
Average	a_j	–0.315	–0.085	0.065	0.01	0.026	–0.069	–0.024
	b_j	–60.828	–30.466	–11.29	–10.712	–5.527	–15.169	–4.251
	c_j	132.107	80.895	43.508	32.958	19.944	22.312	10.292
Lower	a_j	–0.175	–0.037	0.034	0.002	0.014	–0.028	–0.011
	b_j	–64.227	–22.558	–11.799	–8.669	–5.42	–8.66	–2.572
	c_j	117.727	61.853	35.047	24.681	15.056	13.207	6.022

Table 8 Fractal Regression coefficients for LPI on North America

		j						
		0	1	2	3	4	5	6
		1970–	1977–	1984–	1991–	1998–	2005–	2012–
Lower	a_j	0.143	0.19	0.004	–0.042	–0.224	0.168	–0.023
	b_j	–15.857	–4.966	–8.296	–18.684	4.13	–3.509	14.904
	c_j	81.187	78.814	77.953	89.547	42.506	57.573	34.57

This global situation is heavily influenced by the extremely critical cases of freshwater, where the worst-case scenario was observed in five of the last six periods under consideration, and above all in the region Latin America and Caribbean, where the worst-case scenario occurred in all six of the last six periods. This corresponds to a time span of 32 years of permanent, steady, and monotone loss of biodiversity.

The situation in the region of North America is somewhat more promising, as the fractal self-similarity is not found due to a high diversity of factors. In the region of Europe and Central Asia, the situation is somewhat different: after an initial phase (two fractal periods for Asia and Pacific, three fractal periods for Europe and Central Asia) of positive directional coefficients, and thus of positive evolution in biodiversity, the trend was reversed in the last four fractal periods, with negative directional coefficients and the corresponding deterioration in biodiversity.

These cases do not fit exactly into the worst-case scenario and are thus somewhat more favorable than the catastrophic situations relative to the macroregions Freshwater and Latin America and Caribbean. In those regions, an overall negative trend is compounded by very small fractal coefficients, indicating a dangerously monotonic decline in biodiversity. This should be an alert and a clear call for urgent action by policymakers.

We would like to stress the advantages of the fractal regression model used in our approach to the analysis of the LPI with respect to other more classical methods used in the literature ([6, 13, 14, 29]). For these applications, basic time series analysis retains essentially information on trends; analysis of the variations of trends is done, most frequently, in a qualitative way, or via some kind of Fourier analysis like FFT, which is not particularly illuminating in this context.

The fractal regression model used in this paper has several advantages. In the first place, it allows a quantitative measure both of trends (via the directional coefficients) and of oscillations (via the fractal coefficients). In the second place, unlike Fourier-type analysis, the oscillatory component is measured with functions which are themselves determined by the time series through (3), and so are ab initio adapted to it. In the third place, the output of this model is simple and transparent: it consists of two sets of coefficients, which allow a quantitative measure and classification of the dynamics of the time series beyond simple trend analysis.

Appendix

Fractal Real Finder is a software developed by Cristina Serpa to numerically obtain fractal functions that fit real data, performing the so called fractal regression method. The program code is under copyright protection. Therefore, it is not possible to publish it. Here we give a brief description of the algorithm described in the paper [19] containing the theoretical formulation of fractal regression.

The fractal function serving as model is solution of the system of Eq. (1), where the coefficients a_j, b_j and c_j are the parameters to estimate. The explicit solution for $L = 2$ (as in the LPI example) is given in [19] by (3). The algorithm is as following:

1. Write down the expression of the SSR—sum of squared residuals in between the images of the theoretical fractal function (3) and the data.
2. Write down the system of Eq. (4) that equals the partial derivatives of the SSR (3) in order to a_j, b_j and c_j to zero. The explicit equations of this system are (see [19]):

$$
\begin{aligned}
0 = &-c_0 \frac{(a_0 - 1)\, y_0 + c_0}{(a_0 - 1)^3} \\
&- \sum_{\xi_1=1}^{p-1} a_{\xi_1} c_0 \frac{a_{\xi_1} c_0 - (a_0 - 1)\left(c_{\xi_1} - y_{\xi_1}\right)}{(a_0 - 1)^3} \\
&+ \sum_{\xi_1=1}^{p-1} \sum_{\xi_2=1}^{p-1} a_{\xi_1} a_{\xi_2} c_0 \frac{\frac{a_{\xi_1} a_{\xi_2} c_0}{1-a_0} + a_{\xi_1} c_{\xi_2} + b_{\xi_1} \frac{\xi_2}{p} + c_{\xi_1} - y_{\xi_1 \xi_2}}{(1 - a_0)^2} \\
&+ \sum_{\xi_2=1}^{p-1} \left(\frac{a_0 a_{\xi_2} c_0}{(1 - a_0)^2} + \frac{a_{\xi_2} c_0}{1 - a_0} + c_{\xi_2} \right) \left(a_0 c_{\xi_2} + \frac{a_0 a_{\xi_2} c_0}{1 - a_0} + b_0 \frac{\xi_2}{p} + c_0 - y_{0\xi_2} \right),
\end{aligned}
$$

$$
\begin{aligned}
0 = &\, c_0 \frac{a_j c_0 + (1 - a_0)\left(c_j - y_j\right)}{(1 - a_0)^2} \\
&+ \sum_{\substack{\xi_2=1 \\ \xi_2 \neq j}}^{p-1} \left(\frac{a_{\xi_2} c_0}{1 - a_0} + c_{\xi_2} \right) \left(\frac{a_{\xi_2} a_j c_0}{1 - a_0} + b_j \frac{\xi_2}{p} + a_j c_{\xi_2} + c_j - y_{j\xi_2} \right) \\
&+ \sum_{\substack{\xi_1=0 \\ \xi_1 \neq j}}^{p-1} a_{\xi_1} c_0 \frac{\frac{a_{\xi_1} a_j c_0}{1-a_0} + a_{\xi_1} c_j + b_{\xi_1} \frac{j}{p} + c_{\xi_1} - y_{\xi_1 j}}{1 - a_0} \\
&+ \left(2 \frac{a_j c_0}{1 - a_0} + c_j \right) \left(\frac{a_j^2 c_0}{1 - a_0} + b_j \frac{j}{p} + a_j c_j + c_j - y_{jj} \right), \quad j \neq 0,
\end{aligned}
$$

$$0 = \sum_{\xi_2=1}^{p-1} \xi_2 \left(\frac{a_j a_{\xi_2} c_0}{1-a_0} + a_j c_{\xi_2} + c_j + b_j \frac{\xi_2}{p} - y_{j\xi_2} \right) \quad ,$$

$$\begin{aligned} 0 = & \frac{(a_0 - 1)\, y_0 + c_0}{(1-a_0)^2} \\ & + \sum_{\xi_1=1}^{p-1} a_{\xi_1} \frac{a_{\xi_1} c_0 + (1-a_0)\left(c_{\xi_1} - y_{\xi_1}\right)}{(1-a_0)^2} \\ & + \sum_{\xi_1=1}^{p-1} \sum_{\xi_2=1}^{p-1} a_{\xi_1} a_{\xi_2} \frac{\frac{a_{\xi_1} a_{\xi_2} c_0}{1-a_0} + a_{\xi_1} c_{\xi_2} + b_{\xi_1} \frac{\xi_2}{p} + c_{\xi_1} - y_{\xi_1 \xi_2}}{1-a_0} \\ & + \sum_{\xi_2=1}^{p-1} \left(\frac{a_0 a_{\xi_2}}{1-a_0} + 1 \right) \left(a_0 c_{\xi_2} + \frac{a_0 a_{\xi_2} c_0}{1-a_0} + b_0 \frac{\xi_2}{p} + c_0 - y_{0\xi_2} \right), \end{aligned}$$

$$\begin{aligned} 0 = & \left(\frac{a_j c_0}{1-a_0} + c_j - y_j \right) \\ & + \sum_{\substack{\xi_2=1 \\ \xi_2 \neq j}}^{p-1} \left(\frac{a_{\xi_2} a_j c_0}{1-a_0} + a_j c_{\xi_2} + b_j \frac{\xi_2}{p} + c_j - y_{j\xi_2} \right) \\ & + \sum_{\substack{\xi_1=0 \\ \xi_1 \neq j}}^{p-1} a_{\xi_1} \left(\frac{a_{\xi_1} a_j c_0}{1-a_0} + a_{\xi_1} c_j + b_{\xi_1} \frac{j}{p} + c_{\xi_1} - y_{\xi_1 j} \right) \\ & + (a_j + 1) \left(\frac{a_j^2 c_0}{1-a_0} + a_j c_j + b_j \frac{j}{p} + c_j - y_{jj} \right), \quad j \neq 0, \end{aligned}$$

where $0 \leq j \leq p-1$, and $\left\{ y_{\xi_1 \xi_2} \right\}_{\xi_1 \xi_2 \in \{0,1,\ldots,p-1\}}$ is the input data sequence.

3. Find numerical solutions, or approximative solutions, of system given by step 2.
4. From the solutions of step 3, choose the one with less SSR.

Acknowledgements The authors acknowledge partial support from National Funding from FCT—Fundação para a Ciência e a Tecnologia, under the project: UIDB/04561/2020, https://doi.org/10.54499/UIDB/04561/2020

References

1. Alonso JM, Alvarez JA, Riveros CV, Villagra PE (2019) Finite (Hausdorff) dimension of plants and roots as indicator of ontogeny. Rev. FCA UNCUYO 51(2):142–153. ISSN (en línea) 1853–8665
2. Barnsley MF (1993) Fractals everywhere, 2nd edn. Morgan Kaufmann Publishers
3. Barnsley M (1986) Fractal functions and interpolation. Constr Approx 2:303–329
4. Butchart SHM et al (2010) Global biodiversity: indicators of recent declines. Science 328(5982):1164–1168
5. Collen B et al (2008) Monitoring change in vertebrate abundance: the Living Planet Index. Conserv Biol 23(2):317–327
6. Domelas M et al (2018) BioTIME: A database of biodiversity time series for the Anthropocene. Glob Ecol Biogeogr 27:760–786
7. Feng DJ, Wen ZY, Wu J (1999) Some remarks on the box-counting dimensions. Prog Nat Sci 9(6):409–415
8. Fernández-Martínez M, Sánchez-Granero MA (2015) How to calculate the Hausdorff dimension using fractal structures. Appl Math Comput 264(1):116–131
9. Freiberg U, Kohl S (2021) Box dimension of fractal attractors and their numerical computation. Commun Nonlinear Sci Numer Simul 95:105615
10. Geronimo JS, Hardin DP, Massopust PR (1994) Fractal functions and wavelet expansions based on several scaling functions. J Approx Theory 78(3):373–401
11. Khamoush K, Serpa C (2022) Fractal analysis and ferroelectric properties of Nd (Zn1/2 Ti 1/2) O3(NZT). Mod Phys Lett B 36(36):2250167
12. Korner TW (2011) Hausdorff and Fourier dimension. Stud Math 206(1):37–50
13. Ledger S et al Past, present and future of the Living Planet Index. https://doi.org/10.1101/2022.06.20.496803
14. Loh J et al (2005) (2005) The Living Planet Index: using species population time series to track trends in biodiversity. Phil Trans R Soc B 360:289–295
15. Mitić V, Serpa C, Ilić I, Mohr M, Fecht H-J (2021) Fractal Nature of Advanced Ni-Based Superalloys Solidified on Board the International Space Station. Remote Sens 13:1724
16. Newbold T et al (2015) Global effects of land use on local terrestrial biodiversity. Nature 520(7545):45–50
17. Palanivel DA, Natarajan S, Gopalakrishnan S, Jennane R (2019) Trabecular bone texture characterization using regularization dimension and box-counting dimension. In: Proceedings of the 2019 IEEE region 10 conference (TENCON 2019): technology, knowledge, and society, TENCON IEEE region 10 conference proceedings, IEEE345 E 47th ST, New York, NY 10017 USA, pp 1047–1052
18. Singh S (2022) A brief discourse on Hausdorff dimension and self-similarity. Am Math Mon 129(9):831–845
19. Serpa C (2022) Affine fractal least squares regression model. Fractals 30(07):2250138
20. Serpa C, Buescu J (2019) Fractal and Hausdorff dimensions for systems of iterative functional equations. J Math Anal Appl 480(123429):1–19
21. Serpa C, Buescu J (2017) Constructive solutions for systems of iterative functional equations. Constr Approx 45:273–299
22. Serpa C, Buescu J (2015) Explicitly defined fractal interpolation functions with variable parameters. Chaos Solitons Fractals 75:76–83
23. Serpa C, Buescu J (2015) Non-uniqueness and exotic solutions of conjugacy equations. J Differ Equ Appl 21(12):1147–1162
24. Serpa C, Forouharfar A (2021) Fractalization of chaos and complexity: proposition of a new method in the study of complex systems. In: Springer proceedings in complexity, pp 87–105
25. Stajcic I, Stajcic A, Serpa C, Vasiljevic-Radovic D, Randjelovic B, Radojevic V, Fecht H (2022) Microstructure of epoxy-based composites: fractal nature analysis. Fractal Fract 6(12):741
26. van Strien A et al (2016) Global patterns of terrestrial vertebrate diversity and conservation. Proc Natl Acad Sci 113(28):8067–8072

27. Tittensor DP et al (2014) A mid-term analysis of progress toward international biodiversity targets. Science 346(6206):241–244
28. Tsao A, Nardelli P, Waxman AB, Estepar RS, Washko GR, Rahaghi FN (2022) Fractal dimension estimation using box counting method to quantify CT based pulmonary vascular tree simplification. Am J Respir Crit Care Med 205(S):A5436
29. de Vos J et al (2014) Estimating the background rate of species extinction. Conserv Biol 29(2):452–462
30. Our World in Data. https://ourworldindata.org/
31. WWF and ZSL (2020) Living planet report 2020—bending the curve of biodiversity loss. World Wildlife Fund and Zoological Society of London. https://www.worldwildlife.org/pages/living-planet-report-2020

Chapter 3
Constrained Rational Fractal Model Using Function Values

K. Mahipal Reddy and Sana Abdulla

Abstract This research endeavors to create the Rational Cubic Fractal Interpolation Function (RCFIF) by employing an iterated function system that integrates cubic polynomial numerators and linear polynomial denominators within a rational function framework. The objective is to identify appropriate shape parameters and scaling factors, aligning the RCFIF with a predefined piecewise plane. This chapter delivers an in-depth exploration of the theoretical foundations, supplemented by numerical examples to enhance understanding and visualization.

1 Introduction

The foundational work on fractal interpolation functions (FIF) was introduced by Barnsley [1, 2] through the concept of iterated function systems (IFS). Barnsley and Harrington [6] extended this theory to construct m-times differentiable FIFs, assuming knowledge of up to m-th order derivative values at the initial endpoint of the interval. Cubic spline FIFs based on grid slopes were later introduced by Chand and Viswanathan [3], relying on Hermite interpolation data.

Chand and Viswanathan [4] further contributed to the field by constructing cubic FIFs based on Hermite interpolation data. The development of Constrained Fractal Interpolation Functions, positioned between two piecewise lines for restricted IFS parameters, was initiated by the authors [5]. Subsequent works by authors like [7, 10] extended these ideas to construct Constrained Univariate and Bivariate fractal models. Additionally, Chand and Reddy [8] introduced a novel class of fractal interpolation with variable scaling factors.

The authors [9] explored conditions for positive preserving fractal-like B'ezier curves within proposed subdivision matrices. In this paper, we present a novel class of cubic spline rational FIFs utilizing function values with cubic polynomial numerators and linear polynomial denominators. Two shape parameters are introduced, and the

K. Mahipal Reddy (✉) · S. Abdulla
VIT AP, Amaravati, India
e-mail: mahipalnitw@gmail.com

G. Arulprakash et al. (eds.), *Mathematical Modelling of Complex Patterns Through Fractals and Dynamical Systems*, Studies in Infrastructure and Control,
https://doi.org/10.1007/978-981-97-2343-0_3

constrained rational FIF is positioned between two specified piecewise lines to satisfy original data.

The organization of this paper is as follows: Sect. 2 discusses the mathematical background of FIFs based on IFS theory. In Sect. 3, we introduce the new rational FIF with cubic polynomial numerators and linear polynomial denominators. Section 4 establishes sufficient conditions for these interpolants to be constrained rational FIFs, deriving restrictions on the scaling parameters of the associated rational IFS. Finally, Sect. 5 provides numerical examples to illustrate the proposed approach.

2 Basics of FIF

Consider a data set $\{(x_i, y_i) \in \mathbb{R} \times \mathbb{R} : i = 1, 2, \ldots, N+1\}$, where the values of x_i are in ascending order, i.e., $x_1 < x_2 < \cdots < x_N$. Let $I = [x_1, x_{N+1}]$ denote the closed interval of the real line, and $I_i = [x_i, x_{i+1}]$ represent the closed sub-intervals for $i = 1, 2, \ldots, N$.

The contraction maps $w_i : X \to I_i \times \mathbb{R}, i = 1, 2, \ldots, N$, defined as $w_i(x, y) = (L_i(x), F_i(x, y))$, involve $L_i : I \to I_i$ and $F_i : I \times \mathbb{R} \to \mathbb{R}$. These maps satisfy the contraction conditions:

$$
\begin{aligned}
|L_i(x) - L_i(x')| &\leq c_i|x - x'|, \\
|F_i(x, y) - F_i(x, y')| &\leq r_i|y - y'|,
\end{aligned}
$$

for all $x, x' \in I, y, y' \in \mathbb{R}$, where $c_i, r_i \in (0, 1)$. Importantly, F_i is a contraction with respect to the second variable. The maps L_i and F_i also satisfy the join-up conditions:

$$
\begin{aligned}
L_i(x_1) &= x_i, \\
L_i(x_{N+1}) &= x_{i+1}, \\
F_i(x_1, y_1) &= y_i, \\
F_i(x_{N+1}, y_{N+1}) &= y_{i+1},
\end{aligned}
\tag{1}
$$

as specified in [1].

The Iterated Function System (IFS) $\{X, w_1, w_2, \ldots, w_N\}$ possesses a unique attractor $G \subseteq X$, representing the graph of a continuous function f. This function f interpolates the given data set, adhering to $f(x_i) = y_i$ for all $i = 1, 2, \ldots, N+1$. The functional equation for f is expressed as

$$
f(L_i(x)) = F_i(x, f(t)), \quad \text{for } x \in I, \text{ and } i = 1, 2, \ldots, N.
$$

3 Construction of Rational FIFs Using Function Values

Consider the set of interpolation data points $\{(x_i, y_i)\}_{i=1}^{N} = \{(x_1, y_1), (x_2, y_2), \ldots, (x_N, y_N), (x_N, y_N)\}$.

The formulation of the Rational Cubic Fractal Interpolation Function (RCFIF) involves a cubic polynomial as the numerator and function values in the denominator, and it is defined by the equation:

$$f(L_i(x)) = \alpha_i \cdot f(x) + \frac{P_i(x)}{Q_i(x)}. \tag{2}$$

Here, $P_i(x)$ and $Q_i(x)$ are expressed as

$$\begin{aligned} P_i(x) &= (1-\theta)^3 A_i + (1-\theta)^2 \theta B_i + \theta^2 (1-\theta) C_i + D_i \theta^3 \\ Q_i(x) &= (1-\theta) r_i + \theta t_i \end{aligned}$$

where $\theta = \frac{x - x_1}{x_N - x_1}$, and α_i represents vertical scaling factors, while r_i and t_i are shape parameters for $i = 1, 2, \ldots, N-1$.

From Eqs. (1) and (2), we derive the following relationships:

$$\begin{aligned} A_i &= r_i (y_i - \alpha_i y_1) \\ B_i &= (2r_i + t_i) y_i + r_i \Delta_i - \alpha_i \{(2r_i + t_i) y_i + r_i (x_N - x_1) \Delta_i\} \\ C_i &= (r_i + 2t_i) y_{i+1} - \Delta_{i+1} h_i t_i - \alpha_i ((r_i + 2t_i) y_N - \Delta_N (x_N - x_1) t_i) \\ D_i &= t_i (y_{i+1} - \alpha_i y_N) \\ \Delta_i &= \frac{y_{i+1} - y_i}{x_{i+1} - x_i} \\ h_i &= x_{i+1} - x_i \quad \text{for } i = 1, 2, \ldots, N-1 \end{aligned} \tag{3}$$

4 Constrained Rational FIF Using Function Values

In this section, we construct a constrained RCFIF whose graph is between two piecewise straight lines $S_i^l = m_i x + e_i$ and $S_i^u = m_i^* x + e_i^*$, where $i \in \mathbb{N}_{N-1}$, when the given interpolation data set $\{(x_i, y_i)\}_{i=1}^{N+1}$ is between the piecewise straight lines S_i^l and S_i^u.

Now we consider two cases separately:

Case 1: $\alpha_i \geq 0$ for all $i \in \mathbb{N}_{N-1}$.

Case 2: $-a_i < \alpha_i < 0$ for all $i \in \mathbb{N}_{N-1}$.

To prove that the next generation of points satisfies the same constraint, we aim to show that:

$$S_i^l(L_i(x_k)) < f(L_i(x_k)) < S_j^l(L_i(x_k)) \text{ for all } i \in \mathbb{N}_{N-1}.$$

This implies:

$$m_i(L_i(x_k)) + e_i \le \alpha_i f(x_k) + \frac{P_i(\theta_k)}{Q_i(\theta_k)} \le m_i^*(L_i(x_k)) + e_i^*, \quad \theta_k = \frac{x_k - x_1}{x_N - x_1}.$$

Assuming $\alpha_i \ge 0$, the conditions on the rational IFS parameters for the following inequalities to hold are:

$$m_i(a_i x_k + b_i) + e_i \le \alpha_i f(x_k) + \frac{P_i(\theta_k)}{Q_i(\theta_k)} \le m_i^*(a_i x_k + b_i) + e_i^*. \tag{4}$$

The left inequality of Eq. (4) can be written as

$$\alpha_i f(x_k) Q_i(\theta_k) + P_i(\theta_k) - (m_i(a_i x_k + b_i) + e_i) Q_i(x_k) \ge 0. \tag{5}$$

Assuming $\alpha_i \ge 0$, thus

$$f(x_k) \ge m_i x_k + e_i \ \Rightarrow \ \alpha_i f(x_k) Q_i(\theta_k) \ge \alpha_i (m_i x_k + e_i) Q_i(\theta_k). \tag{6}$$

From Eqs. (5) and (6), we have

$$\alpha_i(m_i x_k + e_i) Q_i(\theta_k) + P_i(\theta_k) - (m_i(a_i x_k + b_i) + e_i)\, Q_i(\theta_k) \ge 0$$

where $\theta_k = \frac{x_k - x_1}{x_N - x_1} \Rightarrow x_k = x_1 + \theta_k(x_N - x_1)$. We can get $\alpha_i\ (m_i\ [x_1 + \theta_k(x_N - x_1)] + e_i)\ Q_i(\theta_k) + P_i(\theta_k) - (m_i\ [a_i\ (x_1 + \theta_k(x_N - x_1)) + b_i] + e_i)\ Q_i(\theta_k) \ge 0$. We can be written as

$$\alpha_i\ [m_i x_k + e_i]\ Q_i(\theta_k) + \alpha_i m_i \theta_k (x_N - x_1) Q_i(\theta_k) + P_i(\theta_k) - m_i\ (a_i x_1 + b_i + e_i)\ Q_i(\theta_j)$$
$$- m_i a_i (x_N - x_1) Q_i(\theta_j) \theta_k \ge 0,$$

where

$$Q_i(\theta) = (1 - \theta) r_i + \theta t_i$$

and

$$P_i(\theta) = (1 - \theta)^3 A_i + \theta(1 - \theta)^r B_i + \theta^r (1 - \theta) c_i + D_i \theta^3$$

We can be rewritten as

$$\alpha_i\,(m_i x_1 + e_i)\,\{(1-\theta_k)\,r_i + \theta_k t_i\} + \alpha_i m_i \theta_k (x_N - x_1)\{(1-\theta_k)\,r_i + \theta_k t_i\}$$
$$+ (1-\theta_k)^3 A_i + \theta_k (1-\theta_k)^r A_i + \theta^r (1-\theta_k) c_i + D_i \theta_k^3$$
$$- (m_i (a_i x_1 + b_i + k_i))\,\{(1-\theta_k)\,r_i + \theta_k t_i\} - m_i a_i (x_N - x_1) Q_i(\theta_k)\theta_k \geq 0.$$

Simplify the above inequality

$$A_i(1-\theta)^3 + B_i\theta(1-\theta^2) + c_i\theta^2(1-\theta) + D_i\theta^3 \times \left(S_i^l(x_i) - \alpha_i S_i^l(x_1)\right)((1-\theta)r_i + \theta t_i)$$
$$+ (\alpha_i - a_i) m_i (x_N - x_1) \times \left((1-\theta)\theta r_i + \theta^2 t_i\right) \geq 0$$

Apply the degree evalation technique on $Q_i(\theta_k)$

$$\begin{aligned}(1-\theta_k)r_i + \theta_k t_i &= (1-\theta_k+\theta_k^2)(1-\theta_k)r_i + \theta_k(1-\theta_k)^2 t_i\\ &= (1-\theta_k)^3 r_i + \theta_k^2(1-\theta_k)(2t_i + r_i) + \theta_k(1-\theta_k^2)(2r_i + t_i) + \theta_k^3 t_i\end{aligned}$$

$$(1-\theta_k)\theta_k r_i + \theta_k^2 t_i = (1-\theta_k+\theta_k)\theta_k(1-\theta_k)r_i + \theta_k^2(1-\theta_k+\theta_k)t_i$$

$(1-\theta_k)^2\theta_k r_i + \theta_k^r(1-\theta_k)(r_i+t_i) + \theta_k^2 t_i$ Substitute $A_i,\ B_i,\ C_i$ and D_i expressions using (3), and coefficients of $(1-\theta)^3, \theta(1-\theta)^2, \theta^2(1-\theta),\ \theta^3$ are positive if they satisfies the ifs parameters in the following theorem. Coefficients of $(1-\theta)^3, \theta(1-\theta)^2, \theta^2(1-\theta),\ \theta^3$ are positive then Constrained rational FIF lies between the two piecewise lines.

Theorem 1 *Let $\{(x_i, y_i) : i \in \mathbb{N}_{N+1}\}$ with $y_i \geq S_i^l$ the sufficient condition for the fractal cubic spline f above the piecewise straight line S^l satisfy the following inequality*

1. *For all i,* $0 \leq \alpha_i \leq \min\{\alpha_i, \frac{y_i - S_i^l(x_i)}{y_1 - S_i^l(x_1)}, \frac{y_{i+1} - S_i^l(x_{i+1})}{y_N - S_i^l(x_N)}\}$.
2. *For all i,* $r_i \geq \max\{0, -t_i\{\frac{y_i - S_i^l(x_i) - \alpha_i(y_1 - S_i^l(x_1))}{2y_i + h_i\Delta_i - S_i^l(x_i) - S_i^l(x_{i+1}) - \alpha_i[2y_1 + (x_N - x_1)\Delta_1 - S_i^l(x_1) - S_i^l(x_N)]}\},$
$-t_i\frac{2y_{i+1} - S_i^l(x_i) - S_i^l(x_{i+1}) - h_i\Delta_{i+1} - \alpha_i[2y_N - S_i^L(x_1) - S_i^l(x_N) - (x_N - x_1)\Delta_N])}{y_{i+1} - S_i^l(x_{i+1}) - \alpha_i(y_N - S_i^l(x_N))}$.

Theorem 2 *Let $\{(x_i, y_i) : i \in \mathbb{N}_{N+1}\}$ be Interpolation data, where $y_i < S_i^u(x)$ and $S_i^u(x) = m_i^* x + e_i^*$. Then, the rational cubic fractal interpolation stays below S_i^u if the rational cubic parameters are chosen as follows:*

1. *For all i,* $0 \leq \alpha_i \leq \min\{\frac{S_i^u(x_i) - y_i}{S_1^u(x_i) - y_i}, \frac{S_{i+1}^u(x_{i+1}) - y_{i+1}}{S_i^u(x_N) - y_N}\}$.
2. *For all i,*
$$r_i \geq \max\left\{0, -\frac{t_i[S_i^u(x_{i+1}) + S_i^u(x_i) - 2y_{i+1} + \Delta_{i+1}h_i - \alpha_i(S_i^u(x_N) + S_i^u(x_1) - 2y_N + \Delta_N(x_N - x_1))]}{S_i^u(x_{i+1}) - y_{i+1} - \alpha_i(S_i^u(x_N) - y_N)},\right.$$
$$\left.-t_i\frac{[S_i^u(x_i) - y_i - \alpha_i(S_i^u(x_1) - y_1)]}{S_i^u(x_i) + S_i^u(x_{i+1}) - 2y_i - h_i\Delta_i - \alpha_i[S_i^u(x_1) - S_i^u(x_N) - 2y_1 - (x_N - x_1)\Delta_1]}\right\}.$$

Theorem 3 *Let $\{(x_i, y_i) : i \in \mathbb{N}_{N+1}\}$ be interpolation data, where $S_i^l(x) \leq y_i \leq y_i \leq S_i^u(x)$ for $i \in \mathbb{N}_{N-1}$. Also, $S_i^l(x) = m_i x + e_i$ and $S_i^u(x) = m_i^* x + e_i^*$, then the sufficient condition for the Rational fractal interpolation function (RCFIF) holds between $S_i^l(x)$ and $S_i^u(x)$ if the associated Rational IFS parameters are chosen as*

1. *For all i,*

$$0 \leq \alpha_i < \min\left\{\alpha_i, \frac{y_i - S_i^l(x_i)}{y_1 - S_i^l(x_1)}, \frac{y_{i+1} - S_i^l(x_{i+1})}{y_N - S_i^l(x_N)}, \frac{S^u(x_i) - y_i}{S_i^u(x_1) - y_1}, \frac{S_i^u(x_{i+1}) - y_{i+1}}{S_i^u(x_N) - y_N}\right\}.$$

2. *For all i,*

$$r_i \geq \max\left\{0, \frac{-t_i[y_i - S_i^l(x_i) - \alpha_i(y_1 - S_i^l(x_1))]}{2y_i + h_i\Delta_i - S_i^l(x_i) - S_i^l(x_{i+1}) - \alpha_i(2y_1 + (x_N - x_1)\Delta_1 - S_i^l(x_1) - S_i^l(x_N))}, \right.$$
$$-\frac{t_i[2y_{i+1} - S^l(x_i) - S^l(x_{i+1}) - \Delta_{i+1}h_i - \alpha_i(2y_N - S^l(x_1) - S^l(x_N) - (x_N - x_1)\Delta_N)]}{y_{i+1} - S^l(x_{i+1}) - \alpha_i(y_N - S^l(x_N))},$$
$$-\frac{t_i[S_i^u(x_{i+1}) + S_i^u(x_i) + 2y_{i+1} + \Delta_{i+1}h_i - \alpha_i(S_i^u(x_N) - S_i^u(x_1) + 2y_N + \Delta_N(x_N - x_1))]}{S_i^u(x_{i+1}) - y_{i+1} - \alpha_i(S_i^u(x_N) - y_N)},$$
$$\left. -\frac{t_i[S_i^u(x_i) - y_i - \alpha_i(S_i^u(x_1) - y_1)]}{S_i^u(x_i) + S_i^u(x_{i+1}) - 2y_i - h_i\Delta_i - \alpha_i(S_i^u(x_1) + S_i^u(x_N) - 2y_1 - (x_N - x_1)\Delta_N)}\right\}.$$

Theorem 4 *Let $\{(x_i, y_i) : i \in \mathbb{N}_{N+1}\}$ be an interpolation data. The sufficient conditions for the rational cubic FIF with negative scaling to stay between $S_i^l(x) = m_i x + e_i$ and $S_i^u(x) = m_i x + e_i^*$, if the associated rational IFS parameters are chosen as*

(1) $0 > \alpha_i > \max\{-a_i, \frac{y_i - S_i^l(x_i)}{y_1 - S_1^u(x_1)}, \frac{y_{i+1} - S_i^l(x_{i+1})}{y_N - S_i^u(x_N)}, \frac{S_i^u(x_i) - y_i}{S_i^l(x_1) - y_1}, \frac{S_i^u(x_{i+1}) - y_{i+1}}{S_i^l(x_N) - y_N}\}$

(2) $r_i \geq \max\left\{0, \Lambda_i^1, \Lambda_i^2, \Lambda_i^3, \Lambda_i^4\right\}$

$$\Lambda_i^1 = -\frac{t_i[y_i - S_i^l(x_i) - \alpha_i(y_1 - S_i^u(x_1))]}{2y_i + h_i\Delta_i - S_i^l(x_i) - S_i^l(x_{i+1}) - \alpha_i[2y_1 + (x_N - x_1)\Delta_1 - S_i^l(x_1) - S_i^l(x_N)]},$$
$$\Lambda_i^2 = -\frac{2y_{i+1} - \Delta_{i+1}h_i - S_i^l(x_i) - S_i^l(x_{i+1}) - \alpha_i[2y_N - \Delta_N(x_N - x_1) - S_i^U(x_1) - S_i^u(x_N)]}{y_{i+1} - S_i^l(x_{i+1}) - \alpha_i[y_N - S_i^u(x_N)]},$$
$$\Lambda_i^3 = -\frac{t_i[S_i^u(x_i) - y_i - \alpha_i[S_i^l(x_1) - y_1]}{S_i^u(x_{i+1}) + S_i^u(x_i) - 2y_i - h_i\Delta_i - \alpha_i[S_i^l(x_N) + S_i^l(x_1) - 2y_1 - (x_N - x_1)\Delta_1]},$$
$$\Lambda_i^4 = -\frac{S_i^u(x_{i+1}) - S_i^u(x_i) - 2y_{i+1} + \Delta_{i+1}h_i - \alpha_i[S_i^u(x_N) - S_i^u(x_1) - 2y_N + \Delta_N(x_N - x_1)]}{S_i^u(x_{i+1}) - y_{i+1} - \alpha_i[S_i^l(x_N) - y_N]}.$$

Remark 1 Let $\{(x_i, y_i) : i \in \mathbb{N}_N\}$ be interpolation data where function values are between two straight lines $S_i^l(x_i) = mx_i + e < y_i < S_i^u(x_i) = m^* x_i + e^*$ and $0 < \alpha_i < 1$. The following conditions hold:

(1) For all i,

$$0 \leq \alpha_i < \min\{\alpha_i, \frac{y_i - S^l(x_i)}{y_1 - S^l(x_1)}, \frac{y_{i+1} - S^L(x_{i+1})}{y_N - S^l(x_N)}, \frac{S^u(x_i) - y_i}{S^u(x_1) - y_1}, \frac{S^u(x_{i+1}) - y_{i+1}}{S^u(x_N) - y_N}\}.$$

(2) For all i,

$$r_i \geq \max\left\{0, -\frac{-(y_i - S_i^L - \alpha_i(y_i - S_i^l))t_i}{2y_i + h_i\Delta_i - S^l(x_i) - S^l(x_{i+1}) - \alpha_i(2y_1 + (x_N - x_1)\Delta_1 - S^l(x_1) - S^l(x_N))},\right.$$
$$-\frac{(2y_{i+1} - S^l(x_i) - S^l(x_{i+1}) - \delta_{i+1} - \alpha_i(2y_N - S^l(x_1) - S^l(x_N) - (x_N - x_1)\Delta_N))t_i}{y_{i+1} - S^l(x_{i+1}) - \alpha_i(y_N - S^l(x_N))},$$
$$-\frac{t_i(S_i^u(xi+1) + S_i^u(xi) + 2y_{i+1} + \delta_{i+1} - \alpha_i(S_i^u(x_N) - S_i^u(x_1) + 2y_N + \Delta_N(x_N - x_1)))}{Si^u(x_i+1) - y_{i+1} - \alpha_i(S_i^u(x_N) - y_N)},$$
$$\left.-\frac{t_i(S_i^u(x_i) - y_i - \alpha_i(Si^u(x_1) - y_1))}{S_i^u(x_i) + S_i^u(x_{i+1}) - 2y_i - h_i\delta_i - \alpha_i(Si^u(x_1) + Si^u(x_N) - 2y_1 - (x_N - x_1)\Delta_N)}\right\}.$$

Remark 2 Let $\{(x_i, y_i) : i \in \mathbb{N}_N\}$ be interpolation data where function values are between two straight lines $S^L(x_i) = mx + e < y_i < S^U(x_i) = m^*x + e^*$ and negative scaling vectors. The following conditions hold:

(1) $0 > \alpha_i > \max\{-a_i, \frac{y_i - S^l(x_i)}{y_1 - S^l(x_1)}, \frac{y_{i+1} - S^L(x_i)}{y_N - S^L(x_N)}, \frac{S^u(x_i) - y_i}{S^u(x_1) - y_1}, \frac{S^u(x_{i+1}) - y_{i+1}}{S^u(x_N) - y_N}\}$

(2) $r_i \geq \max\left\{0, \lambda_i^1, \lambda_i^2, \lambda_i^3, \lambda_i^4\right\}$

$$\lambda_i^1 = -\frac{t_i[y_i - S^l(x_i) - \alpha_i(y_1 - S^u(x_1))}{2y_i + h_i\Delta_i - S^l(x_i) - S^l(x_{i+1}) - \alpha_i[2y_1 + (x_N - x_1)\Delta_1 - S^l(x_1) - S^l(x_N)]},$$
$$\lambda_i^2 = -\frac{2y_{i+1} - \Delta_{i+1}h_i - S^l(x_i) - S^l(x_{i+1}) - \alpha_i[2y_N - \Delta_N(x_N - x_1) - S^u(x_1) - S^u(x_N)]}{y_{i+1} - S^l(x_{i+1}) - \alpha_i[y_N - S^u(x_N)]},$$
$$\lambda_i^3 = -\frac{S^u(x_i) - y_i - \alpha_i[S^l(x_1) - y_1}{S^u(x_{i+1}) + S^u(x_i) - 2y_i - h_i\Delta_i - \alpha_i[S^u(x_N) + S^u(x_1) - 2y_1 - (x_N - x_1)\Delta_1]},$$
$$\lambda_i^4 = -\frac{S^u(x_{i+1}) - S^u(x_i) - 2y_{i+1} + \Delta_{i+1}h_i - \alpha_i[S^u(x_N) - S^u(x_1) - 2y_N + \Delta_N(x_N - x_1)}{S^u(x_{i+1}) - y_{i+1} - \alpha_i[S^u(x_N) - y_N]}.$$

Numerical Examples for Curve Constrained Interpolation

In this section, the examples of univariate constrained RCFIF lies between two piecewise functions and lies in straight lines (Table 2).

(i). Constrained α-RCFIF lies between two piecewise lines: Consider a constrained data set $\{(0, 4.5), (2, 6.7), (4, 5), (6, 6.5), (8, 4), (10, 9)\}$ that lies between two piecewise lines. Extended the data set to interpolation data with $S^l(x)$ and $S^u(x)$ by including a data point ('10, 9) so that derivatives at grid point can be approximated by $\Delta_i = \frac{y_{i+1} - y_i}{x_{i+1} - x_i}$.

$$S^l(x) = \begin{cases} x + 4, & \text{if } 0 \leq x \leq 2 \\ -0.75x + 6, & \text{if } 2 \leq x \leq 6.7 \\ 0.75x + 1.5, & \text{if } 4 \leq x \leq 5 \\ -1.25x + 13.5, & \text{if } 6 \leq x \leq 8, \end{cases}$$

$$S^u(x) = \begin{cases} 1.375x + 6, & \text{if } 0 \leq x \leq 2 \\ -1.125x + 11, & \text{if } 2 \leq x \leq 4 \\ x + 2.5, & \text{if } 4 \leq x \leq 6 \\ -1.75x + 19, & \text{if } 6 \leq x \leq 8. \end{cases}$$

Table 1 Scaling vectors and shape parameter r are used in the construction of constrained RCFIF

Scaling vectors α	Figures	Shape parameter vector r	Figures
(0.24, 0.24, 0.24, 0.24)	Figure 1a	(1, 1, 1, 1)	Figure 1a
(0.18, 0.18, 0.18, 0.18)	Figure 1b, d	(201, 1058.5, 101.5, 2156.1)	Figure 1b, c, f
(0.24, 0.24, 0.24, 0.24)	Figure 1c	(500, 500, 500, 500)	Figure 1d
(0.249, 0.249, 0.2, 0.249)	Figure 1e	(200, 200, 200, 200)	Figure 1e
(0, 0, 0, 0)	Figure 1f		

Table 2 Scaling vectors and shape parameter r are used in the construction of constrained straight line RCFIF

Scaling vectors α	Figures	Shape parameter vector r	Figures
(0.245, 0.245, 0.245, 0.245)	Figure 2a	(2, 2, 2, 2)	Figure 2a
(0.18, 0.18, 0.18, 0.18)	Figure 2b	(097.60215.91)	Figure 2b
(0.249, 0.249, 0.2, 0.249)	Figure 2c	(200, 200, 200, 200)	Figure 2c
(0, 97.6, 0, 215.91)	Figure 2d	(0, 0, 0, 0)	Figure 2d

We fix the shape parameters $t = [100]_{1\times5}$. For arbitrary choice of IFS parameters, the α-RCFIF may not preserve the constrained nature of given data, for instance see Fig. 1a. Applying Theorem 3, we have calculated the restrictions on the IFS parameters which satisfy the constrained inequalities, so that the C^1 -RCFIF (2) is bounded between the upper piecewise linear spline S^u and the lower piecewise linear spline S^l. The choice of scaling factors and shape parameters are shown in Table 1. A constrained C^1-RCFIF is generated in Fig. 1b by implementing Theorem 3. Figure 1c is generated by changing the α with respect to IFS parameters in Fig. 1b. Figure 1d is generated by perturbing the shape parameter α with respect to Fig. 1b. Changing the IFS parameters α and r with respect to IFS parameters in Fig. 1b, gives Fig. 1e. The traditional curve is shown in Fig. 1f, by taking $\alpha = 0$.

Here we consider various numerical examples to illustrate the effectiveness of developed constrained FIF schemes and to compare them with the corresponding classical counterparts. Consider the data set $\{(0,\ 4.5), (2,\ 6.7), (4,\ 5), (6,\ 6.5), (8,\ 4), (10,\ 9)\}$. We fix the shape parameters $t = [100]_{1\times5}$. Figure 2a, RCFIF may not be between two straight lines $S^l = 0.1x + 3$ and $S^u = 0.1x + 8.5$. Applying Theorem 1, we have calculated the restrictions on the IFS parameters which satisfy the constrained inequalities, so that the C^1-RCFIF (2) is bounded between the upper piecewise linear spline S^u and the lower piecewise linear spline S^l. The choice of scaling factors and shape parameters are shown in Table 1. A constrained C^1-RCFIF is generated in Fig. 2b by implementing Theorem 3. Figure 2c is generated by changing the α and r with respect to IFS parameters in Fig. 2b. Figure 2d is generated by perturbing the shape parameter $\alpha = (0, 0, 0, 0)$ with respect to Fig. 2b.

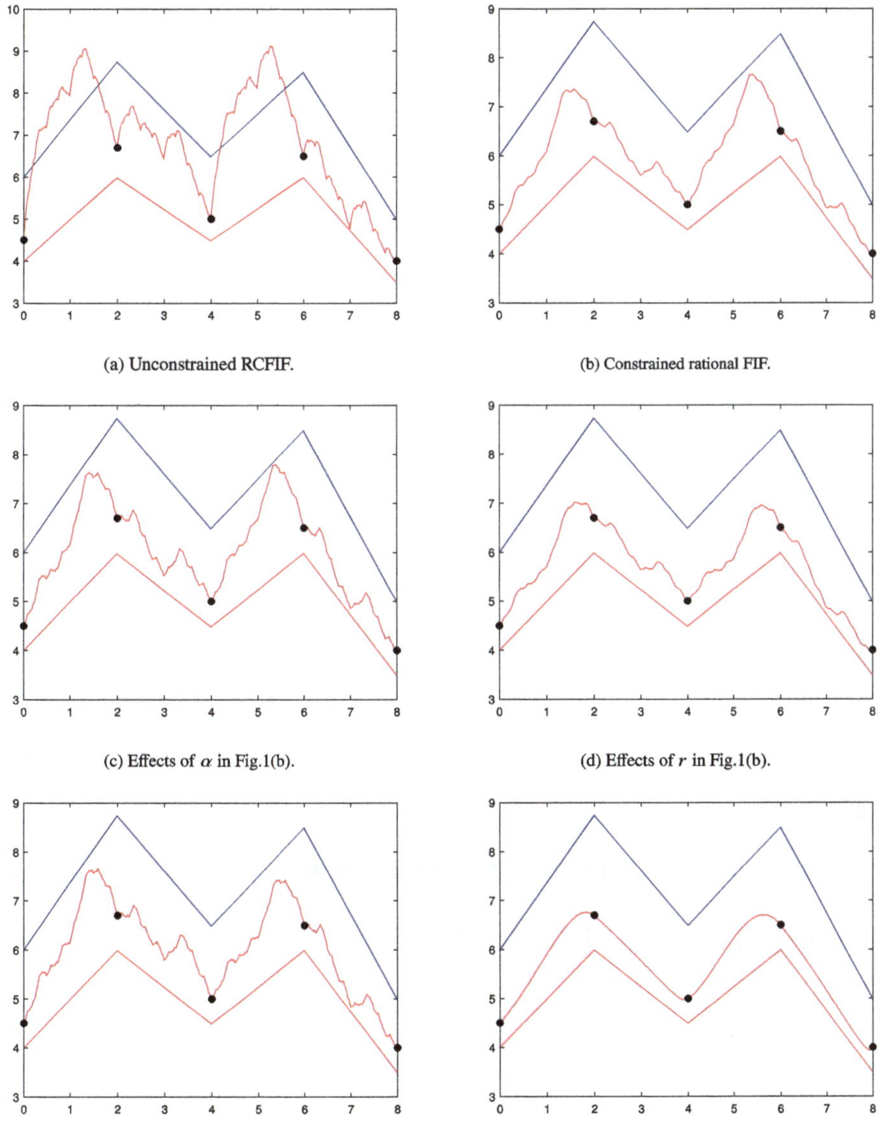

(a) Unconstrained RCFIF.

(b) Constrained rational FIF.

(c) Effects of α in Fig.1(b).

(d) Effects of r in Fig.1(b).

(e) Effects of α and r in Fig.1(b).

(f) Classical constrained interpolant.

Fig. 1 Rational cubic FIF with constrained interpolation

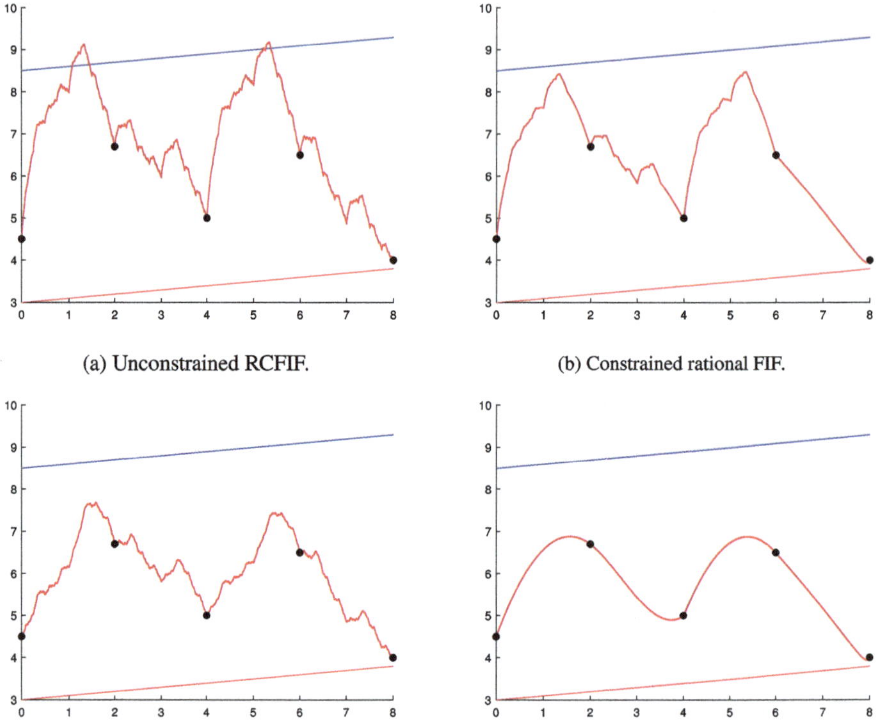

(a) Unconstrained RCFIF.

(b) Constrained rational FIF.

(c) Effects of α in Fig.2(b).

(d) Classical constrained straight line.

Fig. 2 Rational cubic FIF with constrained straight line

Appendix: Fractal Curve Generation Algorithm

Algorithm 1 Fractal Curve Generation

procedure FRACTALCURVEGENERATION(N, P, α)
  Create a 2D N-dimensional grid of values
  Set the number of iterations: P
  Set the vertical scaling factor: $\alpha = (\alpha_1, \alpha_2, \ldots, \alpha_{N-1})$ and $r_n, t_n, n \in \{1, 2, ..., N\}$.
  $j \leftarrow 1$
  while $j \leq P - 1$ **do**
    for $i \leftarrow 1$ to $N - 1$ **do**
      Compute (L_n, F_n) using Eq. (2)
      Generate new interpolation data points (X_d, Y_d)
    end for
    $(X, Y) \leftarrow (X_d, Y_d)$ ▷ Input for the next iteration
    $j \leftarrow j + 1$
  end while
  Draw the graph of generating points (X, Y)
  Output: Fractal interpolation curve G_g
end procedure

References

1. Barnsley MF (1986) Fractal functions and interpolation. Constr Approx 2:303–329
2. Barnsley MF (1988) Fractals everywhere. Academic Press, Orlando
3. Chand AKB, Kapoor GP (2006) Generalized cubic spline fractal interpolation functions. SIAM J Numer Anal 44(2):655–676
4. Chand AKB, Viswanathan P (2013) A constructive approach to cubic Hermite fractal interpolation function and its constrained aspects. BIT Numer Math 53:841–865
5. Chand AKB, Viswanathan P, Reddy KM (2015) Towards a more general type of univariate constrained interpolation with fractal splines. Fractals 23(4). Article ID 1550040
6. Barnsley MF, Harrington AN (1989) The calculus of fractal interpolation functions. J Approx Theory 57(1):14–34
7. Reddy KM, Chnad AKB (2019) Constrained univariate and bivariate rational fractal interpolation. Taylor and Francis 20(5):404–422
8. Chand AKB, Reddy KM (2018) Constrained fractal interpolation functions with variable scaling. Sib Èlektron Mat Izv 15:50–73
9. Reddy KM, Saravana Kumar G, Chand AKB (2022) Family of shape perserving fractal-like Bezier curves. Fractals 28(6):2050105
10. Reddy KM, Vijender N (2023) A fractal model for constrained curve and surface interpolation. Eur Phys J Spec Top 232:1015–1025
11. Böhm W (1984) A survey of curve and surface methods in CAGD. Comput Aided Geom Des 1:1–60

Chapter 4
Defining Coherent Upper Conditional Previsions in a General Metric Space Using Distinct Dimensional Fractal Outer Measures

Serena Doria and Bilel Selmi

Abstract This article presents a novel approach to constructing models of CUCP for both bounded and unbounded random variables in a metric space. The models are built upon the concept of different dimensional outer measures, and the relationships between these measures are thoroughly examined. By utilizing a dimensional measure, it becomes possible to determine the dimensional outer measure of the conditioning event, even in cases where the metric space is not second countable. This metric serves as an indicator of whether the conditioning event possesses a measure of zero, positive, finite, or infinite dimension within its specific context. In cases where the conditioning event exhibits a positive and finite-dimensional measure in its dimension, the coherent upper conditional forecast is established through the utilization of the Choquet integral. Additionally, this definition aligns with an expanded version of the monotone convergence theorem for non-linear integrals, ensuring the maintenance of equivalence among unbounded random variables with identical distributions. On the contrary, if the conditioning event has an outer measure dimensionally equal to 0 or $+\infty$, the coherent upper conditional forecast is delineated about a probability that is finitely but not countably additive, taking on values of 0 or 1.

S. Doria
Department of Engineering and Geology, University G. d'Annunzio, Via dei Vestini, 31, Chieti-Pescara, Italy
e-mail: serena.doria@unich.it

B. Selmi (✉)
Analysis, Probability & Fractals Laboratory: LR18ES17, Department of Mathematics, Faculty of Sciences of Monastir, University of Monastir, 5000 Monastir, Tunisia
e-mail: bilel.selmi@fsm.rn.tn; bilel.selmi@isetgb.rnu.tn

G. Arulprakash et al. (eds.), *Mathematical Modelling of Complex Patterns Through Fractals and Dynamical Systems*, Studies in Infrastructure and Control,
https://doi.org/10.1007/978-981-97-2343-0_4

1 Introduction

One of the aims of *Artificial Intelligence* and knowledge representation is to propose mathematical models able to describe human decision-making and human behaviors that are guided also by unconscious aspects of the mind. A recent publication [15] introduced a novel approach to coherent upper conditional previsions (CUCP), utilizing HOMs (HOMs) to address inherent biases in human reasoning. The study highlighted the model's ability to elucidate a specific facet of unconscious brain activity: the adept handling of unexpected events, akin to selective attention. Within the discussed framework, various metric spaces were introduced to capture diverse human reactions to unforeseen circumstances [16]. This current paper proposes an innovative model of CUCP, building upon distinct fractal outer measures (FOM) that attribute varying measures to different events. Notably, this model enables the representation of diverse reactions to unexpected events within a unified metric space. It is crucial to emphasize that certain coherent upper probabilities, as demonstrated in [13, Example 8], assign zero probability solely to empty sets and consequently cannot effectively depict unexpected events.

CUCP are functionals defined on the linear space of all bounded random variables, adhering to coherence axioms outlined in [37]. Unlike linear conditional expectations, as defined by the Radon–Nikodym derivative in the axiomatic approach [2], these previsions do not always emerge through such extensions. The discrepancy stems from a conflict between the defining property of the Radon–Nikodym derivative, necessitating measurability with respect to the σ-field of conditioning events, and a crucial condition for coherence [9, Theorem 1]. If it is possible to determine which events have positive probability and which ones are null events (i.e., events with probability zero), coherent upper conditional probability (CUCPr) can be expressed as a ratio of coherent upper unconditional probabilities of these events. However, this determination is only possible in a topological space that is second countable. In cases where the space is not second countable, the support of a probability measure may not exist, making it impossible to identify all the null sets of the space. A metric space is considered second countable if it is separable. However, even in a metric space, which is not separable, it is possible to define dimensional outer measures and calculate the outer measure for each event. In this case, all events have Hausdorff outer measure (HOM) equal to infinity and so a CUCP is defined by a finitely, but not countably, additive 0–1 valued probability. The objective of this research paper is to establish a connection between the concept of CUCP and the notions of fractal sets and fractal dimensional outer measures. In many cases, information is conveyed through an extensive amount of data, which is represented by the conditioning set. In a metric space, this data can be aggregated by considering the fractal dimension of the conditioning set. If the fractal dimension is non-integer, it indicates that the conditioning set is a fractal set. The fractal dimension, denoted by s, provides a measure of the complexity of the available information. By defining CUCP and probabilities based on the s-dimensional fractal measure, it becomes possible to establish a link between the updating of partial knowledge and the complexity of the given information. This approach allows for incorporating the fractal dimension as a means of capturing the intricacies inherent in the available data.

This research has introduced a novel model of CUCP for bounded random variables within a metric space. The properties of this model have been rigorously demonstrated in several works [8–12], and its extension to the linear space of Choquet integrable bounded and unbounded random variables has been proposed in [13].

CUCP are established through the Choquet integral with respect to HOMs, specifically when the conditioning event exhibits a positive and finite Hausdorff outer measure in its Hausdorff dimension. In instances where the conditioning event fails to meet this criterion, the previsions are determined by a 0–1 valued finitely additive probability. This deliberate choice ensures that the limitation to a field of events for the given CUCP fulfills the criteria of being a complete conditional probability, as defined by Dubins and demonstrated in [13]. This distinction does not hold true for all coherent upper probabilities, as illustrated in [13, Example 1].

In the realm of unbounded random variables, Seidenfeld et al. [33] have observed that coherent previsions can result in an arrangement of random variables that contravenes the conventional Archimedean principle and the monotone convergence theorem. However, it is crucial to highlight that when the conditioning event possesses a positive and finite HOM in its Hausdorff dimension, the CUCP under consideration adheres to the monotone convergence theorem. This adherence is attributed to the continuity from below exhibited by HOM, enabling the preservation of the monotone convergence property in this specific scenario. This outcome is facilitated by the countable additive nature of the measures involved. Our research results ensure that CUCP, established through HOMs, maintain the equivalence between unbounded random variables that share identical distributions. In this investigation, we introduce models of CUCP applicable to both bounded and unbounded random variables within a linear space. These models incorporate dimensional outer measures with continuity from below, and we delve into exploring the interconnections among these measures. By employing this diverse set of CUCP models, we expand the scope of previsions, ensuring continuity from below across a broader spectrum of scenarios. In cases where a conditioning event exhibits a positive and finite HOM in its dimension, we establish the conditional probability utilizing the HOM. However, when the HOM of the conditioning event in its Hausdorff dimension is zero, we investigate the possibility of identifying an alternative dimensional outer measure within the same metric space. This alternative measure should assign a positive and finite outer measure to the conditioning event. This approach allows us to explore different measures and dimensions, providing a means to determine the suitable coherent upper conditional precision. As an illustration, if a conditioning event possesses an HOM in its Hausdorff dimension that is zero, yet exhibits a positive and finite packing outer measure in the same dimension, we have the option to define a CUCP using the packing outer measure. This provides an alternative to depending on a 0–1 valued finitely additive probability that is not countable. To ascertain the most appropriate model for defining a CUCP that ensures continuity from below, it is essential to establish a framework for assessing whether a given set possesses a zero HOM in its dimension. Extending this methodology to a probability space necessitates evaluating whether each event has zero probability or not. However, this determination isn't always feasible without defining the measure's support, a task achievable in a second countable topological space. Consequently, it becomes imperative to define conditional probability in a metric space by employing dimensional measures, facilitating the computation of a dimensional outer measure for each event.

This paper presents an extension of the model of CUCP, previously introduced in [13], by incorporating alternative FOM. We derive new models of CUCP using the packing s-dimensional outer measure (Theorems 3 and 4), the lower H–S s-dimensional outer measure (Theorems 5 and 6), the lower average H–S s-dimensional outer measure (Theorem 7), and the lower n'th-order average Hölder Hewitt–Stromberg s-dimensional outer measure. The lower H–S dimension is obtained through interpolation between the original Hausdorff and packing dimensions, allowing us to explore cases where the classical dimensions diverge from each other.

The novelty of these proposed models of CUCP, based on different dimensional outer measures, lies in the fact that they exemplify outer measures defined within the same metric space that do not share the same null sets. In previous works [14, 15], the introduction of outer measures with distinct null sets required considering a different metric space with a non-bi-Lipschitz metric in relation to the initial metric space. Additionally, applications of these different outer measures within the same metric space for representing unexpected events are discussed in [16, 17].

The paper's outline is as follows: In Sect. 2, we discuss the representation of sets with zero probability within a metric space. Section 3 presents the proposed models of coherent conditional upper previsions based on different FOM. Examples illustrating sets with different Hausdorff and packing dimensions are constructed in Sect. 4, providing the motivation for constructing CUCP using average dimensional H–S measures in Sect. 5 and HCA of dimensional H–S measures in Sect. 6.

2 Representation of Sets with Zero Probability in a Metric Space

Within a probability space, the quantification of belief in the occurrence of an event, represented by a set, doesn't necessitate a metric. However, if the aim is to identify sets with zero probability under a specific measure, defining the measure's support becomes essential. Typically, support is defined in a second countable topological space. Defining coherent conditional previsions in any metric space using dimensional measures offers an advantage enabling the computation of the probability measure for any set. When the metric space is separable, it is also second countable, facilitating the determination of the probability measure's support. In such instances, sets with zero probability can always be pinpointed, and coherent conditional previsions are established using the dimensional measure if and only if the measure of the conditioning event is positive and finite. However, when the metric space is non-separable, each set possesses a fractal dimension and a fractal dimensional measure equal to infinity. In such scenarios, the identification of specific sets with zero probability becomes more challenging.

The concept of null sets, which refers to sets with zero probability, is closely related to the notion of the support of a probability measure. To provide a better understanding, we present the following definitions as a reminder.

Definition ([15]) A topological space $\mathcal{T}$ is deemed second countable when there exists a countable collection $\mathcal{U} = \{U_i\}_{i=1}^{\infty}$ comprising open subsets of $\mathcal{T}$ in such a way that any open subset of $\mathcal{T}$ can be represented as the union of elements from a subfamily of $\mathcal{U}$. Essentially, this implies the presence of a countable basis for the topology of $\mathcal{T}$ that comprehensively covers all open sets within the space. □

Definition ([15]) The *support* of a measure ν defined on a topological space (Ω, τ) is the minimal closed set X for which $\nu(\Omega - X) = 0$. □

Defined by its nature, the complement of the support of a probability measure ν is identified as the most extensive open set E for which $\nu(E) = 0$. Let us consider the class of sets $\mathbf{F}_0 = \{F \subset \Omega : F \text{ is open}, \nu(F) = 0\}$. The set $E = \bigcup_{F \in \mathbf{F}_0} F$ is open, and it includes every open set with measure 0.

To establish $\nu(E) = 0$, the topological space must possess second countability. This enables the representation of U as the union of a countable subfamily $\mathbf{V}_0$ derived from $\mathbf{V}$. Thus, we can write $E = \bigcup_{F \in \mathbf{F}_0} F$, and it follows that

$$\nu(E) = \nu\left(\bigcup_{F \in \mathbf{F}_0} F\right) \leq \sum_{F \in \mathbf{F}_0} \nu(F) = 0.$$

Example ([15]) Consider Ω as an uncountable set equipped with the discrete topology, formed by the topology generated by the discrete metric

$$d(n, m) = \begin{cases} 1 \; if \; n \neq m, \\ 0 \; if \; n = m \end{cases}$$

and let

$$\nu(A) = \begin{cases} 0, & \text{if } A \text{ is finite or countable,} \\ +\infty, & \text{if } A \text{ is uncountable.} \end{cases}$$

In this case, the measure ν lacks support, as for every (closed) subset $A \subset \Omega$ with $\nu(\Omega - A) = 0$, there exists a proper subset $T \subset A$ such that $\nu(\Omega - T) = 0$. For instance, one can choose $T = A - \{\omega\}$ for any $\omega \in A$. □

3 CUCP Defined by Dimensional Outer Measures

Consider a metric space (Ω, d), and let $\mathbf{B}$ be a partition of Ω. In this context, a random variable is defined as a function $X : \Omega \to \mathbb{R}^* = \mathbb{R} \cup \{-\infty, +\infty\}$. The set $\mathscr{L}^*(\Omega)$ denotes the linear space of bounded random variables defined on Ω. However, it is crucial to emphasize that $\mathscr{L}^*(\Omega)$ itself is not a linear space. To address this limitation, we introduce $\mathscr{L}(\Omega)$ as a linear space contained within $\mathscr{L}^*(\Omega)$. For any $B \in \mathbf{B}$, the restriction of X to B is denoted as $X|B$. Additionally, we define $\sup(X|B)$ as the supremum value that X assumes on B. Let I_A represent the indicator function of any event $A \in \wp(B)$, where $\wp(B)$ denotes the set of all subsets of B. Specifically, $I_A(\omega) = 1$ if $\omega \in A$, and $I_A(\omega) = 0$ if $\omega \in A^c$. For each $B \in \mathbf{B}$, CUCP $\overline{C}(\cdot|B)$ are functionals defined on a linear space $\mathscr{L}(B)$.

Definition ([16]) CUCP, denoted as functionals $\overline{C}(\cdot|B)$, are defined on $\mathscr{L}(B)$. These previsions adhere to the axioms of coherence for every X and Y in $\mathscr{L}(B)$, as well as for every strictly positive constant λ

(1) $\overline{C}(X|B) \leq \sup(X|B)$;

(2) $\overline{C}(\lambda X|B) = \lambda\overline{C}(X|B)$ (positive homogeneity);

(3) $\overline{C}(X + Y|B) \leq \overline{C}(X|B) + \overline{C}(Y|B)$ (subadditivity). □

If we regard $\mathscr{L}(B)$ as the linear space encompassing all bounded random variables, Definition 1 aligns with the definition presented in [37]. Assuming that $\overline{C}(X|B)$ represents a coherent upper conditional forecast established on $\mathscr{L}(B)$, its corresponding conjugate coherent lower conditional forecast is defined as follows:

$$\underline{C}(X|B) = -\overline{C}(-X|B).$$

If, for every random variable X within a linear space $\mathscr{L}(B)$, the equality $C(X|B) = \underline{C}(X|B) = \overline{C}(X|B)$ holds, then $C(X|B)$ is identified as a coherent linear conditional forecast [5, 6, 18, 28, 29]. This coherent linear conditional forecast exhibits linearity, positivity, and positive homogeneity as functional properties on $\mathscr{L}(B)$ [37, Corollary 2.8.5]. The unconditional coherent upper forecast, denoted as $\overline{C} = \overline{C}(\cdot|\Omega)$, represents a special case where the conditioning event is the entire sample space Ω. CUCPr is derived when considering only 0–1 valued random variables. Based on axioms (1)–(3) and utilizing the conjugacy property, it follows that

$$\underline{C}(I_B|B) = 1 = \overline{C}(I_B|B).$$

3.1 CUCP Defined by HOMs

Given a metric space (Ω, d), the diameter of a non-empty set U in Ω is defined as $|U| = \sup\{d(x, y) : x, y \in U\}$. If there exists a subset A of the space Ω such that A is contained in the union of sets U_i for all i, and each U_i has a size less than δ (where $0 < |U_i| < \delta$ for each i), then the collection $\{U_i\}$ is termed a δ-cover of A. Consider a non-negative real number s. For a given positive value of δ, we define $\hbar_{s,\delta}(A)$ as the infimum of the sum $\sum_{i=1}^{\infty} |U_i|^s$ taken over all possible countable δ-covers $\{U_i\}$. The *s-dimensional HOM* of A, denoted by $\hbar^s(A)$, is then defined as per the following expression:

$$\hbar^s(A) = \lim_{\delta \to 0} \hbar_{s,\delta}(A).$$

This limit is well defined, although it has the potential to be infinite, given that $\hbar_{s,\delta}(A)$ grows as δ decreases. The *Hausdorff dimension* of a set A, denoted as $\mathscr{D}_H(A)$, is characterized as the singular value for which

$$\hbar^s(A) = \infty \text{ if } 0 \leq s < \mathscr{D}_H(A),$$

$$\hbar^s(A) = 0 \text{ if } \mathscr{D}_H(A) < s < \infty.$$

Remark 1 In the situation where Ω lacks separability, choosing a sufficiently small value for $\delta > 0$ makes it impractical to find a countable cover of Ω comprising sets with diameters smaller than δ. Consequently, the infimum in the definition of HOM is taken over an empty set, yielding a value of $+\infty$. As δ diminishes, the limit of the outer measure also tends toward $+\infty$. Therefore, for a non-separable set Ω, for any $s \in [0, +\infty)$, the HOM $\hbar^s(\Omega)$ attains a value of $+\infty$. This signifies that the Hausdorff dimension of Ω is likewise $+\infty$.

The work by Doria in [9] introduces a probability conditioning model grounded in HOMs. This model characterizes conditional probability as the HOM of order s, also referred to as the s-dimensional HOM, particularly when the conditioning event possesses a Hausdorff dimension equal to s. The concept of conditional upper probability is conventionally interpreted as an assessment of the likelihood of an event happening, given the occurrence of another event, which is assumed, presumed, asserted, or supported by evidence.

Theorem 1 ([9]) *Consider a metric space (Ω, d) and suppose $\mathbf{B}$ forms a partition of Ω. For each $B \in \mathbf{B}$, we denote the Hausdorff dimension of the conditioning event B as s and the corresponding s-dimensional HOM as $\hbar^s$. Let m_B represent a 0–1 valued finitely additive, though not countably additive, probability defined on the power set $\wp(B)$. Thus, for every $B \in \mathbf{B}$, we possess a function defined on $\wp(B)$ given by*

$$\overline{C}(A|B) = \begin{cases} \dfrac{h^s(A \cap B)}{h^s(B)} & if\ 0 < h^s(B) < +\infty, \\ m_B & if\ \ h^s(B) \in \{0, +\infty\} \end{cases}$$

is a CUCPr.

Consider a set $B \in \mathbf{B}$ with a positive and finite HOM in its Hausdorff dimension s. Under this circumstance, the monotone set function ν_B^* is established for each $A \in \wp(B)$ as $\nu_B^*(A) = \frac{h^s(A \cap B)}{h^s(B)}$. This function constitutes a CUCPr and exhibits several significant properties. In particular, it demonstrates submodularity, signifying that for any $A, C \in \wp(B)$, the inequality $\nu_B^*(A \cap C) + \nu_B^*(A \cup C) \leq \nu_B^*(A) + \nu_B^*(C)$ holds. Moreover, it exhibits continuity from below, and when restricted to the σ-field of all sets measurable with respect to ν_B^*, it transforms into a Borel regular countably additive probability.

In Theorem 1, for every conditioning set B where the HOM equals 0 or $+\infty$, the definition of conditional probability involves a 0–1 valued finitely additive probability that is not countably additive. This ensures that, under such conditions, the constrained conditional probability on the Borel σ-field serves as a complete conditional probability, following Dubins' concept [18]. Specifically, it adheres to the general compound rule for every Borelian set A, B, C.

$$C(A \cap B|C) = C(A|B \cap C)C(B|C).$$

Remark 2 It is noteworthy that in a discrete metric space, the condition stated in Theorem 1, which mandates both Ω and the conditioning event B to possess positive and finite HOM, is equivalent to requiring that these sets are compact. In a different formulation, within this framework, it is possible to obtain a finite δ-cover of the entire space Ω from any given δ-cover of Ω. This equivalence stems from the fact that, in a discrete metric space, the diameter of each set consistently remains 1. When the δ-covers in the definition of HOMs are countable, the HOM of the set becomes infinite. Furthermore, in a discrete metric space, a set is compact if and only if it is finite. Consequently, in the discrete metric space, the definition of conditional probability involves the use of the counting measure h^0 because the conditioning events B with finite and positive HOM precisely correspond to finite sets with a Hausdorff dimension of 0. Therefore, for each conditioning event with a positive and finite HOM in its Hausdorff dimension, the conditional probability is established on the power set, as per Theorem 1. This entails the application of the counting measure, aligning with the 0-dimensional HOM. It is crucial to highlight that even though the concept of support for any probability measure does not exist in this metric space, the definition of conditional probability remains applicable.

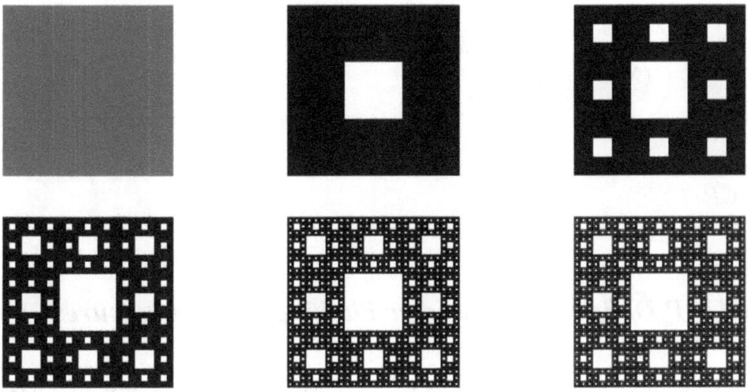

Fig. 1 The first six steps of the construction of a Sierpiński carpet

If the conditioning event B is a fractal set, characterized by a non-integer Hausdorff dimension s as depicted in Fig. 1, we can infer from Theorem 1 that

$$\overline{C}(A|B) = \frac{h^s(A \cap B)}{h^s(B)} \text{ if } 0 < h^s(B) < +\infty.$$

The Choquet integral concerning the s-dimensional HOM h^s is expressed as

$$\int X dh^s = \int_{-\infty}^{0} h^s\left(\{\omega \in \Omega : X(\omega) \geq x\} - h^s(\Omega)\right) dx + \int_{0}^{+\infty} h^s\left(\{\omega \in \Omega : X(\omega) \geq x\}\right) dx.$$

The integral resides in the real numbers $\Re$, or it takes on values of $-\infty$ or $+\infty$, or it may not exist. If

$$-\infty < \int X dh^s < +\infty$$

then the random variable X is called Choquet integrable.

In [13], a theorem is introduced that broadens the scope of the CUCPr, as defined in Theorem 1, to encompass all random variables—both bounded and unbounded—with a finite Choquet integral relative to h^s. This extension is applicable when the conditioning event B possesses a positive and finite HOM in its Hausdorff dimension. In such instances, the coherent upper prevision is established through the application of the Choquet integral [4]. Further details on this theorem can be found in [13].

Theorem 2 *Consider a metric space (Ω, d) and assume **B** forms a partition of Ω. For each set B within **B**, let s represent the Hausdorff dimension of the conditioning event B, and let h^s denote the s-dimensional HOM. Let m_B be a finitely additive probability, taking values of 0 or 1, and defined on the power set $\wp(B)$. For every B in **B**, we establish the functional $\overline{C}(X|B)$ on $\mathscr{L}(B)$ in the following manner:*

$$\overline{C}(X|B) = \begin{cases} \dfrac{1}{\hbar^s(B)} \displaystyle\int_B X d\hbar^s \; if \; 0 < \hbar^s(B) < +\infty, \\ \\ \displaystyle\int_B X dm_B \quad\quad if \;\; \hbar^s(B) \in \{0, +\infty\} \end{cases}$$

is a CUCP.

3.2 CUCP Defined by Diameter Packing Outer Measures

Packing measures exhibit similarities with HOM. Nevertheless, research has shown that specifically for geometric limit sets generated by Kleinian groups [35], there are instances where the HOM proves to be a more fitting measure to consider, whereas, in other scenarios, the packing measure is the more appropriate choice. Specifically, it has been observed that there are instances where the HOM of the limit set is zero, while the packing measure is positive and finite (see Examples 5 (Sect. 4.5) and 6 (Sect. 4.6)). In the study conducted by Taylor and Tricot in [35], they focused on Brownian trajectories and demonstrated that by using a specific gauge function, the packing measure is positive and finite. This finding highlights the significance of the packing measure in certain contexts, where it provides a more appropriate measure for capturing the properties of the system under investigation. The packing measure and the HOM can be seen as duals of each other in a sense. The HOM is defined based on efficient coverings, where the goal is to cover the set with a minimal number of subsets of small diameter. On the other hand, the packing measure is defined in terms of bountiful packings, aiming to densely pack the set with subsets of small diameter. Therefore, the packing measure can be considered as dual to the HOM, as they capture different aspects of the spatial distribution and structure of sets.

Let $E \subset \Omega$ and $\delta > 0$, we say that a collection of closed balls $\big(B(x_i, r_i)\big)_i$ is a centered δ-packing of E if

$$\forall i, \; 0 < r_i < \delta, \; x_i \in E, \text{ and } B(x_i, r_i) \cap B(x_j, r_j) = \emptyset, \; \forall \, i \neq j.$$

Similarly, we say that $\big(B(x_i, r_i)\big)_i$ is a centered δ-covering of E if

$$\forall i, \; 0 < r_i < \delta, \quad x_i \in E, \quad \text{and} \quad E \subset \bigcup_i B(x_i, r_i).$$

For $s > 0$, $E \subseteq \Omega$ and $\delta > 0$, we define

$$\overline{\mathcal{P}}^s_\delta(E) = \sup\left\{\sum_i (2r_i)^s\right\},$$

where the highest value is determined among all centered δ-packings of E. The packing pre-measure is expressed as

$$\overline{\mathscr{P}}^s(E) = \inf_{\delta>0} \overline{\mathscr{P}}^s_\delta(E).$$

To compare the packing measure with the HOM we consider the centered HOM defined considering only centered δ coverings made by balls

$$\overline{\mathscr{H}}^s_\delta(E) = \inf \left\{ \sum_i (2r_i)^s \right\},$$

where the smallest value is calculated among all centered δ-coverings of E. The centered Hausdorff pre-measure is defined as

$$\overline{\mathscr{H}}^s(E) = \sup_{\delta>0} \overline{\mathscr{H}}^s_\delta(E).$$

We implement the following adjustments to the Hausdorff and packing measures,

$$\mathscr{H}^s(E) = \sup_{F\subseteq E} \overline{\mathscr{H}}^s(F) \text{ and } \mathscr{P}^s(E) = \inf_{E\subseteq\bigcup_i E_i} \sum_i \overline{\mathscr{P}}^s(E_i).$$

The functions $\mathscr{H}^s$ and $\mathscr{P}^s$ serve as metric outer measures, thus qualifying as measures on the family of Borel subsets of Ω. An important characteristic of the Hausdorff and packing measures is the relationship $\mathscr{H}^s \le \mathscr{P}^s \le \overline{\mathscr{P}}^s$ and $2^{-s}h^s \le \mathscr{H}^s \le h^s$. Consequently, h^s and $\mathscr{H}^s$ share identical null sets and sets with a measure equal to $+\infty$.

The measure $\mathscr{P}^s$ assigns in a usual way a fractal dimension to each subset E of Ω, denoted by $\mathscr{D}_P(E)$ and defined as follows:

$$\mathscr{P}^s(E) = \infty \text{ if } 0 \le s < \mathscr{D}_P(E),$$

$$\mathscr{P}^s(E) = 0 \text{ if } \mathscr{D}_P(E) < s < \infty.$$

It directly ensues from the given definitions that

$$\mathscr{D}_H(E) \le \mathscr{D}_P(E).$$

Theorem 3 *Take a metric space (Ω, d) and assume that $\mathbf{B}$ forms a partition of Ω. For each set $B \in \mathbf{B}$, let s represent the packing dimension of the event B, and let $\mathscr{P}^s$ denote the packing s-dimensional outer measure. Define m_B as a finitely additive probability, with values restricted to 0 or 1, and defined on the power set $\wp(B)$. For each $B \in \mathbf{B}$, we establish the functional $\overline{C}(X|B)$ on $\mathscr{L}(B)$ in the following manner:*

$$\overline{C}(X|B) = \begin{cases} \dfrac{1}{\mathcal{P}^s(B)} \displaystyle\int_B X d\mathcal{P}^s \ if\ 0 < \mathcal{P}^s(B) < +\infty, \\ \displaystyle\int_B X dm_B \quad if\ \ \mathcal{P}^s(B) \in \{0, +\infty\} \end{cases}$$

is a CUCP.

Proof To demonstrate that $\overline{C}(X|B)$ adheres to conditions (1), (2), and (3) outlined in Definition 3 for each $B \in \mathbf{B}$, it is necessary to examine various scenarios based on the packing outer measure of B within its dimension s. If B possesses a finite and positive packing outer measure in its dimension s, then $\overline{C}(X|B) = \frac{1}{\mathcal{P}^s(B)} \int_B X d\mathcal{P}^s$. In these cases, the fulfillment of properties (1) and (2) is assured, as these properties hold true for the Choquet integral according to [7, Proposition 5.1]. The establishment of Property (3) is achieved by invoking the Subadditivity Theorem from [7, Theorem 6.3], as packing outer measures demonstrate monotonicity, submodularity, and continuity from below. In cases where B possesses a packing outer measure with a dimension equal to 0 or $+\infty$, the set of all coherent (upper) previsions on $\mathcal{L}(B)$ is synonymous with the set of 0–1 valued additive probabilities defined on $\wp(B)$, and we observe that $\overline{C}(X|B) = m_B$. Under these circumstances, properties (1), (2), and (3) are fulfilled, given that m_B is a 0–1 valued finitely additive probability on $\wp(B)$. By confining CUCP to the set of indicator functions, we derive coherent upper conditional probabilities. This leads to a framework for coherent upper conditioning prevision grounded in packing outer measures. In this framework, conditional probability, typically interpreted as the likelihood of an event happening given the occurrence of another event, is defined by the packing measure of order s or the s-dimensional packing measure when the conditioning event possesses a packing dimension equal to s.

Theorem 4 *Take a metric space (Ω, d) and assume that $\mathbf{B}$ constitutes a partition of Ω. For each set $B \in \mathbf{B}$, let s represent the packing dimension of the event B, and let $\mathcal{P}^s$ indicate the packing s-dimensional outer measure. Define m_B as a finitely additive probability, taking values of 0 or 1, although it is not countably additive, and defined on the power set $\wp(B)$. For each $B \in \mathbf{B}$, we define a function on $\wp(B)$ as follows:*

$$\overline{C}(A|B) = \begin{cases} \dfrac{\mathcal{P}^s(A \cap B)}{\mathcal{P}^s(B)} \ if\ 0 < \mathcal{P}^s(B) < +\infty, \\ m_B \quad if\ \ \mathcal{P}^s(B) \in \{0, +\infty\} \end{cases}$$

is a CUCPr.

Proof Utilizing Theorem 4, it is determined that when the conditioning event B has a positive and finite packing outer measure, the CUCP defined by the packing outer measure can be confined to the category of indicator functions. In such instances, the constrained functions thus obtained qualify as coherent upper conditional probabilities. Conversely, if the packing outer measure of the conditioning event B equals 0 or $+\infty$, the CUCP are synonymous with a finitely additive probability that is not countably additive.

3.3 CUCP Defined by the Lower and Upper Dimensional H–S Measures

The two primary fractal measures widely recognized are the HOM and the packing measure. Felix Hausdorff introduced the HOM in 1919, while Tricot introduced the packing measure in the early 1980s [36]. Additionally, there exists a less familiar category of fractal measures known as the Hewitt–Stromberg (H–S) measures. Generally, both Hausdorff and packing measures are formulated by considering coverings and packings of sets with diameters less than a specified positive value r. In contrast, H–S measures are defined by considering packings of balls with a fixed diameter r. Subsequently, we present the definitions for the H–S measures and dimensions. Consider a (totally) bounded subset E of Ω and let $t \geq 0$. The definitions for H–S pre-measures are outlined as follows:

$$\overline{\mathscr{V}}^s(E) = \limsup_{r \to 0} \mathscr{M}_r(E)\ (2r)^s$$

and

$$\overline{\mathscr{U}}^s(E) = \liminf_{r \to 0} \mathscr{N}_r(E)\ (2r)^s,$$

where

$$\mathscr{N}_r(E) = \inf\left\{ \sharp\{I\} \;\middle|\; \Big(B(x_i, r)\Big)_{i \in I} \text{ is a family of closed balls with } x_i \in \Omega \text{ and } E \subseteq \bigcup_i B(x_i, r) \right\}$$

and

$$\mathscr{M}_r(E) = \sup\left\{ \sharp\{I\} \;\middle|\; \Big(B(x_i, r)\Big)_{i \in I} \text{ is a family of closed balls with } x_i \in E \text{ and } B(x_i, r) \cap B(x_j, r) = \emptyset \text{ for } i \neq j \right\},$$

where $\sharp\{I\}$ denotes the cardinal of the set I. The lower and upper s-dimensional H–S measures, denoted as $\mathscr{u}^s(E)$ and $\mathscr{v}^s(E)$, respectively, are defined as follows:

$$\mathscr{u}^s(E) = \inf_{E \subseteq \bigcup_i E_i} \sum_i \overline{\mathscr{u}}^s(E_i)$$

and

$$\mathscr{v}^s(E) = \inf_{E \subseteq \bigcup_i E_i} \sum_i \overline{\mathscr{v}}^s(E_i).$$

Specific foundational inequalities apply to the H–S, Hausdorff, and packing measures, which can be expressed as follows:

$$\overline{\mathscr{u}}^s(E) \leq \overline{\mathscr{v}}^s(E) \leq \overline{\mathscr{p}}^s(E)$$

and

$$\mathscr{H}^s(E) \leq \mathscr{u}^s(E) \leq \mathscr{v}^s(E) \leq \mathscr{p}^s(E) \leq \overline{\mathscr{p}}^s(E).$$

The lower and upper dimensions of H–S, denoted as $\underline{\mathscr{D}}_{MB}(E)$ and $\overline{\mathscr{D}}_{MB}(E)$, are defined as follows:

$$\underline{\mathscr{D}}_{MB}(E) = \sup\left\{s \geq 0 \;\middle|\; \mathscr{u}^t(E) = +\infty\right\} = \inf\left\{s \geq 0 \;\middle|\; \mathscr{u}^t(E) = 0\right\}$$

and

$$\overline{\mathscr{D}}_{MB}(E) = \sup\left\{s \geq 0 \mid \mathscr{v}^t(E) = +\infty\right\} = \inf\left\{s \geq 0 \mid \mathscr{v}^t(E) = 0\right\}.$$

The dimensions mentioned above satisfy the following inequalities (see, for example, [31, 32])

$$\mathscr{D}_H(E) \leq \underline{\mathscr{D}}_{MB}(E) \leq \overline{\mathscr{D}}_{MB}(E) = \mathscr{D}_P(E).$$

Remark 3 Derived from the construction referred to as Method I [30, Theorem 4], it can be inferred that $\mathscr{u}^s$ and $\mathscr{v}^s$ constitute outer measures, establishing them as measures on the Carathéodory-measurable algebra. Specifically, $\mathscr{u}^s$ qualifies as a metric outer measure, making it a measure on the Borel algebra. However, it is important to highlight that $\mathscr{v}^s$ does not possess the characteristics of a metric outer measure, as discussed in [19, 31, 32]. Consequently, this implies that $\mathscr{v}^s$ does not function as a Borel measure, as articulated in Theorem 1.7 of [24].

Similarly, symmetrical results hold for the lower and upper H–S measures, and their proofs are analogous to those presented in the aforementioned theorems.

Theorem 5 *Consider a metric space (Ω, d) and suppose $\boldsymbol{B}$ forms a partition of Ω. Let s represent the lower H–S dimension of each conditioning event $B \in \boldsymbol{B}$, and let*

u^s denote the lower H–S s-dimensional outer measure. Furthermore, let m_B be a finitely additive probability, taking 0 or 1 values, defined on the power set $\wp(B)$- though not countably additive. For each $B \in \mathbf{B}$, we establish a functional $\overline{C}(X|B)$ on $\mathscr{L}(B)$ in the following manner:

$$\overline{C}(X|B) = \begin{cases} \dfrac{1}{u^s(B)} \displaystyle\int_B X du^s \ if \ 0 < u^s(B) < +\infty, \\ \\ \displaystyle\int_B X dm_B \quad \ if \ \ u^s(B) \in \{0, +\infty\} \end{cases}$$

is a CUCP.

Theorem 6 *Take a metric space (Ω, d) and have $\mathbf{B}$ form a partition of Ω. For each $B \in \mathbf{B}$, designate s as the lower H–S dimension of the conditioning event B, and let u^s denote the lower H–S s-dimensional outer measure. Furthermore, let m_B be a 0–1 valued finitely additive probability, which is not countably additive, defined on the power set $\wp(B)$. For every $B \in \mathbf{B}$, the function defined on $\wp(B)$ is expressed as follows:*

$$\overline{C}(A|B) = \begin{cases} \dfrac{u^s(A \cap B)}{u^s(B)} \ if \ 0 < u^s(B) < +\infty, \\ \\ \quad m_B \quad \ if \ \ u^s(B) \in \{0, +\infty\} \end{cases}$$

is a CUCPr.

Next, we try to study a strictly intermediate case, where the classical measures and dimensions diverge from each other. The motivations and the importance of this theory come from several examples, see, for example, [31]: In the subsequent sections, we will delve into further motivations and provide examples that are closely associated with these concepts.

4 Examples

We illustrate our main results with some examples.

4.1 Example 1

Note that the equality $\mathscr{H}^s(E) = p^s(E)$ holds only rarely unless they are both 0 or both $+\infty$. Also, it is shown in [23] that the 1-dimensional Hausdorff, packing, and H–S measures have the same value for p-regular sets in $\mathbb{R}^d$. In [34], Taylor and

Tricot construct an example of a compact set E with $\mathcal{H}^s(E) = 0$ and $\mathcal{p}^s(E) = \infty$, i.e., the HOM of a subset E is as small as possible, and the packing measure of E is as big as possible. We give the constriction of the set E for the reader's convenience. The closed nowhere dense set E in the closed disk $D_0 = B(\frac{1}{2}, 0)$ is constructed as follows. For $k \geq 1$, let a_k be a sequence of positive integers and put

$$c_k = \sum_{i=1}^{k} a_i, \quad b_k = \sum_{i=1}^{k} a_{2i}.$$

Consider a closed disk D with radius r. We have two operations that we can perform on D:
* Operation O_1: Replace D with a single concentric disk of radius $\frac{r}{3}$.
* Operation O_2: Replace D with 7 equal disjoint disks of radius $\frac{r}{3}$ packed inside D. Starting with the initial disk D_0, we apply operation O_1 a total of a_1 times, resulting in a closed disk with diameter $\frac{1}{3^{a_1}}$. We then apply operation O_2 to this disk a_2 times, giving us a set E_2 consisting of 7^{a_2} disks, each with diameter $\frac{1}{3^{c_2}}$. Continuing this process, if E_{2k} is a set containing 7^{b_k} closed disks of diameter $\frac{1}{3^{c_{2k}}}$, we obtain E_{2k+1} by applying operation O_1 a_{2k+1} times, resulting in 7^{b_k} disks of diameter $\frac{1}{3^{c_{2k+1}}}$. Similarly, E_{2k+2} is obtained by applying operation O_2 a_{2k+2} times to each disk in E_{2k+1}, resulting in $7^{b_{k+1}}$ disks of diameter $\frac{1}{3^{c_{2k+2}}}$. The set E is defined as the intersection of all the sets E_i for $i = 1, 2, 3, \ldots$. For each k, it can be shown that E is covered by E_{2k+1}.

$$\mathcal{H}^1(E) \leq \liminf_{k \to \infty} 7^{b_k - c_{2k+1}}$$

and the set E can be represented by a packing consisting of the disks in E_{2k}. This implies that

$$\mathcal{p}^1(E) \geq \limsup_{k \to \infty} 7^{b_k - c_{2k}}.$$

Now, fix an integer $n \geq 2$ and put $a_k = N^{k-1}$ then

$$b_k = \frac{N^{2k+2} - 1}{N^2 - 1}, \quad c_k = \frac{N^{k+1} - 1}{N - 1},$$

which implies that $\mathcal{H}^1(E) = 0$ and $\mathcal{p}^1(E) = \infty$. Also, if we can choose a rapidly increasing sequence a_k, for example, $a_k = k!$, we can show that

$$\mathcal{D}_H(E) = 0 \quad \text{and} \quad \mathcal{D}_P(E) = \frac{\log 7}{\log 3}.$$

A more complex variation of the construction, in which the operation O_2 is replaced by an operation involving an increasing number of disks, would result in a set E with

$$\mathscr{D}_H(E) = 0 \quad \text{and} \quad \mathscr{D}_P(E) = 2.$$

4.2 Example 2

Let E_0 be a closed equilateral triangle with a side length of 1. We define E_k as a set consisting of 3^k equal triangles, each with a side length of 3^{-k}. To obtain E_{k+1} from E_k, we replace each triangle in E_k with 3 equal triangles, each with a side length of 3^{-k+1}, as illustrated in Fig. 2. Finally, we define the set E as follows:

$$E = \bigcap_{k=0}^{\infty} E_k.$$

In other words, E can be considered as the complement set of the osculatory packing of E_0 using open hexagons. It has been shown in [34] that

$$\mathscr{H}^1(E) = 1 < 4 = p^1(E).$$

4.3 Example 3

Let $\mathscr{F}_n$ be the set of the 3-adic intervals of the nth generation of $[0, 1]$ and let $a_1, a_2, a_3 \in (0, 1)$ such that $a_1 + a_2 + a_3 = 1$. Let $(T_k)_{k \geq 1}$ be a sequence of integers such that

$$T_1 = 1, \quad T_k < T_{k+1}, \quad \text{and} \quad \lim_{k \to +\infty} \frac{T_{k+1}}{T_k} = +\infty.$$

Fig. 2 The constriction of the set E

Now, if $\left(I_{\varepsilon_1\cdots\varepsilon_n}\right)_{\varepsilon_1,\ldots,\varepsilon_n\in\{0,1,2\}}$ are the 3^n intervals of $\mathscr{F}_n$, we will define the measure ν with the following transitions:

1. $\nu\left(I_0\right)=a_1$, $\nu\left(I_1\right)=a_2$, and $\nu\left(I_2\right)=a_3$.
2. If $T_{2n-1}\leq k+1<T_{2n}$, then

$$\frac{\nu\left(I_{\varepsilon_1\cdots\varepsilon_{k+1}}\right)}{\nu\left(I_{\varepsilon_1\cdots\varepsilon_k}\right)}=\begin{cases}a_1 \text{ if } \varepsilon_{k+1}=\varepsilon_k,\\ a_2 \text{ if } \varepsilon_{k+1}=\varepsilon_k+1,\\ a_3 \text{ if } \varepsilon_{k+1}=\varepsilon_k+2\ [\text{mod}.3].\end{cases}$$

3. If $T_{2n}\leq k+1<T_{2n+1}$, then

$$\frac{\nu\left(I_{\varepsilon_1\cdots\varepsilon_{k+1}}\right)}{\nu\left(I_{\varepsilon_1\cdots\varepsilon_k}\right)}=\begin{cases}1/2 \text{ if } \varepsilon_{k+1}=0 \text{ or } 2,\\ 0 \quad \text{ if } \varepsilon_{k+1}=1.\end{cases}$$

Denote $E:=\operatorname{supp}(\nu)$. By using similar techniques as [3], we obtain that

$$\mathscr{D}_H(E)=\frac{\log 2}{\log 3}<\mathscr{D}_P(E)=1.$$

4.4 Example 4

In the ensuing discussion, we form a metric space Ω intending to minimize its Hausdorff dimension while maximizing its packing dimension.

For $n=1,2,3,\ldots$ let $k_n\in\{2,3,4,5,\ldots\}$ and $r_n\in(0,1/2)$. Let T_n be a set with k_n elements (with the discrete topology). Let

$$K_n=k_1k_2\ldots k_n,\quad R_n=r_1r_2\ldots r_n$$

such that $K_n\nearrow\infty$ and $R_n\searrow 0$ as $n\to+\infty$. Also, let $R_0=1$. Our space is $\Omega=\prod_{n=1}^{\infty}T_n$ with the product topology. So Ω is separable and compact. Define metric d on Ω as follows:

Let $x=(x_1,x_2,\ldots)$, $y=(y_1,y_2,\ldots)\in\Omega$. For $x=y$, define $d(x,y)=0$. For $x\neq y$, let $m\in\mathbb{N}$ be such that

$$x_j=y_j\quad\text{for } 1\leq j\leq m,\quad\text{and } x_{m+1}\neq y_{m+1}$$

and define $d(x,y)=R_m$. Then Ω has diameter R_0. Ω is the disjoint union of k_1 closed subsets $\Omega\left[a_1\right]$, $a_1\in T_1$,

$$\Omega\left[a_1\right]=\{(x_1,x_2,\ldots)\in\Omega: x_1=a_1\}\text{ has diameter } R_1.$$

Each set $\Omega[a_1]$ is the disjoint union of k_2 closed subsets $\Omega[a_1, a_2], a_2 \in T_2$,

$$\Omega[a_1, a_2] = \{(x_1, x_2, \ldots) \in \Omega : x_1 = a_1, x_2 = a_2\} \text{ has diameter } R_2.$$

Continue in this way, so that each $\Omega[a_1, \ldots, a_m]$ is the disjoint union of k_{m+1} closed subsets $\Omega\left[a_1, \ldots, a_m, a_{m+1}\right], a_{m+1} \in T_{m+1}$, and $\Omega[a_1, \ldots, a_m]$ has diameter R_m. For each m, the space Ω is the disjoint union of K_m closed sets of diameter R_m. Define a "uniform" measure ν on T so that

$$\nu\left(\Omega[a_1, \ldots, a_m]\right) = \frac{1}{K_m}.$$

Now, let $s \in (0, \infty)$. The upper s-density of ν at a point $x \in E$ is

$$\begin{aligned}\overline{d}_\nu^s(x) &= \limsup_{\eta \to 0} \frac{\nu\left(B(x, \eta)\right)}{\left(\operatorname{diam} B(x, \eta)\right)^s} \\ &= \limsup_{n \to \infty} \frac{1/K_n}{R_n^s} \\ &= \limsup_{n \to \infty} \frac{1}{K_n R_n^s}.\end{aligned}$$

The lower density is

$$\underline{d}_\nu^s(x) = \liminf_{n \to \infty} \frac{1}{K_n R_n^s}.$$

Now, what we need to do is select k_n, r_n at the beginning so that, for all $s \in (0, \infty)$ we have

$$\limsup_{n \to \infty} \frac{1}{K_n R_n^s} = +\infty, \quad \liminf_{n \to \infty} \frac{1}{K_n R_n^s} = 0.$$

Then the metric space (Ω, d) will have Hausdorff dimension 0 and packing dimension ∞.

4.5 Example 5

Consider the following set

$$\mathcal{A}_u^r = \left\{ \sum_{n=0}^{\infty} a_n r^n : a_n \in \{0, 1, u\} \right\}.$$

Let us examine a constant value $u \in \mathbb{R}$, using r as a parameter. As we are specifically interested in sets with Hausdorff dimensions below one, we confine our investigation to scenarios where $r < \frac{1}{3}$. Without loss of generality, let's assume that $u \geq 2$. In accordance with [27, Theorem 2.1], the ensuing result holds

$$\mathscr{H}^{s(r)}\left(\mathcal{A}_u^r\right) = 0 \quad \text{and} \quad 0 < \mathscr{P}^{s(r)}\left(\mathcal{A}_u^r\right) < \infty,$$

with $s(r) = \frac{\log 3}{|\log r|}$, for all $u \in (2, 4)$ and a.e. $r \in \left(\frac{1}{1+u}, \frac{1}{3}\right)$.

4.6 Example 6

Consider the set $C_r^2 = C_r \times C_r$, where C_r represents the middle-α Cantor set with $\alpha = 1 - 2r$. It can be observed that the family of projections $\{\text{proj}_\theta C_r^2\}_{0 \leq \theta < \pi}$ is affine-equivalent to the self-similar sets generated by the iterated function systems $\{r_x, r_x + 1, r_x + u, r_x + 1 + u\}_{u>0}$. To ensure $\mathscr{D}_\text{H}(C_r^2) < 1$, we assume that $r < \frac{1}{4}$. Without loss of generality, we may consider $u \geq 1$, as demonstrated in Example 5 (Sect. 4.5). According to [27], for every $r \in \left(\frac{1}{6}, \frac{1}{4}\right)$ and for almost every $\theta \in \left[\arctan \frac{1-2r}{r}, \arctan \frac{2}{1-3r}\right]$, we obtain the following result:

$$\mathscr{H}^s\left(\text{proj}_\theta C_r^2\right) = 0 \quad \text{and} \quad 0 < \mathscr{P}^s\left(\text{proj}_\theta C_r^2\right) < \infty,$$

where s is the similarity dimension $\frac{\log 4}{\log r^{-1}}$.

Remark 4 As established by [1, 20], the upper H–S dimension aligns with the packing dimension. Nevertheless, it should be noted that, in general, the packing measure $\mathscr{P}^s$ and the upper H–S measure $\mathscr{v}^s$ do not exhibit concurrence. The lower H–S dimension essentially serves as an interpolation between the original Hausdorff and packing dimensions. So, whenever, these two dimensions are equal, there is somehow no motivation to investigate the interpolations as they will be automatically the same. We are interested in studying a strictly intermediate case, where the classical dimensions diverge from each other. The rationale and significance of this study stem from various instances found in the literature where one or more of the mentioned inequalities are distinct. In other words, the lower H–S dimension differs from both the Hausdorff and packing dimensions. We refer, for example, [21, Proposition 2], [22, Theorem 2], [23, Example 20] and [25, Theorems 1.1 and 6.1].

4.7 Example 7

Consider the probability space $\Omega = [0, 1]$, with the Lebesgue measure P serving as the unconditional probability measure. Let $A = \left[0, \frac{1}{4}\right]$ be an event, and $B = \left\{0, \frac{1}{2}, \frac{1}{3}, \frac{1}{4}, \frac{1}{5}\right\}$ represent a finite set, characterizing an unexpected event related to C. Remarkably, the fractal dimensions of B are all uniformly zero. The associated fractal measures of order 0 are described by the counting measure, which, in this case, amounts to 5. Consequently, we have $h^0(B) = u^0(B) = p^0(B) = 5$. Based on the models proposed in Theorems 1, 4, and 6, we can compute the conditional probability $C(A|B)$ using the counting measures h^0, p^0, and u^0 so that

$$C(A|B) = \frac{h^0(A \cap B)}{h^0(B)} = \frac{u^0(A \cap B)}{u^0(B)} = \frac{p^0(A \cap B)}{p^0(B)} = \frac{3}{5}.$$

4.8 Example 8

Consider the probability space $\Omega = [0, 1]$ with the Lebesgue measure C as the unconditional probability measure. Consider the event $A = \left[0, \frac{1}{4}\right]$ and let B denote the set of rational numbers within the interval $[0, 1]$. Due to B being an infinite set with a probability of zero under C, it manifests as an unforeseen event. The fractal dimensions of B are all equal to 0, indicating that B is a set with zero-dimensional measure. The corresponding measures of order 0 are given by the counting measure. However, in this case, we have $h^0(B) = u^0(B) = p^0(B) = +\infty$, as the set B is infinite. Utilizing the frameworks presented in Theorems 1, 4, and 6, the conditional probability $C(A|B)$ can be computed using a finitely additive probability measure m_B that takes values in the range of 0 to 1 but is not countably additive.

5 CUCP Defined by Average Dimensional H–S Measures

For a subset E of a metric space Ω and $s > 0$, the s-box-counting function $\psi_E^s : (0, \infty) \to [0, \infty]$ of E is defined as $\psi_E^s(t) = \mathscr{M}_{e^{-t}}(E)(2e^{-t})^s$, where $\mathscr{M}_{e^{-t}}(E)$ denotes the e^{-t}-packing number of E. The H–S pre-measures of E can be expressed as follows:

$$\overline{u}^s(E) = \liminf_{t \to \infty} \psi_E^s(t)$$

and by

$$\overline{v}^s(E) = \limsup_{t \to \infty} \psi_E^s(t).$$

Definition ([26, 31]) An averaging system is a family $\Xi = (\Xi_t)_{t \geq t_0 > 0}$ such that

1. Ξ_t constitutes a finite Borel measure over the interval $[t_0, \infty)$.
2. The support of Ξ_t is compact.
3. The consistency condition holds when a positive measurable function $\psi : [t_0, \infty) \to [0, \infty)$ exists, and there is a real number α such that $\psi(t) \to \alpha$ as $t \to \infty$, then $\int \psi d\Xi_t \to \alpha$ as $t \to \infty$.

Consider a positive measurable function $\psi : [t_0, \infty) \to [0, \infty)$. The lower and upper Ξ-averages of ψ are defined as follows:

$$\underline{A}_{\Xi}\psi = \liminf_{t\to\infty} \int \psi d\Xi_t$$

and

$$\overline{A}_{\Xi} f = \limsup_{t\to\infty} \int \psi d\Xi_t.$$

By utilizing average systems on the box-counting function $\psi_E^s(t)$, we arrive at the concept of average H–S measures, which was introduced in [26, 31]. Consider $s > 0$ and $\Xi = (\Xi_t)_{t\geq t_0}$ as an averaging system. In the context of a metric space Ω and a subset $E \subseteq \Omega$, we establish the lower and upper Ξ-average s-dimensional H–S pre-measures of E in the following manner:

$$\overline{\mathcal{U}}_{\Xi}^s(E) = \underline{A}_{\Xi}\psi_E^s = \liminf_{t\to\infty} \int \psi_E^s d\Xi_t$$

and

$$\overline{\mathcal{V}}_{\Xi}^s(E) = \overline{A}_{\Xi}\psi_E^s = \limsup_{t\to\infty} \int \psi_E^s d\Xi_t.$$

The lower and upper s-dimensional H–S measures of E with respect to the Ξ-average are defined, respectively, in [26, 31] as follows:

$$\mathcal{U}_{\Xi}^s(E) = \inf\left\{\sum_{i=1}^{\infty} \overline{\mathcal{U}}_{\Xi}^s(E_i) \;\middle|\; E \subseteq \bigcup_{i=1}^{\infty} E_i\right\}$$

and

$$\mathcal{V}_{\Xi}^s(E) = \inf\left\{\sum_{i=1}^{\infty} \overline{\mathcal{V}}_{\Xi}^s(E_i) \;\middle|\; E \subseteq \bigcup_{i=1}^{\infty} E_i\right\}.$$

Remark 5 It is worth noting that the H–S measures can be considered as special cases of average H–S measures. In particular, if we define the average system Ξ as $\Xi = (\delta_t)_{t\geq 1}$, where δ_t represents the Dirac measure concentrated at t, then for any subset E of Ω we have the following relationship:

$$\overline{\mathcal{U}}^s(E) = \overline{\mathcal{U}}_{\Xi}^s(E), \quad \overline{\mathcal{V}}^s(E) = \overline{\mathcal{V}}_{\Xi}^s(E)$$

and

$$\mathcal{U}^s(E) = \mathcal{U}_{\Xi}^s(E), \quad \mathcal{V}^s(E) = \mathcal{V}_{\Xi}^s(E).$$

Subsequently, we outline the essential inequalities applicable to the H–S measures, average H–S measures, HOMs, and packing measures. These inequalities hold for any metric space Ω and any subset $E \subseteq \Omega$, as presented in [31],

$$\mathcal{H}^s(E) \leq u^s(E)$$

and

$$\overline{u}^s(E) \leq \overline{u}^s_\Xi(E) \leq \overline{v}^s_\Xi(E) \leq \overline{v}^s(E) \leq \overline{\mathcal{P}}^s(E)$$
$$\vee| \qquad \vee| \qquad \vee| \qquad \vee| \qquad \vee|$$
$$u^s(E) \leq u^s_\Xi(E) \leq v^s_\Xi(E) \leq v^s(E) \leq p^s(E).$$

Now, applying the average systems to the box-counting function $\psi^s_E(t)$ to obtain the definition of the average H–S dimensions. Let $\Xi = (\Xi_t)_{t \geq t_0}$ be an averaging system. For a subset $E \subseteq \Omega$, the lower and upper Ξ-average H–S dimensions of E are defined by

$$\underline{\mathcal{D}}_{\Xi,MB}(E) = \inf \left\{ s \geq 0 \mid u^s_\Xi(E) = 0 \right\} = \sup \left\{ s \geq 0 \mid u^s_\Xi(E) = +\infty \right\}$$

and

$$\overline{\mathcal{D}}_{\Xi,MB}(E) = \inf \left\{ s \geq 0 \mid v^s_\Xi(E) = 0 \right\} = \sup \left\{ s \geq 0 \mid v^s_\Xi(E) = +\infty \right\}.$$

It is clear from [31] that

$$\mathcal{D}_H(E) \leq \underline{\mathcal{D}}_{MB}(E) \leq \underline{\mathcal{D}}_{\Xi,MB}(E) \leq \overline{\mathcal{D}}_{\Xi,MB}(E) \leq \overline{\mathcal{D}}_{MB}(E) = \mathcal{D}_P(E).$$

Remark 6 Note that the H–S dimensions are, in fact, average H–S dimensions. Specifically, considering a metric space X and the average system denoted by $\Xi = (\delta_t)_{t \geq 1}$ (where δ_t signifies the Dirac measure concentrated at t), it follows that, for all subsets E of Ω

$$\underline{\mathcal{D}}_{\Xi,MB}(E) = \underline{\mathcal{D}}_{MB}(E)$$

and

$$\overline{\mathcal{D}}_{\Xi,MB}(E) = \overline{\mathcal{D}}_{MB}(E).$$

The following results provide an extension of Theorems 5 and 6, and show that not only is the function $\psi^{s(t)}_X = \mathcal{M}_{e^{-t}}(X)\left(2e^{-t}\right)^s$ defines a CUCPr as $t \to \infty$. More precisely, we have

Theorem 7 *Take a metric space (Ω, d) and a partition $\mathbf{B}$ of Ω. For each $B \in \mathbf{B}$, designate s as the lower average H–S dimension of the conditioning event B, and let u^s_Ξ denote the lower average H–S s-dimensional outer measure.*

1. *Suppose m_B is a finitely additive probability, taking values of 0 or 1, but not countably additive, defined on the power set $\wp(B)$. For every $B \in \mathbf{B}$, the functional $\overline{C}(X|B)$ defined on the linear space $\mathscr{L}(B)$ is given by*

$$\overline{C}(X|B) = \begin{cases} \dfrac{1}{\mathcal{u}_{\Xi}^{s}(B)} \displaystyle\int_B X d\mathcal{u}_{\Xi}^{s} \; if \; 0 < \mathcal{u}_{\Xi}^{s}(B) < +\infty, \\ \displaystyle\int_B X dm_B \; if \; \mathcal{u}_{\Xi}^{s}(B) \in \{0, +\infty\} \end{cases}$$

is a CUCP.

2. *Consider m_B, a finitely additive probability taking values of 0 or 1, and not being countably additive, defined on the power set $\wp(B)$. Consequently, for every $B \in \mathbf{B}$, the function defined on $\wp(B)$ is given by*

$$\overline{C}(A|B) = \begin{cases} \dfrac{\mathcal{u}_{\Xi}^{s}(A \cap B)}{\mathcal{u}_{\Xi}^{s}(B)} \; if \; 0 < \mathcal{u}_{\Xi}^{s}(B) < +\infty, \\ m_B \quad if \; \mathcal{u}_{\Xi}^{s}(B) \in \{0, +\infty\} \end{cases}$$

is a CUCPr.

3. *The same results hold for the upper average H–S measure and dimension.*

6 CUCP Defined by HCA of Dimensional H–S Measures

We revisit the concept of Hölder and Cesaro averages (HCA) of H–S measures, as introduced in [26, 31]. These averages utilize two popular averaging systems: HCA. We will provide the definitions of these systems and employ them with the function $\psi_E^s(t) = \mathscr{M}_{e^{-t}}(E)\left(2e^{-t}\right)^s$, where E is a subset of a metric space Ω. Let $a > 0$ and $\psi : (a, \infty) \to [0, \infty)$ be a positive measurable function, the function $M\psi : (a, \infty) \to [0, \infty)$ is defined by

$$(M\psi)(t) = \frac{1}{t}\int_a^t \psi(s)ds.$$

Let $n \in \mathbb{N} - \{0\}$. Subsequently, the lower and upper nth-order Hölder averages of ψ are established as follows:

$$\underline{H}_n\psi = \liminf_{t\to\infty} \left(M^n\psi\right)(t)$$

and

$$\overline{H}_n\psi = \limsup_{t\to\infty} \left(M^n\psi\right)(t).$$

In a similar way, we will define the Cesaro averages as follows: we define $I\psi : (a,\infty) \to [0,\infty)$ by

$$(I\psi)(t) = \int_a^t \psi(s)ds.$$

For $n \in \mathbb{N} - \{0\}$, we proceed to define the lower and upper n'th-order Cesaro averages of ψ as follows:

$$\underline{C}_n\psi = \liminf_{t\to\infty} \frac{n!}{t^n}\left(I^n\psi\right)(t)$$

and

$$\overline{C}_n\psi = \limsup_{t\to\infty} \frac{n!}{t^n}\left(I^n\psi\right)(t).$$

The HCA exhibit the following inequalities (as stated in [26]):

$$\liminf_{t\to\infty}\psi(t) = \underline{H}_0\psi \le \underline{H}_1\psi \le \underline{H}_2\psi \le \cdots \le \overline{H}_2\psi \le \overline{H}_1\psi \le \overline{H}_0\psi = \limsup_{t\to\infty}\psi(t)$$

and

$$\liminf_{t\to\infty}\psi(t) = \underline{C}_0\psi \le \underline{C}_1\psi \le \underline{C}_2\psi \le \cdots \le \overline{C}_2\psi \le \overline{C}_1\psi \le \overline{C}_0\psi = \limsup_{t\to\infty}\psi(t).$$

The average H–S measures, employing HCA, are delineated in [26, 31] as follows: For a positive integer n, the averaging system is defined as $\Xi_n^{\mathrm{H}} = \left(\Xi_{n,t}^{\mathrm{H}}\right)_{t\ge a}$

$$\Xi_{n,t}^{\mathrm{H}}(A) = \frac{1}{(n-1)!t}\int_{[a,t]\cap A} (\log t - \log s)^{n-1} ds$$

for any Borel subsets A of $[a,\infty)$, which implies that

$$\underline{H}_n\psi = \liminf_{t\to} \int \psi d\Xi_{n,t}^{\mathrm{H}}$$

and

$$\overline{H}_n\psi = \limsup_{t\to\infty} \int \psi d\Xi_{n,t}^{\mathrm{H}}.$$

Similarly, for $n \in \mathbb{N} - \{0\}$ we define the averaging system $\Xi_n^{\mathrm{C}} = \left(\Xi_{n,t}^{\mathrm{C}}\right)_{t\ge a}$, for any Borel subsets A of $[a,\infty)$, by

$$\Xi_{n,t}^{C}(A) = \frac{n}{t^n}\int_{[a,t]\cap A} (t-s)^{n-1} ds.$$

This gives that

$$\underline{C}_n \psi = \liminf_t \int \psi d\Xi^{\mathrm{C}}_{n,t}$$

and

$$\overline{C}_n \psi = \limsup_t \int \psi d\Xi^{\mathrm{C}}_{n,t}.$$

Consider Ω as a metric space, and let n be an integer with $n \geq 0$. The introduction of average Hölder and Cesaro H–S pre-measures and measures can be accomplished by employing the definitions of HCA on $\psi^s_X(t) = \mathscr{M}_{e^{-t}}(X)\left(2e^{-t}\right)^s$. These measures are denoted as $\overline{\mathcal{U}}^{\mathrm{H},s}_n(E)$, $\overline{\mathcal{V}}^{\mathrm{H},s}_n(E)$, $\mathcal{U}^{\mathrm{H},s}_n(E)$, and $\mathcal{V}^{\mathrm{H},s}_n(E)$, and they are defined as follows:

$$\overline{\mathcal{U}}^{\mathrm{H},s}_n(E) = \overline{\mathcal{U}}^s_{\Xi^H_n}(E), \quad \mathcal{U}^{\mathrm{H},s}_n(E) = \mathcal{U}^s_{\Xi^H_n}(E)$$

and

$$\overline{\mathcal{V}}^{\mathrm{H},s}_n(E) = \overline{\mathcal{V}}^s_{\Xi^H_n}(E), \quad \mathcal{V}^{\mathrm{H},s}_n(E) = \mathcal{V}^s_{\Xi^H_n}(E).$$

We can also define the lower and upper n'th-order average Cesaro H–S pre-measures and measures of a subset E of Ω, which are denoted as $\overline{\mathcal{U}}^{\mathrm{C},s}_n(E)$, $\overline{\mathcal{V}}^{\mathrm{C},s}_n(E)$, $\mathcal{U}^{\mathrm{C},s}_n(E)$, and $\mathcal{V}^{\mathrm{C},s}_n(E)$, respectively. These measures are defined as

$$\overline{\mathcal{U}}^{\mathrm{C},s}_n(E) := \overline{\mathcal{U}}^s_{\Xi^C_n}(E), \quad \mathcal{U}^{\mathrm{C},s}_n(E) := \mathcal{U}^s_{\Xi^C_n}(E)$$

and

$$\overline{\mathcal{V}}^{\mathrm{C},s}_n(E) := \overline{\mathcal{V}}^s_{\Xi^C_n}(E), \quad \mathcal{V}^{\mathrm{C},s}_n(E) := \mathcal{V}^s_{\Xi^C_n}(E).$$

The hierarchical structure of higher order average Hölder and Cesaro H–S pre-measures and measures lies between the classical fractal measures, comprising countably infinite levels. In simpler terms, as outlined in [32]

$$\mathscr{H}^s(E) \leq \mathcal{U}^s(E) = \mathcal{U}^{\mathrm{H},s}_0(E) \leq \mathcal{U}^{\mathrm{H},s}_1(E) \leq \cdots \leq \mathcal{V}^{\mathrm{H},s}_1(E) \leq \mathcal{V}^{\mathrm{H},s}_0(E) = \mathcal{V}^s(E) \leq \mathcal{P}^s(E),$$
$$\mathscr{H}^s(E) \leq \mathcal{U}^s(E) = \mathcal{U}^{C,s}_0(E) \leq \mathcal{U}^{C,s}_1(E) \leq \cdots \leq \mathcal{V}^{C,s}_1(E) \leq \mathcal{V}^{C,s}_0(E) = \mathcal{V}^s(E) \leq \mathcal{P}^s(E),$$

and

$$\overline{\mathcal{U}}^s(E) = \overline{\mathcal{U}}^{\mathrm{H},s}_0(E) \leq \overline{\mathcal{U}}^{\mathrm{H},s}_1(E) \leq \cdots \leq \overline{\mathcal{V}}^{\mathrm{H},s}_1(E) \leq \overline{\mathcal{V}}^{\mathrm{H},s}_0(E) = \overline{\mathcal{V}}^s(E) \leq \overline{\mathcal{P}}^s(E),$$
$$\overline{\mathcal{U}}^s(E) = \overline{\mathcal{U}}^{C,s}_0(E) \leq \overline{\mathcal{U}}^{C,s}_1(E) \leq \cdots \leq \overline{\mathcal{V}}^{C,s}_1(E) \leq \overline{\mathcal{V}}^{C,s}_0(E) = \overline{\mathcal{V}}^s(E) \leq \overline{\mathcal{P}}^s(E).$$

Subsequently, we introduce the lower and upper nth-order average Hölder H–S dimensions of E, denoted, respectively, by $\underline{\mathscr{D}}^{\mathrm{H}}_{\mathrm{MB},n}(E)$ and $\overline{\mathscr{D}}^{\mathrm{H}}_{\mathrm{MB},n}(E)$, as follows:

$$\underline{\mathscr{D}}^{\mathrm{H}}_{\mathrm{MB},n}(E) = \inf\left\{s \geq 0 \;\middle|\; \mathcal{U}^{\mathrm{H},s}_n(E) = 0\right\} = \sup\left\{s \geq 0 \;\middle|\; \mathcal{U}^{\mathrm{H},s}_n(E) = +\infty\right\}$$

and

$$\overline{\mathscr{D}}^{\mathrm{H}}_{\mathrm{MB},n}(E) = \inf\left\{s \geq 0 \,\middle|\, \upsilon_n^{\mathrm{H},s}(E) = 0\right\} = \sup\left\{s \geq 0 \,\middle|\, \upsilon_n^{\mathrm{H},s}(E) = +\infty\right\}.$$

Now, we proceed to define the lower and upper nth-order average Cesaro H–S dimensions of E, denoted, respectively, by $\underline{\mathscr{D}}^{\mathrm{C}}_{\mathrm{MB},n}(E)$ and $\overline{\mathscr{D}}^{\mathrm{C}}_{\mathrm{MB},n}(E)$, as follows:

$$\underline{\mathscr{D}}^{\mathrm{C}}_{\mathrm{MB},n}(E) = \inf\left\{s \geq 0 \,\middle|\, u_n^{\mathrm{C},s}(E) = 0\right\} = \sup\left\{s \geq 0 \,\middle|\, u_n^{\mathrm{C},s}(E) = +\infty\right\}$$

and

$$\overline{\mathscr{D}}^{\mathrm{C}}_{\mathrm{MB},n}(E) = \inf\left\{s \geq 0 \,\middle|\, \upsilon_n^{\mathrm{C},s}(E) = 0\right\} = \sup\left\{s \geq 0 \,\middle|\, \upsilon_n^{\mathrm{C},s}(E) = +\infty\right\}.$$

The hierarchical structure of higher order average Hölder and Cesaro H–S dimensions establishes a continuum between the classical fractal dimensions, encompassing countably infinite levels. In simpler terms, this relationship is outlined in [32] as

$$\underline{\mathscr{D}}_{\mathrm{MB}}(E) = \underline{\mathscr{D}}^{\mathrm{H}}_{\mathrm{MB},0}(E) \leq \underline{\mathscr{D}}^{\mathrm{H}}_{\mathrm{MB},1}(E) \leq \cdots \leq \overline{\mathscr{D}}^{\mathrm{H}}_{\mathrm{MB},1}(E) \leq \overline{\mathscr{D}}^{\mathrm{H}}_{\mathrm{MB},0}(E) = \overline{\mathscr{D}}_{\mathrm{MB}}(E)$$

and

$$\underline{\mathscr{D}}_{\mathrm{MB}}(E) = \underline{\mathscr{D}}^{\mathrm{C}}_{\mathrm{MB},0}(E) \leq \underline{\mathscr{D}}^{\mathrm{C}}_{\mathrm{MB},1}(E) \leq \cdots \leq \overline{\mathscr{D}}^{\mathrm{C}}_{\mathrm{MB},1}(E) \leq \overline{\mathscr{D}}^{\mathrm{C}}_{\mathrm{MB},0}(E) = \overline{\mathscr{D}}_{\mathrm{MB}}(E).$$

Utilizing the insights from Theorem 7, we demonstrate the subsequent findings.

Theorem 8 *Consider a metric space* (Ω, d)*, and let* $\mathbf{B}$ *constitute a partition of* Ω*.*

1. *Let* $\mathbf{B}$ *be a partition of* Ω*, and for each* $B \in \mathbf{B}$ *and* $n \in \mathbb{N} \cup 0$*, let* s *denote the lower n'th-order average Hölder H–S dimension of the conditioning event* B*, and let* $u_n^{H,s}$ *denote the lower n'th-order average Hölder H–S* s*-dimensional outer measure.*

   a. *Suppose* m_B *is a finitely additive probability, taking values of 0 or 1 and not being countably additive, defined on* $\wp(B)$*. For every* $B \in \mathbf{B}$*, the functional* $\overline{C}(X|B)$ *is defined on* $\mathscr{L}(B)$ *as follows:*

$$\overline{C}(X|B) = \begin{cases} \dfrac{1}{u_n^{H,s}(B)} \displaystyle\int_B X du_n^{H,s} & \textit{if } \ 0 < u_n^{H,s}(B) < +\infty, \\ \displaystyle\int_B X dm_B & \textit{if } \ u_n^{H,s}(B) \in \{0, +\infty\} \end{cases}$$

   is a CUCP.

b. *Consider m_B, a finitely additive probability with values of 0 or 1, but not being countably additive, defined on $\wp(B)$. Therefore, for each $B \in \mathbf{B}$, the function defined on $\wp(B)$ is given by*

$$\overline{C}(A|B) = \begin{cases} \dfrac{u_n^{H,s}(A \cap B)}{u_n^{H,s}(B)} & \text{if } 0 < u_n^{H,s}(B) < +\infty, \\ m_B & \text{if } u_n^{H,s}(B) \in \{0, +\infty\} \end{cases}$$

is a CUCPr.

2. *For every $B \in \mathbf{B}$ and $n \in \mathbb{N} \cup 0$, let s denote the lower n'th-order average Cesaro H–S dimension of the conditioning event B, and let $u_n^{C,s}$ denote the lower n'th-order average Cesaro H–S s-dimensional outer measure.*

   a. *Consider a 0–1 valued finitely additive probability, denoted as m_B, which is not countably additive, defined on the power set $\wp(B)$. For every $B \in \mathbf{B}$, the functional $\overline{C}(X|B)$ is defined on $\mathcal{L}(B)$ as follows:*

$$\overline{C}(X|B) = \begin{cases} \dfrac{1}{u_n^{C,s}(B)} \displaystyle\int_B X du_n^{C,s} & \text{if } 0 < u_n^{C,s}(B) < +\infty, \\ \displaystyle\int_B X dm_B & \text{if } u_n^{C,s}(B) \in \{0, +\infty\} \end{cases}$$

   is a CUCP.

   b. *Consider a 0–1 valued finitely additive probability, denoted as m_B, which lacks countable additivity and is defined on the power set $\wp(B)$. Therefore, for every $B \in \mathbf{B}$, the function defined on $\wp(B)$ is given by*

$$\overline{C}(A|B) = \begin{cases} \dfrac{u_n^{C,s}(A \cap B)}{u_n^{C,s}(B)} & \text{if } 0 < u_n^{C,s}(B) < +\infty, \\ m_B & \text{if } u_n^{C,s}(B) \in \{0, +\infty\} \end{cases}$$

   is a CUCPr.

3. *The same results hold for the upper higher order average Hölder and Cesaro H–S measures and dimensions for all $n \in \mathbb{N} \cup \{0\}$.*

Consider a partition **B** of Ω. The CUCP $\overline{C}(X|\mathbf{B})$ is a random variable defined on Ω as $\overline{C}(X|\mathbf{B}) = \overline{C}(X|B)$ if $\omega \in B$. The disintegration is valid on the partition **B** if

$$\overline{C}(\overline{C}(X|\mathbf{B})) = \overline{C}(X).$$

In the work by Doria [13], it has been demonstrated that when CUCP are established using HOMs and **B** represents an $\hbar^s$-measurable partition, where s denotes the Hausdorff dimension of Ω, then the CUCP, defined as per Theorem 1, adheres to the disintegration property. This finding holds true when coherent conditional previsions are constructed using the centered HOM, packing outer measure, and lower H–S s-dimensional outer measure. Notably, in these scenarios, the disintegration property is satisfied for every Borel partition due to the metric nature of these outer measures. However, the disintegration property does not hold for all Borel partitions of Ω when the CUCP is defined by the upper H–S s-dimensional outer measure. This discrepancy arises because the latter measure is not metric, rendering Borel sets not measurable with respect to it.

7 Conclusions

This paper introduces models for CUCP and probabilities utilizing various dimensional outer measures. It investigates the interrelations among these models, encompassing the application of HOMs, centered HOMs, packing outer measures, lower H–S s-dimensional outer measures, and Hölder and Cesàro averages of dimensional H–S measures. These outer measures are metric, enabling them to be treated as countably additive measures on the Carathéodory field. As a result, when the conditioning event possesses a positive and finite-dimensional outer measure in its dimension, the CUCP exhibit continuity from below and adhere to the monotone convergence theorem [7]. This facilitates the extension of CUCP to unbounded random variables within the class of all Choquet integrable random variables. Moreover, these models do not preclude indifference between equivalent random variables, as evidenced by instances involving random variables with a geometric distribution. An intriguing observation emerges when the conditioning event possesses an HOM of zero in its Hausdorff dimension, yet maintains a positive and finite packing outer measure in its dimension. This scenario is illustrated in Examples 5 (Sect. 4.5) and 6 (Sect. 4.6). In such instances, instead of defining CUCP through a 0–1 valued finitely additive, but non-countably additive probability, we can opt for CUCP defined in relation to the packing measure. This choice ensures the continuity from below of the CUCP. In summarizing the motivations behind proposing coherent forecasting models based on FOM, we can affirm that they enable the handling of null events, exhibit continuity from below, and demonstrate linearity on the Borel σ-field. These properties collectively contribute to the formulation of a model of conditional previsions capable of capturing certain facets of the unconscious activity of the human brain, often interpreted as bias.

References

1. Attia N, Selmi B (2021) A multifractal formalism for Hewitt-Stromberg measures. J Geom Anal 31:825–862
2. Billingsley P (1986) Probability and measure. Wiley
3. Ben Nasr F, Bhouri I, Heurteaux Y (2002) The validity of the multifractal formalism: results and examples. Adv Math 165:264–284
4. Choquet G (1953) Theory of capacities. Annales de l'Institut Fourier 5:131–295
5. de Finetti B (1970) Probability induction and statistics. Wiley, New York
6. de Finetti B (1974) Theory of probability. Wiley
7. Denneberg D (1994) Non-additive measure and integral. Kluwer Academic Publishers
8. Doria S (2007) Probabilistic independence with respect to upper and lower conditional probabilities assigned by Hausdorff outer and inner measures. Int J Appr Reason 46:617–635
9. Doria S (2012) Characterization of a coherent upper conditional prevision as the Choquet integral with respect to its associated Hausdorff outer measure. Ann Oper Res 195:33–48
10. Doria S (2014) Symmetric coherent upper conditional prevision by the Choquet integral with respect to Hausdorff outer measure. Ann Oper Res 229:377–396
11. Doria S (2017) On the disintegration property of a coherent upper conditional prevision by the Choquet integral with respect to its associated Hausdorff outer measure. Ann Oper Res 256:253–269
12. Doria S, Dutta B, Mesiar R (2018) Integral representation of coherent upper conditional prevision with respect to its associated Hausdorff outer measure: a comparison among the Choquet integral, the pan-integral and the concave integral. Int J Gen Syst 216:569–592
13. Doria S (2021) Disintegration property of coherent upper conditional previsions with respect to Hausdorff outer measures for unbounded random variables. Int J General Syst 50:262–280
14. Doria S (2022) Coherent upper conditional previsions with respect to outer Hausdorff measures and the mathematical representation of the selective attention. In: Information processing and management of uncertainty in knowledge-based systems, IPMU 2022. Communications in Computer and Information Science, vol 1602, pp 667–680
15. Doria S (2023) Coherent conditional previsions with respect to inner and outer Hausdorff measures to represent conscious and unconscious human brain activity. Int J Approx Reas 156:134–146
16. Doria S, Selmi B (2023) Coherent upper conditional previsions defined by fractal outer measures to represent the unconscious activity of human brain. Lecture Notes Comput Sci MDAI 2023(13890):70–82
17. Doria S, Selmi B (2024) Conditional aggregation operators defined by the Choquet integral and the Sugeno integral with respect to general fractal measures. Fuzzy Sets Syst 477:108811. https://doi.org/10.1016/j.fss.2023.108811
18. Dubins LE (1975) Finitely additive conditional probabilities, conglomerability and disintegrations. Ann Probab 3:89–99
19. Edgar GA (2023) Errata for "Integral, probability, and fractal measures". Springer, New York
20. Falconer KJ (1990) Fractal geometry: mathematical foundations and applications. Wiley
21. Haase H (1985) A contribution to measure and dimension of metric spaces. Math Nachr 124:45–55

22. Haase H (1988) The dimension of analytic sets. Acta Universitatis Carolinae. Mathematica et Physica 29:15–18
23. Lee HH, Baek I (1992) The comparison of d-meuasure with packing and Hausdorff measures. Kyungpook Math J 32:523–53
24. Mattila P (1995) Geometry of sets and measures in Euclidian Spaces: fractals and rectifiability. Cambridge University Press, Cambridge
25. Mitchell A, Olsen L (2018) Coincidence and noncoincidence of dimensions in compact subsets of [0, 1]. arXiv:1812.09542v1
26. Olsen L (2019) On average Hewitt-Stromberg measures of typical compact metric spaces. Math Z 293:1201–1225
27. Pöyhtäri T (2013) Self-similar sets of zero Hausdorff measure and positive packing measure. Masters Thesis, University of Oulu
28. Regazzini E (1987) De Finetti's coherence and statistical inference. Ann Stat 15:845–864
29. Regazzini E (1985) Finitely additive conditional probabilities. Rendiconti del Seminario Matem-atico e Fisico di Milano 55:69–89
30. Rogers CA (1970) Hausdorff measures. Cambridge University Press, Cambridge
31. Selmi B (2022) Average Hewitt-Stromberg and box dimensions of typical compact metric spaces. Quaest Math 46:411–444
32. Selmi B (2023) A review on multifractal analysis of Hewitt-Stromberg measures. J Geom Anal 32:1–44
33. Seidenfeld T, Schervish M, Kadane JB (2009) Preferences for equivalent random variables: a price for unbounded utilities. J Math Econ 45:329–340
34. Taylor SJ, Tricot C (1986) The packing measure of rectifiable subsets of the plane. Math Proc Camb Philos Soc 99:285–296
35. Taylor SJ, Tricot C (1985) Packing measure and its evaluation for Brownian paths. Trans Amer Math Soc 288:679–699
36. Tricot C (1982) Two definitions of fractional dimension. Math Proc Camb Philos Soc 91:54–74
37. Walley P (1991) Statistical reasoning with imprecise probabilities. Chapman and Hall, London

Chapter 5
Linear and Non-linear Stretching Surfaces of MHD Casson Nanofluid with Heat and Mass Transfer Analysis

K. Varatharaj, R. Tamizharasi, and N. A. Bakar

Abstract The chapter discusses the magnetohydrodynamic heat and mass transfer flow of a Casson nanofluid over linear and non-linear stretching surfaces embedded in a porous medium. The study takes into account the effects of Brownian motion, thermophoresis, thermal radiation, and chemical reaction. Proper transformations are employed to convert the non-linear partial differential systems into ordinary differential equations. The Runge–Kutta method, in combination with the shooting technique, is utilized for finding the numerical solutions to the momentum, temperature, and concentration equations, the subject to the boundary conditions. The impact of different relevant parameters on the velocity, temperature, and concentration profiles of the fluid is analyzed and the findings are presented in graphical form through plots. The study highlights the intricate dynamics of nanofluids, where the Casson, magnetic, buoyancy ratio, and mixed convection parameters collectively influence behavior. Notably, contrasting effects emerge in the realm of mixed convection, adding depth to our understanding of these influential parameters. It is noted that as the Brownian motion parameter increases, the velocity and temperature profiles of the fluid increase, whereas the opposite behavior is observed in the concentration profile. Moreover, a detailed investigation is conducted on the skin-friction coefficient, the Nusselt number and the Sherwood number and the results are presented in tabular format.

K. Varatharaj · R. Tamizharasi (✉)
Department of Mathematics, School of Advanced Sciences, Vellore Institute of Technology, Vellore 632 014, Tamil Nadu, India
e-mail: tamizharasi.r@vit.ac.in

N. A. Bakar
Department of Science and Mathematics, Centre for Diploma Studies, Universiti Tun Hussein Onn Malaysia, Parit Raja, Johor, Malaysia
e-mail: norhaliza@uthm.edu.my

G. Arulprakash et al. (eds.), *Mathematical Modelling of Complex Patterns Through Fractals and Dynamical Systems*, Studies in Infrastructure and Control,
https://doi.org/10.1007/978-981-97-2343-0_5

1 Introduction

1.1 Introduction on Stretching Sheet

In this chapter, an extensive literature survey has been conducted, focusing on various problem domains that researchers have addressed in the past. The study was carefully done with the goal of understanding how a mix of elements, including linear stretching surfaces and non-linear stretching sheets, might impact the quality of the final product in a typical manufacturing process. Consequently, the research work related to these factors has been thoroughly reviewed and analyzed.

The chapter is divided into six sections, each dedicated to a specific problem identified in the research literature. These problems are as follows:

1. Stretching sheet problem involving Newtonian Fluid.
2. MHD stretching sheet issue using Newtonian and non-Newtonian fluids.
3. Stretching sheet problem involving heat and mass transfer.
4. Stretching sheet problem involving nanofluid.
5. Flow over a linearly stretching sheet.
6. Flow over a non-linear stretching sheet.
7. Practical applications of flows over a stretching surface.

Throughout the survey, the focus was on understanding the existing knowledge and findings related to these problem domains. By exploring the literature on these subjects, the chapter aims to provide a comprehensive understanding of the state of research and the various approaches that have been employed to address these challenging problems.

1.1.1 Stretching Sheet Problem Involving Newtonian Fluid

Various researchers have studied the stretching sheet problem involving different types of liquids, particularly those with non-linear stress-strain relationships. Crane [1] explored the dynamics of a steady, boundary layer flow in an incompressible and viscous fluid, triggered by the linear expansion of a stretchable, flat sheet. Gupta and Gupta [2] explored the heat and the mass exchange occurring in fluid flow over an expanding sheet, which included the effects of suction or injection through a slit, with the scenario based on a moving sheet that does not maintain a constant temperature. Banks [3] analyzed similarity solutions related to fluid flow caused by a stretching surface, identifying a set of solutions that form a one-parameter family akin to the solutions described by the Falkner–Skan equation. Grubka and Bobba [4]. Dutta et al. [5] delved into the distribution of temperature in fluid flow over a uniformly heated stretching surface, extending research on thermal transfer where surface temperature varies according to a power law. Dutta and Gupta [6] addressed the interconnected issues of heat transfer associated with a stretching sheet, specifically examining how the sheet temperature changes with the distance from the slit. Dutta [7] an analytical approach to cool a thin, stretching sheet immersed in a viscous flow, incorporating effects of suction (or) blowing. Chen and Char [8] explored how variations in surface temperature and heat flux, both following a power-law, influence heat transfer within a linearly stretching sheet, also considering the impact of suction and blowing. Soewono et al. [9] delved into non-linear boundary value issues related to flow and the heat transfer over a sheet, where both the thermal conductivity and the heat sink (or) source vary with temperature. Karahalios et al. [10] achieved an exact similarity solution for the time-dependent Navier–Stokes equations in the case of a flat surface undergoing radial stretching. Vajravelu [11] investigated the convective motion and heat transmission characteristics of a viscous liquid producing heat around a vertically stretched surface, taking into consideration the impacts of free convection, suction, or injection.

Kumaran and Ramanaiah [12] investigated the flow of the boundary layer over a sheet stretching in a quadratic manner, and Magyari and Keller [13–17] explored various aspects of boundary layer flows caused by continuous stretching surfaces. Mahapatra et al. [18] analyzed a similarity solution for the steady, asymmetric flow at a stagnation point directed towards a stretching surface. Partha et al. [19] conducted an analysis of mixed convection flow and heat transfer emanating from a vertical surface that stretches exponentially. Liao [20, 21], Liao and Pop [22], and Xu [23] applied the HAM to tackle non-linear issues associated with stretching sheets. They furnished explicitly analytic solutions, including recursive formulas for determining the coefficients. It's important to note that all the above researchers focused on Newtonian liquids. Now, in the following sections, we will review the literature specifically concerning liquids with non-linear stress–strain relationships.

1.1.2 MHD Stretching Sheet Issue Using Newtonian and Non-Newtonian Fluids

In the literature review, several researchers have investigated the problem of MHD boundary layer flow over a stretching sheet with different variations and conditions. Pavlov [24] formulated a precise solution for the unidirectional, steady flow of an electrically charged fluid caused by stretching an elastic sheet, set against a uniform magnetic field crossing the flow. Chakrabarti et al. [25] expanded upon Pavlov's research, exploring the temperature profiles within an MHD boundary layer flow incorporating uniform suction. They employed a similarity solution approach to examine both velocity and heat transfer features, specifically focusing on the effects of uniform wall suction. Kumari et al. [26] investigated how an induced electromagnetic field and the existence of a source (or) sink influence the flow dynamics and thermal transfer properties over a stretching surface. Anderson [27] delved into the MHD flow of Walters B fluid over an extending surface, providing an analytical resolution for the complex non-linear boundary layer equations that characterize this specific flow dynamic.

Vajravelu and Nayfeh [28] examined MHD convective flow as well as heat transmission in a viscous temperature-generating liquid around an infinite vertical extending surface. They included the effect of free convection and heat production/absorption effects. Char [29] an investigation into the MHD flow of a Walters B liquid across a stretching sheet, deriving exact solutions and numerically assessing the heat and mass transfer properties for various parameters. Dandapat et al. [30] delved into the stability of MHD flow over a stretching sheet, discovering that the magnetic field plays a stabilizing role in the flow dynamics. Gupta and Mahapatra [31] researched the steady, 2D stagnation-point motion of a fluid with electrical conductivity across a flat, flexible sheet, demonstrating the variations in velocity at a point due to the magnetic field and the stretching velocity of the sheet. Liao [32] introduced the homotopy analysis method as a solution technique for non-linear MHD viscosity flows of power-law fluids across a stretched sheet, offering explicit analytic solutions for various scenarios. Other researchers, such as Datti et al. [33], Afify [34], Liu [35], and Cortell [36], also contributed to the study of MHD flow and thermal transfer over-stretching sheets with various considerations, such as radiation, chemical reactions and second-grade fluid behavior. Massoudi and Manechy [37] employed numerical methods, including the quasi-linearization technique, to investigate the flow of a second-grade liquid over a stretching sheet.

In summary, the reviewed literature demonstrates the extensive research conducted by various researchers on MHD boundary layer flow over-stretching sheets, considering different fluid behaviors and external influences like magnetic fields, chemical reactions and heat generation.

1.1.3 Stretching Sheet Problem Involving Heat and Mass Transfer

A series of research works conducted by various authors have investigated different aspects of fluid flow, heat transfer and mass transfer over-stretching surface. Layek and colloquies [38] explored the influence of suction/blowing on flow separation and thermal transfer of an incompressible viscous fluid toward a porous stretching surface in a porous medium. Anuar Ishak and co-authors [39] examined the MHD stagnation-point flow towards a variable-temperature stretching surface, employing the Keller-box method to explore the effects on flow and heat transfer. Mahapatra et al. [40] explored the MHD stagnation-point flow of power-law fluids over a stretching sheet using the Homotopy Analysis Method, assessing the impact of different parameters on flow, mass and heat transfer. Kuznetsov et al. [41] investigated nanofluid flow, heat, and mass transfer past a vertical plate, using analytical methods to assess the impact of various parameters on flow and thermal characteristics. Hayat and his teams [42] explored the stagnation point flow over a stretching surface with radiation effects, applying the Homotopy Analysis Method to solve the transformed non-linear ordinary differential equations analytically.

These works were followed by Abbas et al. [43], Mustafa et al. [44], Makinde and Aziz [45], Hassani et al. [46], Kandasamy et al. [47], Bhattacharyya et al. [48], Noghrehabadi et al. [49], Bhattacharyya [50], Turkyilmazoglu and Pop [51] and Ibrahim et al. [52], all of whom investigated various aspects of flow, heat transfer, and mass transfer in different scenarios, using analytical or numerical methods, and studying the effects of different parameters on the flow and heat transfer characteristics. These studies have contributed significantly to the understanding of fluid dynamics and heat transfer phenomena in complex situations, providing valuable insights for practical applications.

1.1.4 Stretching Sheet Problem Involving Nanofluid

Nanofluids are highly valued for their exceptional thermal performance, distinct physical and chemical characteristics, and wide range of industrial and technical applications. As is already widely recognized, nanofluid is a mixture of small (diameter 100nm) nanoparticles in base fluids like water, ethylene glycol, kerosene, etc. that include chemically stable metals like Al, Cu, Ag, Au and Fe as well as oxides like Al_2O_3, CuO, SiO_2, carbides (SiC), nitrides (AIN, SiN), and oxides like SiO_2. Researchers are particularly interested in the heat transfer of boundary layer flow of nanofluids across a stretched sheet due to its important applications in manufacturing. Moreover, nanofluids find widespread use in contemporary science and technology, including nuclear reactors, electronics, biomedicine, transportation, and crystal growth. In particular, they are employed in the heat removal process of industrial applications like the cooling system in electronic devices and the extrusion of plastic and rubber sheets. Choi [53] coined the term "nanofluid" while presenting a groundbreaking model addressing inherent thermal property limitations in base fluids. His research demonstrated that the incorporation of nanoparticles effectively enhances the thermal properties of these fluids, offering a comprehensive solution to the challenge. Xuan and Li [54] outlined a method for crafting a nanofluid suspension, incorporating nanophase powders into a base liquid. Eastman et al. [55] revealed that nanofluids with copper nanoparticles in ethylene glycol outperform in thermal conductivity compared to pure ethylene glycol or those with oxide nanoparticles. Das et al. [56] explored the impact of temperature on enhancing thermal conductivity in nanofluids through experimental analysis. Buongiorno [57] clarified the abnormal heat transfer enhancement in nanofluids, emphasizing the pivotal role of Brownian diffusion and thermophoresis effects, which notably improve nanofluid thermal conductivity. Das and Choi [58] examined convection, boiling, critical heat flux in pool boiling, and applications of nanofluids in their comprehensive discussion. Additionally, a foreign mass and the working fluid that moves as a result of the surface stretching react chemically in many chemical engineering processes. Numerous factors influence the chemical reaction's sequence. One of the most basic types of chemical reactions is first-order reaction, in which the species concentration directly determines the reaction's rate. A numerical solution for studying boundary layer flow and heat transfer of nanofluid over a non-linear stretching sheet, incorporating the effects of thermophoresis and Brownian motion by Rana et al. [59]. Noghrehabadi et al. [60] explored how nanoparticle volume and slip effects influence nanofluid flow and heat transfer over a stretched sheet, noting increased thermal boundary layer thickness with higher slip parameters. By accounting for convective boundary conditions, Rahman et al. [61] investigated radiative flow and heat transfer analysis of nanofluid over a stretched sheet. The mass and heat transport study of nanofluid across a stretched sheet under the effect of thermal radiation and a magnetic field was examined by Rashidi et al. [62].

1.1.5 Flow over a Linearly Stretching Sheet

Early studies on flow behavior over a flat sheet were conducted and reported by Sakiadis [63, 64], where boundary layer flow equations for a moving surface were derived, and the equations were solved and compared with results for stationary surfaces. Tsou et al. [65] studied the thermal behavior of a boundary layer flow on a continuous moving surface. Since then, several studies have been carried out on boundary layer problems on continuously moving surfaces. Boundary layer formation over a moving solid surface is observed in various industrial processes like polymer production, tempering of metallic and glass plates, and sheeting material production. The study of heat and flow dynamics over-stretching surfaces has captured researchers' interest because of its relevance to processes like polymer extrusion, wire drawing, and paper production [66]. Studies on heat transfer over-stretching surfaces have expanded upon initial examinations of flow dynamics [67–69].

While most early works assumed quiescent fluid, practical applications involve fluids in motion. Mahapatra and Gupta [70] studied the flow at a stagnation point against a stretching sheet with free stream velocity, identifying the emergence of a boundary layer. Lok et al. [71]explored the non-orthogonal flow at a stagnation point across a stretching sheet. Attia [72] and Rosali et al. [73] examined the stagnation-point flow in porous medium. Magnetic fields significantly impact the final product in industrial processes. The effect of magnetic fields has been investigated and analyzed by Andersson [74], and reported that viscoelastic fluid has similar effects to a magnetic field over a stretching sheet. The influence of magnetic fields on the flow dynamics and heat transfer properties investigated in [75–79]. The quality of the end product is determined by the heating and cooling rates, controllable via magnetic fields, affecting velocity and temperature profiles of the fluid under diverse conditions. Enhancing the magnetic field leads to a decrease in fluid velocity and a rise in its temperature. Research has delved into how changes in surface temperature and heat flux affect the heat transfer characteristics of a linearly stretching surface [80–82]. The effect of radiation near stagnation-points has been discussed in [83–85], whereas [86–88] investigated the radiation effect on heat transfer characteristics under different physical conditions. The non-linear radiation effect over a stretching surface has been analyzed by Cortell [89]. These studies reported that temperature of the fluid increases with an increase in the radiation parameter. Overall, research on flow and heat transfer characteristics over moving and stretching surfaces is crucial for understanding industrial processes and optimizing the quality of final products.

1.1.6 Flow over Non-linear Stretching Sheet

In various manufacturing processes, fluid flow over non-linearly stretching surfaces is common, such as in plastic sheet extrusion, wire drawing, and aerodynamic shaping. Chiam [90] studied the magnetic effect on flow characteristics over a non-linearly stretching surface with a power law velocity profile of the fluid. The combined effect of magnetohydrodynamics and thermal radiation on flow and heat transfer characteristics over an exponentially stretching surface has been investigated in [91, 92]. Previous studies assumed constant thermo-physical properties of the fluid. However, in many processes, these properties vary with temperature, leading to changes in flow and heat transfer rates. This aspect of flow has been studied on stretching surfaces [93]. For lubricating fluids, internal friction generates heat, resulting in a rise in temperature and a corresponding change in the physical properties of the fluid. This temperature increase affects the heat transport rate by varying the fluid properties across the thermal boundary layer. Therefore, considering variable fluid properties is crucial in the analysis of flow and heat transfer rates. In magnetohydrodynamic (MHD) flows, the effect of variable fluid properties on the flow and heat transfer rates over non-linearly stretching surfaces has been analyzed in [94, 95]. It was reported that the magnetic field tends to decrease the velocity and increase the temperature of the fluid. Taking into account the variable fluid properties and the influence of magnetic fields and thermal radiation provides a more comprehensive understanding of the flow and heat transfer characteristics over non-linearly stretching surfaces, which is essential for optimizing various industrial processes.

1.1.7 Practical Applications of Flows over a Stretching Surface

This chapter focuses on studying a specific type of flow where a sheet is stretched within its own plane, and its velocity is related to the distance from a reference point on the sheet. This flow scenario, known as flow over a stretching surface, has significant relevance in various engineering processes with industrial applications. Some of these applications include polymer extrusion, copper wire drawing, continuous casting of metals, glass blowing, hot rolling, and production of plastic and rubber sheets. In the context of manufacturing flat plastic sheets, understanding cooling and heat transfer is essential for improving the quality of the final product. While conventional cooling fluids like water and air are commonly used in industrial settings, certain sheet materials may require more efficient heat exchange methods. Researchers have explored the application of a magnetic field as a potential means to enhance the rate of heat exchange for such materials.

Furthermore, the momentum and heat transfer characteristics of boundary layer flow over a stretching sheet have practical implications in various chemical engineering processes. These processes may involve cooling molten liquids by stretching them into a cooling system, such as drawing continuous strips or filaments through a quiescent fluid. Understanding the flow and heat transfer in such scenarios is crucial for optimizing industrial processes and improving product quality.

This chapter unveils an innovative analysis of the effects of thermal radiation and chemical reactions on MHD boundary layer flow, heat, and mass transfer of Casson nanofluid across linear and non-linear stretching surfaces in a porous medium. The novelty of this research lies in the utilization of appropriate similarity transformations to convert the non-linear partial differential equations into a system of ordinary differential equations. Moreover, the numerical solution using the highly validated Runge–Kutta method with shooting technique offers a unique and robust approach to analyze the problem.

To the best of our knowledge, the specific problem addressed in this article, involving the simultaneous consideration of thermal radiation, chemical reaction, magnetohydrodynamics, and nanofluid flow over porous media with Casson fluid model and stretching surfaces, has not been previously reported in the existing literature. The comprehensive examination of velocity, temperature, and concentration profiles, along with other relevant parameters, provides valuable insights into the impacts of thermal radiation and chemical reaction on the flow and heat/mass transfer characteristics. The article's findings are likely to contribute significantly to the understanding of such complex phenomena and may have potential applications in various engineering and industrial processes.

2 Formulation of the Problem

The coordinate system is established such that the $x\text{-}axis$ is aligned with the motion of the stretching surface, originating from the origin. Meanwhile, the $y\text{-}axis$ is positioned normal to the surface of the sheet, as depicted in Fig. 1. The fundamental equations are derived to account for the physical conditions mentioned above, including the governing equation and the boundary-layer are proposed by Darcy–Boussinesq [96].

$$\frac{\partial v}{\partial y} + \frac{\partial u}{\partial x} = 0 \tag{1}$$

$$v\frac{\partial u}{\partial y} + u\frac{\partial u}{\partial x} = \nu\frac{\partial^2 u}{\partial y^2}\left(1 + \frac{1}{\beta}\right) - \frac{\mu}{K}u + \frac{g}{\rho}[(1 - C_\infty)\rho_{f\infty}\beta(\mathcal{T} - \mathcal{T}_\infty(C - C_\infty))] - \frac{\sigma B_0^2}{\rho} \tag{2}$$

$$v\frac{\partial \mathcal{T}}{\partial y} + u\frac{\partial \mathcal{T}}{\partial x} = \alpha_{nf}\left(\frac{\partial^2 \mathcal{T}}{\partial y^2}\right) + \tau\left(\left(\frac{\partial C}{\partial y}\right)\mathcal{D}_B\left(\frac{\partial \mathcal{T}}{\partial y}\right) + \left(\frac{\mathcal{D}_{\mathcal{T}}}{\mathcal{T}_\infty}\right)\left(\frac{\partial \mathcal{T}}{\partial y}\right)^2\right) - \frac{1}{\rho C_p}\left(\frac{\partial q_r}{\partial y}\right) \tag{3}$$

$$v\frac{\partial C}{\partial y} + u\frac{\partial C}{\partial x} = \mathcal{D}_B\left(\frac{\partial^2 C}{\partial y^2}\right) + \frac{\mathcal{D}_T}{\mathcal{T}_\infty}\left(\frac{\partial^2 \mathcal{T}}{\partial y^2}\right) - (C - C_\infty)\mathcal{K}r. \tag{4}$$

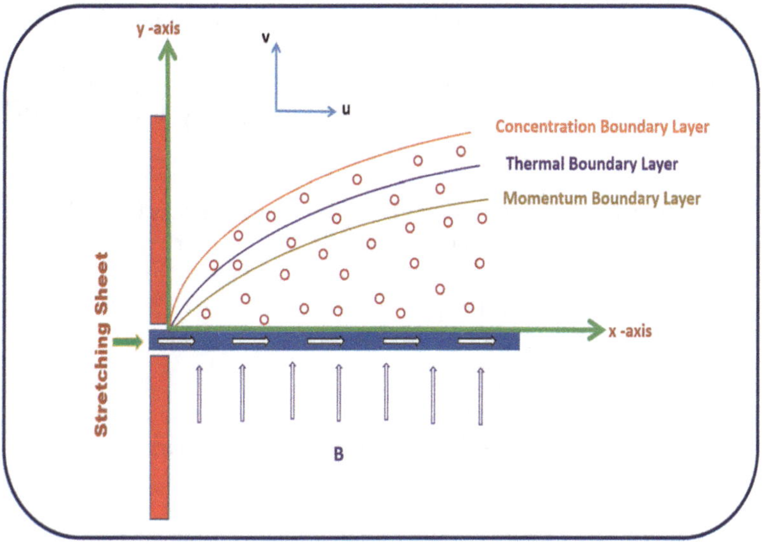

Fig. 1 Physical configuration of the problem

For the relevant boundary conditions are: [96]

$$\text{as } y = 0 \;\; u = ax^n, v = 0, \mathcal{T} = \mathcal{T}_{wall}, C = C_{wall} \tag{5}$$

$$\text{as } y \to \infty \;\; u \to 0, \mathcal{T} \to \mathcal{T}_\infty, C \to C_\infty. \tag{6}$$

The Rosseland approximation is a useful approach for estimating the radiative heat flux vector (q_r) in optically thick fluids. In such fluids, self-absorption plays a significant role alongside emission. Since the absorption coefficient varies with wavelength and can be considerable, employing the Rosseland approximation is appropriate. Consequently, we can define q_r in a manner that captures both emission and self-absorption [96].

$$q_r = -\left(\frac{4}{3}\right)\left(\frac{\sigma_1}{K}\right)^* \frac{\partial \mathcal{T}^4}{\partial y}. \tag{7}$$

In the equation mentioned, the symbol K^* represents the Rosseland mean absorption coefficient, while σ_1 refers to the $Stefan-Boltzmann$ constant value.

According to the condition of insignificant thermal variations within the flow, an approximation for $\mathcal{T}^4$ ($\mathcal{T}$ raised to the power of 4) can be obtained by extending it around the stream's free temperature $\mathcal{T}$ by extending Taylor's series. By neglecting higher-order terms, the following approximation is derived (AM).

$$\mathcal{T}^4 \simeq -3\mathcal{T}_\infty^4 + 4\mathcal{T}_\infty^3\mathcal{T}. \tag{8}$$

By substituting Eqs. (7) and (8) into Eq. (3), the energy equation is simplified to:

$$v\frac{\partial \mathcal{T}}{\partial y}+u\frac{\partial \mathcal{T}}{\partial x}=\alpha_{nf}\left(\frac{\partial^2 \mathcal{T}}{\partial y^2}\right)+\tau\left(\left(\frac{\partial C}{\partial y}\right)\left(\frac{\partial \mathcal{T}}{\partial y}\right)\mathcal{D}_{\mathcal{B}}+\frac{\mathcal{D}_{\mathcal{T}}}{\mathcal{T}_\infty}\left(\frac{\partial \mathcal{T}}{\partial y}\right)^2\right)+\frac{16\sigma_1\mathcal{T}_\infty^3}{3\rho C_p\mathcal{K}^*}\frac{\partial^2 \mathcal{T}}{\partial y^2}. \tag{9}$$

The similarity transformation converts PDEs into dimensionless ODEs by introducing dimensionless variables and parameters [97].

$$\begin{aligned}
\eta &= \sqrt{\frac{a(n+1)}{2\nu}}x^{\frac{n-1}{2}}y\\
u &= ax^n f'(\eta)\\
v &= \sqrt{\frac{\nu a\,(n+1)}{2}}x^{\frac{n-1}{2}}\left(f(\eta)+\frac{n-1}{n+1}\eta f'(\eta)\right)\\
\phi &= \frac{C_\infty - C}{C_\infty - C_w}\\
\theta &= \frac{\mathcal{T}_\infty - \mathcal{T}}{\mathcal{T}_\infty - \mathcal{T}_w}.
\end{aligned} \tag{10}$$

In dimensionless equations, important parameters are defined as non-dimensional variables or coefficients.

$$\begin{aligned}
\mathcal{M} &= \frac{2\sigma B_0^2}{a\rho(n+1)},\\
\mathcal{P}r &= \frac{\gamma}{\alpha},\\
\mathcal{S}C &= \frac{\gamma}{D_B},\\
Cr &= \frac{\mathcal{K}_r}{\mathcal{D}_B},\\
\mathcal{R}a &= \frac{2g_\beta(\mathcal{T}_W - \mathcal{T}_\infty)(1-C_\infty)\rho f_\infty}{\rho a^2(n+1)},\\
\mathcal{N}b &= \frac{\mathcal{D}_B}{\gamma}\tau\left((C_w - C_\infty)\right),\\
\mathcal{N}t &= \frac{\mathcal{D}_T}{\gamma N_r} = \tau\left((\mathcal{T}_w - \mathcal{T}_\infty)\right),\\
\mathcal{N}r &= \frac{\left(\rho_p - \rho_{f\infty}\right)(C_W - C_\infty)}{\rho_{f\infty}\beta(\mathcal{T}_W - \mathcal{T}_\infty)(1-C_\infty)},\\
\mathcal{R} &= \frac{16}{3}\frac{\sigma_1 T_\infty^3}{k^* k},\\
\mathcal{K} &= \left(\frac{2\sigma\beta_o^2}{a(n+1)\rho_k}\right).
\end{aligned} \tag{11}$$

Now Eq. (1) is automatically realized and Eqs. (2) to (4) were derived utilizing Eqs. (10) and (11)to yield the following non-linear ordinary differential equations.

$$(1+\frac{1}{\beta})\mathcal{F}''' + \mathcal{F}\mathcal{F}'' - \frac{2n}{n+1}\mathcal{F}'^2 + \mathcal{R}a\,(\theta - \mathcal{N}r\phi) - (\mathcal{M}+\mathcal{K})\mathcal{F}' = 0 \tag{12}$$

$$\frac{1}{\mathcal{P}r}\,(1+\mathcal{R})\,\Theta'' + \mathcal{N}b\Theta'\Phi' + \mathcal{F}\Theta' + \mathcal{N}t\Theta'^2) = 0 \tag{13}$$

$$\Phi'' + \mathcal{SCF}\Phi' + \frac{\mathcal{N}t}{\mathcal{N}b}\Theta'' - Cr\Phi = 0. \tag{14}$$

Dimensionless boundary conditions that exhibit correlation between variables.

$$\begin{aligned} \text{at } \eta = 0 \quad \mathcal{F}'(0) &= 1, \\ \mathcal{F}(0) &= 0, \\ \Phi(0) &= 1, \\ \Theta(0) &= 1, \\ \text{at } \eta \to \infty \quad \Theta(\eta) &\to 0, \\ \Phi(\eta) &\to 0, \\ \mathcal{F}'(\eta) &\to 0. \end{aligned} \tag{15}$$

3 Physical Quantities

The physical quantities of interest, namely the skin friction coefficients (Cf_x) the local Nusselt number (Nu_x) and the local Sherwood number (Sh_x), are defined as follows:

$$Cf_x = \frac{2\tau_{wx}}{\rho}, \tag{16}$$

$$Nu_x = \frac{xq_w}{k(\mathcal{T}_w - \mathcal{T}_\infty)}, \tag{17}$$

$$Sh_x = \frac{xj_m}{\mathcal{D}_B(C_w - C_\infty)}. \tag{18}$$

The following expressions provide the wall skin friction coefficients (τ_{wx}) and (τ_{wy}), wall heat flux (q_w), and wall mass flux (j_w):

$$
\begin{aligned}
\tau_{wx} &= \left(\frac{\partial u}{\partial y}\right)_{y=0} \mu, \\
q_w &= -k\left(\frac{\partial \mathcal{T}}{\partial y}\right)_{y=0}, \\
j_m &= -\left(\frac{\partial C}{\partial y}\right)_{y=0} \mathcal{D}_B.
\end{aligned} \tag{19}
$$

The non-dimensional versions of the coefficients of Cf_x, Nu_x and Sh_x expressed in terms of the similarity variable are given as follows,

$$
\begin{aligned}
Cf_x &= \sqrt{\frac{n+1}{2Re_x}}\mathcal{F}''(0), \\
Nu_x &= -\sqrt{\frac{n+1}{2Re_x}}\Theta'(0), \\
Sh_x &= -\sqrt{\frac{n+1}{2Re_x}}\Psi'(0).
\end{aligned} \tag{20}
$$

4 Result and Discussion

Comprehensive numerical simulations were carried out for various values of the relevant parameters in order to give a physical understanding of the flow problem produced by both linear and non-linear stretching surfaces. The findings are shown both visually and tabulator. The following parameters were held constant during the analysis [96]: $\mathcal{M} = 0.5, \mathcal{N}r = 0.5, \beta = 0.5, \mathcal{R}a = 0.5, \mathcal{P}r = 1.0, \mathcal{R} = 0.1, Cr = 0.1, \mathcal{N}b = 1.0, \mathcal{N}t = 1.0$. An analysis is conducted on the impact of these factors on the fluid's velocity ($\mathcal{F}'$), temperature (Θ), and concentration (Φ).

4.1 *Effect of Casson Parameter (β)*

The study's results, presented in Fig. 2a–c under both linear and non-linear stretching conditions, reveal a consistent trend: as the Casson parameter increases, flow velocity near the nanofluid surface decreases due to heightened viscous forces. This behavior is evident in both stretching scenarios, indicating the universal impact of the Casson parameter on flow behavior. Moreover, higher Casson parameter values lead to elevated temperature and concentration levels, attributed to enhanced heat and mass transfer within the fluid. This effect is observed in both linear and non-linear stretching cases, emphasizing the Casson parameter's significant influence on

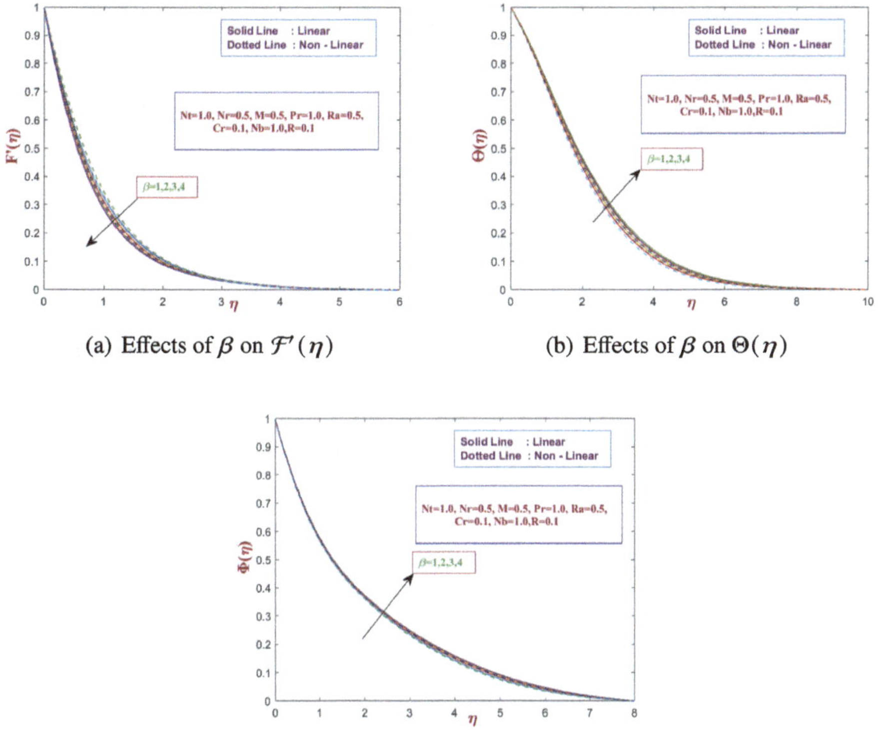

(a) Effects of β on $\mathcal{F}'(\eta)$

(b) Effects of β on $\Theta(\eta)$

(c) Effects of β on $\Phi(\eta)$

Fig. 2 Effects of Casson parameter (β) on $\mathcal{F}'(\eta)$, $\Theta(\eta)$ and $\Phi(\eta)$

thermal and concentration characteristics. In conclusion, the Casson parameter plays a pivotal role in shaping flow behavior and thermal and concentration profiles of the nanofluid over the stretching surface, irrespective of the stretching condition being linear or non-linear.

4.2 *Effect of Magnetic Parameter ($\mathcal{M}$)*

In Fig. 3a–c, the investigation examines the impact of the magnetic field parameter $\mathcal{M}$ on $\mathcal{F}'(\eta)$, $\Theta(\eta)$, and $\Phi(\eta)$ profiles under both linear and non-linear stretching conditions. With higher $\mathcal{M}$ values, the hydrodynamic boundary layer thickness reduces, particularly pronounced in the non-linear case (Fig. 3a), attributed to the influence of the Lorentz force, which decelerates fluid particle movement. Meanwhile, the thermal and solutal boundary layer thicknesses increase with higher $\mathcal{M}$ values in both cases, with a more significant enhancement observed in the non-linear stretching scenario. These results reveal that the magnetic field parameter plays a crucial role in

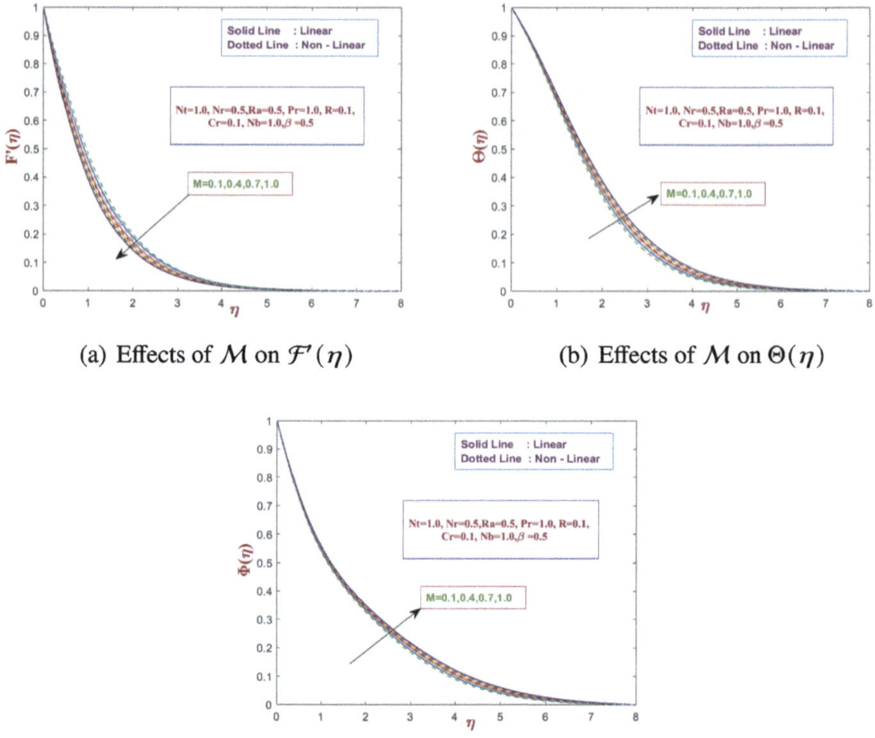

(a) Effects of $\mathcal{M}$ on $\mathcal{F}'(\eta)$

(b) Effects of $\mathcal{M}$ on $\Theta(\eta)$

(c) Effects of $\mathcal{M}$ on $\Phi(\eta)$

Fig. 3 Effects of magnetic field ($\mathcal{M}$) on $\mathcal{F}'(\eta)$, $\Theta(\eta)$ and $\Phi(\eta)$

shaping thermal and concentration profiles, leading to an increase in their boundary layer thickness, impacting fluid behavior over the stretching surface under both linear and non-linear stretching conditions.

4.3 Effect of Buoyancy Ratio Parameter ($\mathcal{N}r$)

Figure 4a–c shows the influence of the $\mathcal{N}r$ on $\mathcal{F}'(\eta)$, $\Theta(\eta)$, and $\Phi(\eta)$ profiles for both linear and non-linear stretching conditions. As $\mathcal{N}r$ increases, the fluid velocity decreases in both cases, with a more significant reduction observed in the non-linear stretching case (Fig. 4a). This behavior is attributed to the impact of buoyancy, which impedes fluid motion. Moreover, the $\Theta(\eta)$ and $\Phi(\eta)$ profiles rise with higher $\mathcal{N}r$ values, and this increase is notably more pronounced in the non-linear stretching case comparing to the linear case. The elevated $\mathcal{N}r$ values enhance the fluid's temperature and concentration levels, resulting in greater thermal and solutal gradients. These

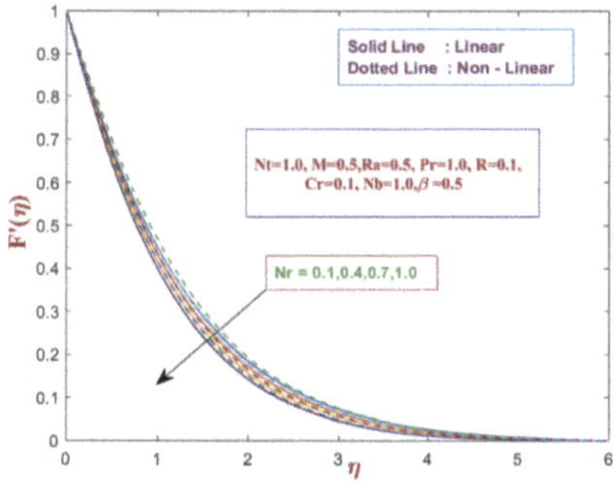

(a) Effects of $\mathcal{N}r$ on $\mathcal{F}'(\eta)$

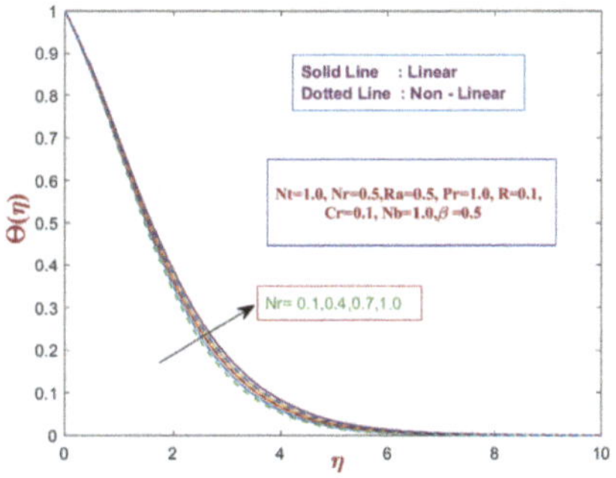

(b) Effects of $\mathcal{N}r$ on $\Theta(\eta)$

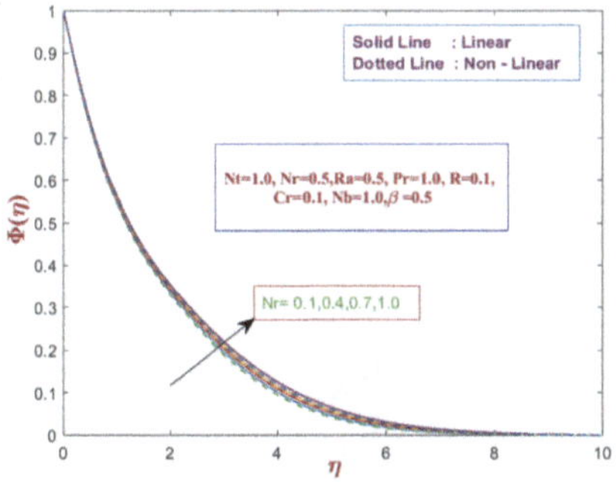

(c) Effects of $\mathcal{N}r$ on $\Phi(\eta)$

Fig. 4 Effects of Buoyancy ratio parameter ($\mathcal{N}r$) on $\mathcal{F}'(\eta)$, $\Theta(\eta)$ and $\Phi(\eta)$

findings highlight the significant role of the buoyancy ratio parameter in shaping the flow behavior and thermal and concentration characteristics of the fluid over the stretching surface, under both linear and non-linear stretching conditions.

4.4 Effect of Mixed Convection Parameter ($\mathcal{R}a$)

In Fig. 5a–c, the impact of the $\mathcal{R}a$ on $\mathcal{F}'(\eta)$, $\Theta(\eta)$ and $\Phi(\eta)$ profiles is illustrated for both linear and non-linear stretching conditions. Notably, in the boundary layer regime, higher $\mathcal{R}a$ values result in increased velocity profiles, observed in both stretching cases. In contrast, the temperature and concentration profiles show a decrease as $\mathcal{R}a$ increases, with this reduction more pronounced in the linear stretching scenario compared to the non-linear stretching case. These findings demonstrate the significant influence of the mixed convection parameter on the fluid's velocity, temperature, and concentration behaviors, underlining the distinct effects observed in the linear and non-linear stretching conditions.

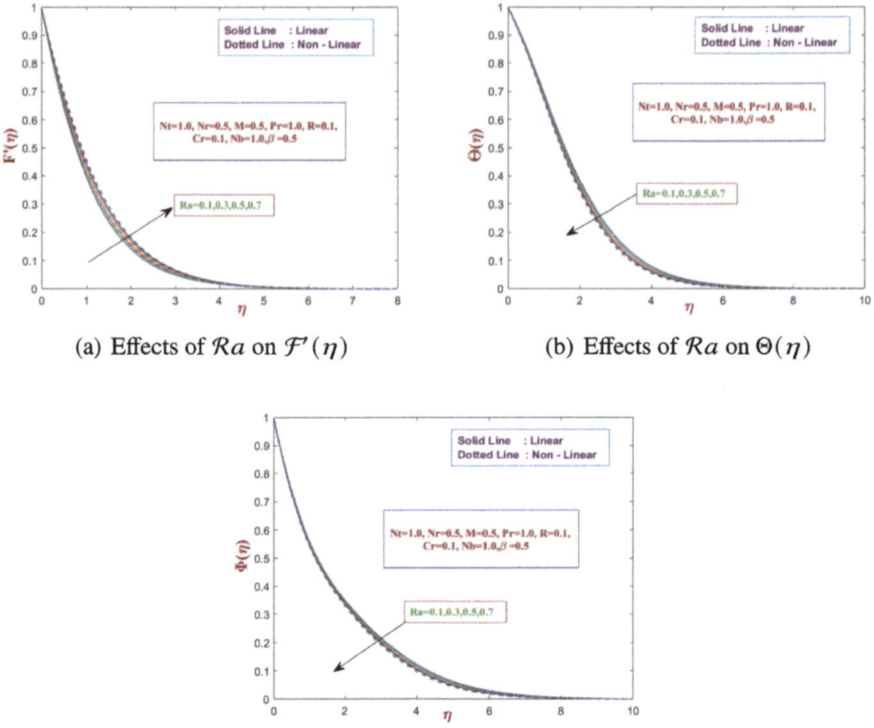

(a) Effects of $\mathcal{R}a$ on $\mathcal{F}'(\eta)$

(b) Effects of $\mathcal{R}a$ on $\Theta(\eta)$

(c) Effects of $\mathcal{R}a$ on $\Phi(\eta)$

Fig. 5 Effects of Buoyancy ratio parameter ($\mathcal{R}a$) on $\mathcal{F}'(\eta)$, $\Theta(\eta)$ and $\Phi(\eta)$

4.5 Effect of Radiation Parameter ($\mathcal{R}$)

In Fig. 6a–c, the $\mathcal{F}'(\eta)$, $\Theta(\eta)$ and $\Phi(\eta)$ distributions are presented for various radiation parameter values $\mathcal{R}$ under both linear and non-linear stretching conditions. Increasing $\mathcal{R}$ enhances the velocity and temperature profiles in both cases, with the velocity increment more prominent in the linear stretching scenario (Fig. 6a). In contrast, the temperature increment shows an opposite trend, as higher thermal radiation leads to elevated fluid temperature and an increase in the thickness of hydrodynamic and thermal boundary layers (Fig. 6b). Regarding the $\Phi(\eta)$ profiles, they decrease with increasing $\mathcal{R}$ values, with this reduction slightly more pronounced in the linear stretching case compared to the non-linear stretching case (Fig. 6c). These results reveal the contrasting effects of the radiation parameter on velocity, temperature, and concentration behaviors under both linear and non-linear stretching conditions.

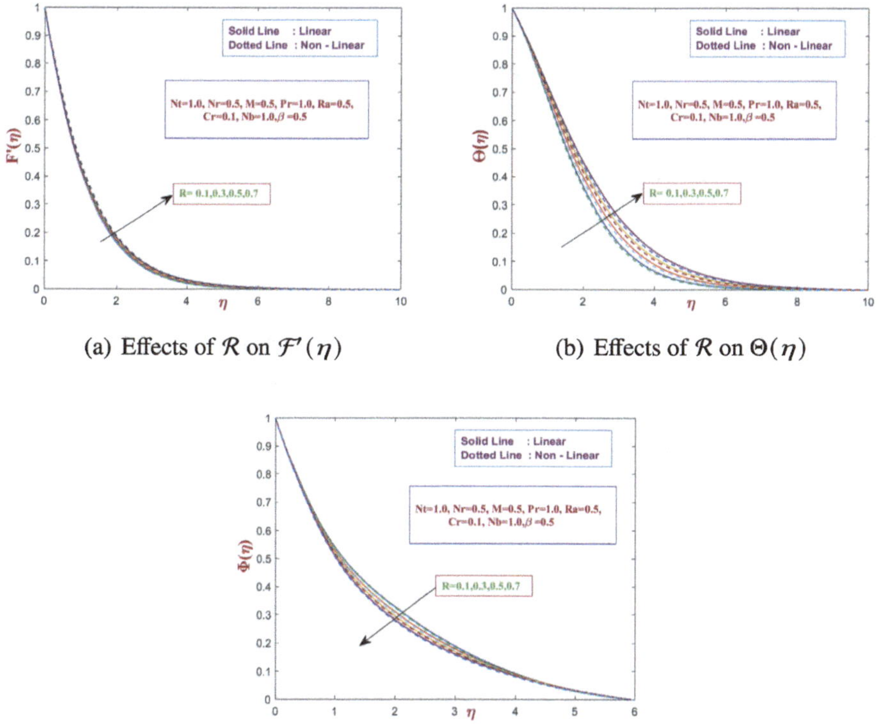

(a) Effects of $\mathcal{R}$ on $\mathcal{F}'(\eta)$

(b) Effects of $\mathcal{R}$ on $\Theta(\eta)$

(c) Effects of $\mathcal{R}a$ on $\Phi(\eta)$

Fig. 6 Effects of radiation parameter ($\mathcal{R}$) on $\mathcal{F}'(\eta)$, $\Theta(\eta)$ and $\Phi(\eta)$

4.6 Effect of Prandtl Number ($\mathcal{P}r$)

In Fig. 7a–b, the influence of the $\mathcal{P}r$ on $\mathcal{F}'(\eta)$, $\Theta(\eta)$ and $\Phi(\eta)$ profiles is demonstrated. As $\mathcal{P}r$ increases, both $\mathcal{F}'(\eta)$ and $\Theta(\eta)$ distributions decrease. Higher $\mathcal{P}r$ values correspond to higher momentum diffusivity or lower thermal diffusivity. This results in a reduction in the thermal boundary layer thickness, with a more pronounced effect observed in the linear stretching case compared to the non-linear stretching case. However, in contrast to velocity and temperature, the concentration profiles increase with higher $\mathcal{P}r$ values. These findings illustrate the significance of the Prandtl number in shaping $\mathcal{F}'(\eta)$,$\Theta(\eta)$ and $\Phi(\eta)$ behaviors under both linear and non-linear stretching conditions.

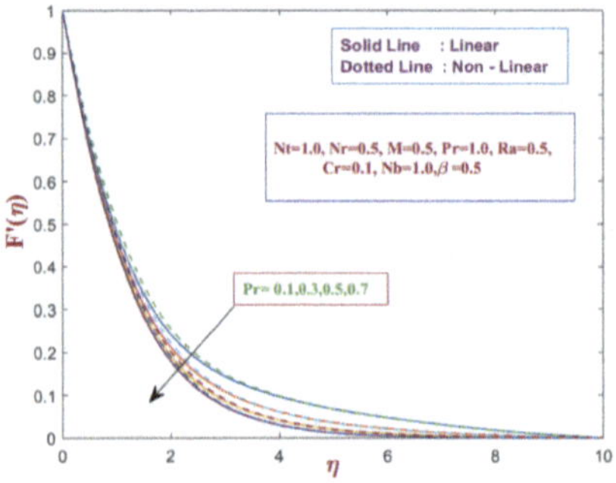

(a) Effects of $\mathcal{P}r$ on $\mathcal{F}'(\eta)$

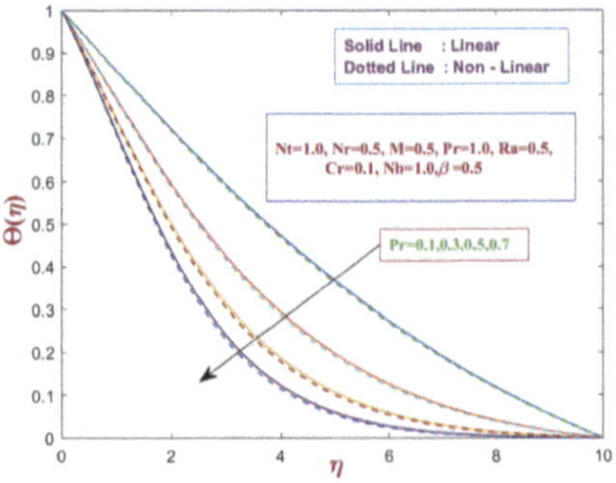

(b) Effects of $\mathcal{P}r$ on $\Theta(\eta)$

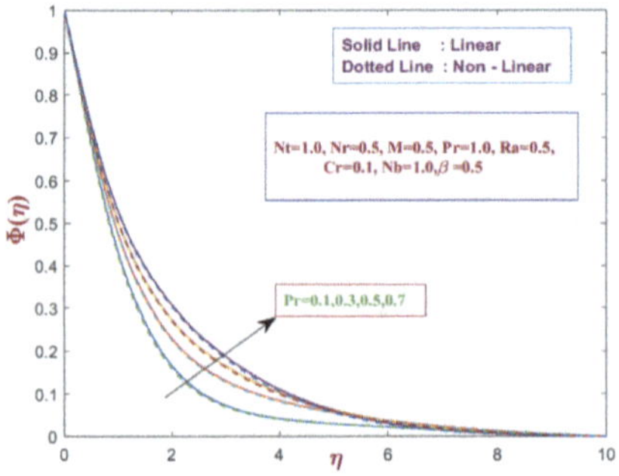

(c) Effects of $\mathcal{P}r$ on $\Phi(\eta)$

Fig. 7 Effects of Prandtl number ($\mathcal{P}r$) on $\mathcal{F}'(\eta)$, $\Theta(\eta)$ and $\Phi(\eta)$

4.7 Effect of Brownian Motion ($\mathcal{N}b$)

In Fig. 8a–c, the velocity, temperature, and concentration profiles are presented for different values of the Brownian motion parameter ($\mathcal{N}b$) under both linear and non-linear stretching conditions. In both cases, increasing ($\mathcal{N}b$) values result in higher velocity and temperature profiles. This behavior is a consequence of the enhanced random movement of nanoparticles within the base fluid, which becomes more pronounced with higher ($\mathcal{N}b$) values. Simultaneously, the concentration profiles decrease as ($\mathcal{N}b$) increases. The observed effects stem from increased nanoparticle randomness, diminishing concentration profiles and thickening the thermal boundary layer, underscoring the Brownian motion parameter's impact on $\mathcal{F}'(\eta)$, $\Theta(\eta)$ and $\Phi(\eta)$ in stretching scenarios.

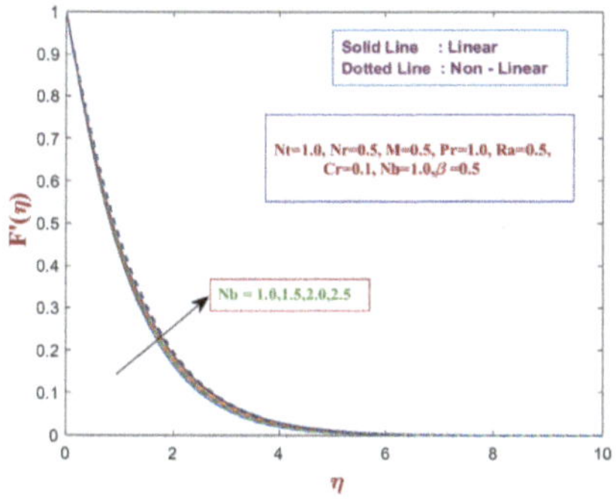

(a) Effects of $\mathcal{N}b$ on $\mathcal{F}'(\eta)$

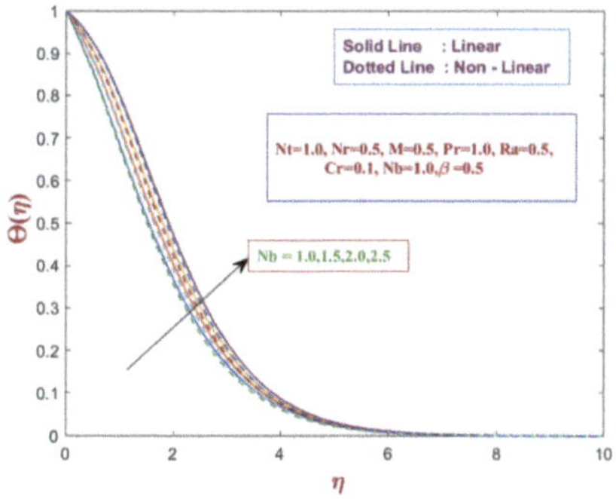

(b) Effects of $\mathcal{N}b$ on $\Theta(\eta)$

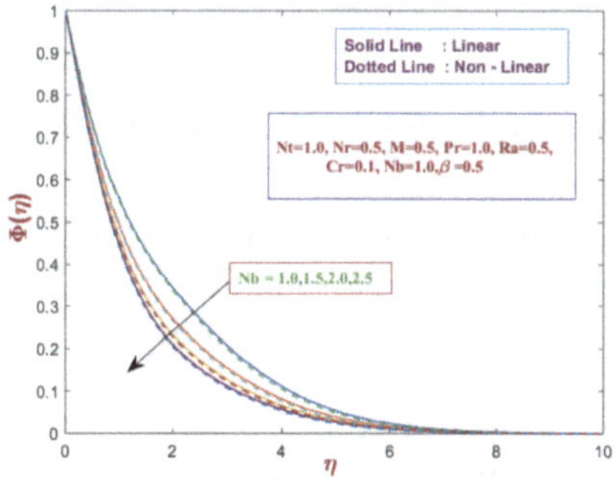

(c) Effects of $\mathcal{N}b$ on $\Phi(\eta)$

Fig. 8 Effects of Brownian motion ($\mathcal{N}b$) on $\mathcal{F}'(\eta)$, $\Theta(\eta)$ and $\Phi(\eta)$

4.8 Effect of Thermophoretic Parameter ($\mathcal{N}t$)

In the visuals presented in Fig. 9a and b, the role of the thermophoretic parameter ($\mathcal{N}$t) on shaping the fluid's temperature and concentration is depicted. An escalation in $\mathcal{N}$t correlates with heightened levels of temperature and concentration within the fluid, a consequence attributed to the action of thermophoresis, which moves nanoparticles from regions of higher to lower temperature, thereby enhancing the boundary layers for both temperature and solute concentration. This effect is notably more accentuated in scenarios involving non-linear stretching as opposed to linear stretching, underscoring the pivotal influence of thermophoresis on modulating the fluid's thermal and concentration profiles under varied stretching conditions.

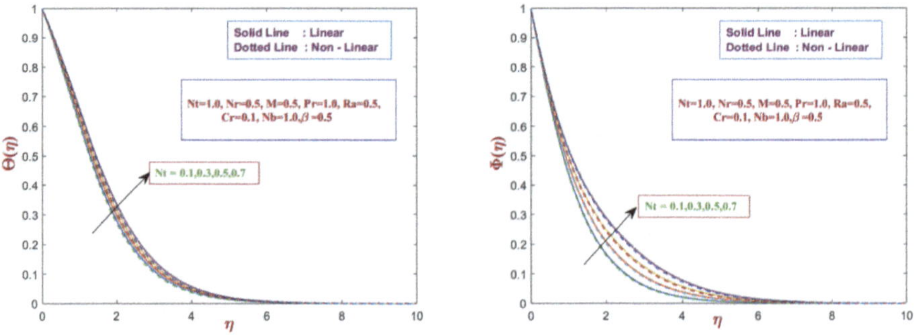

(a) Effects of $\mathcal{N}t$ on $\Theta(\eta)$

(b) Effects of $\mathcal{N}t$ on $\Phi(\eta)$

Fig. 9 Effects of thermophoretic parameter $(\mathcal{N}t)$ on $\mathcal{F}'(\eta)$, $\Theta(\eta)$ and $\Phi(\eta)$

4.9 Effect of Chemical Reaction Parameter (Cr)

Figure 10a depict the influence of the Cr on the concentration profiles. As Cr increases, the concentration profiles are significantly affected and decrease in the flow region. This reduction is more pronounced in the linear stretching case compared to the non-linear stretching case.

The study meticulously evaluated the variations in dimensionless velocity, as well as the rates of heat and mass transfer, across both linear and non-linear stretching conditions, incorporating an extensive range of influential parameters. The findings, concisely summarized in Table 1, reveal distinct patterns. As the parameter $\mathcal{M}$ increases, there is a noticeable rise in the local skin-friction coefficient, coupled with a decrease in dimensionless heat and mass transfer rates. Interestingly, dimensionless velocity rates exhibit an overall rise across the fluid spectrum, while the trend is reversed for heat and mass transfer rates with increasing $\mathcal{N}r$ values. Furthermore, $(\mathcal{R})$ values correspond to a decline in velocity and thermal transfer rates, but an upsurge

Fig. 10 Effect of chemical reaction parameter (Cr) and Activation Energy $(\mathcal{E})$ on $\Phi(\eta)$

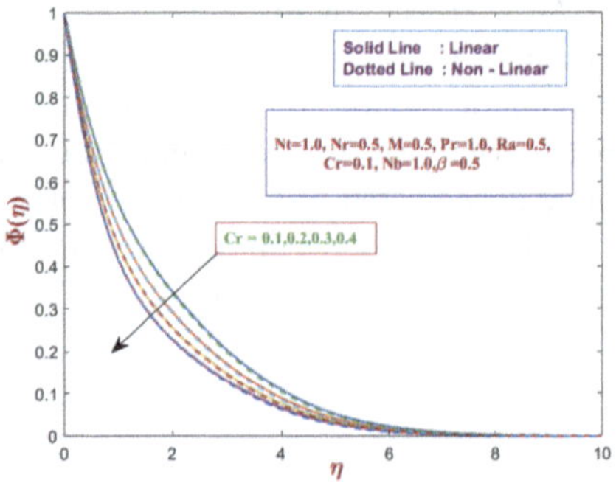

(a) Effect of Cr on $\Phi(\eta)$

Table 1 Skin friction, Nusselt number and Sherwood Number

Parameters							Cf_x		Nu_x		Sh_x	
$\mathcal{M}$	Nr	$\mathcal{R}$	Nb	Nt	Cr	β	Linear	Non-linear	Linear	Non-linear	Linear	Non-linear
0.1	0.5	0.1	1.0	1.0	0.1	0.5	−0.434756	−0.451383	0.27653	0.402098	0.694999	1.002602
0.4	0.5	0.1	1.0	1.0	0.1	0.5	−0.522037	−0.597313	0.269062	0.389865	0.681932	0.979516
0.7	0.5	0.1	1.0	1.0	0.1	0.5	−0.600029	−0.723469	0.262331	0.379111	0.671256	0.961258
1.0	0.5	0.1	1.0	1.0	0.1	0.5	−0.670960	−0.835604	0.256201	0.369486	0.662385	0.946453
0.5	0.1	0.1	1.0	1.0	0.1	0.5	−0.495215	−0.557768	0.267962	0.387956	0.680159	0.976191
0.5	0.4	0.1	1.0	1.0	0.1	0.5	−0.535468	−0.620315	0.271503	0.393254	0.686202	0.985832
0.5	0.7	0.1	1.0	1.0	0.1	0.5	−0.575949	−0.683087	0.264253	0.382413	0.674108	0.966541
0.5	1.0	0.1	1.0	1.0	0.1	0.5	−0.619503	−0.749801	0.261553	0.378299	0.645622	0.929233
0.5	0.5	0.1	1.0	1.0	0.1	0.5	−0.548937	−0.641216	0.266744	0.386136	0.678141	0.972973
0.5	0.5	0.3	1.0	1.0	0.1	0.5	−0.543881	−0.632566	0.267579	0.387572	0.677054	0.970835
0.5	0.5	0.5	1.0	1.0	0.1	0.5	−0.539500	−0.624984	0.266497	0.386054	0.677994	0.971866
0.5	0.5	0.7	1.0	1.0	0.1	0.5	−0.535687	−0.618338	0.264484	0.383097	0.679950	0.974509
0.5	0.5	0.1	1.0	1.0	0.1	0.5	−0.548937	−0.641216	0.266744	0.386136	0.678141	0.972973
0.5	0.5	0.1	1.5	1.0	0.1	0.5	−0.535662	−0.619382	0.200145	0.290050	0.733911	1.053890
0.5	0.5	0.1	2.0	1.0	0.1	0.5	−0.525746	−0.603023	0.147852	0.214489	0.756041	1.086135
0.5	0.5	0.1	2.5	1.0	0.1	0.5	−0.517559	−0.589452	0.107736	0.156448	0.765601	1.100182
0.5	0.5	0.1	1.0	0.1	0.1	0.5	−0.551446	−0.644441	0.356620	0.515631	0.719662	1.034182
0.5	0.5	0.1	1.0	0.3	0.1	0.5	−0.55129	−0.644347	0.333940	0.482913	0.700916	1.007105
0.5	0.5	0.1	1.0	0.5	0.1	0.5	−0.550901	−0.643893	0.312912	0.452601	0.688179	0.988562
0.5	0.5	0.1	1.0	0.7	0.1	0.5	−0.550284	−0.643084	0.293407	0.424505	0.680732	0.977501
0.5	0.5	0.1	1.0	1.0	0.1	0.5	−0.548937	−0.641216	0.266744	0.386136	0.678141	0.972973
0.5	0.5	0.1	1.0	1.0	0.2	0.5	−0.545445	−0.635779	0.266744	0.376923	0.770309	1.101250
0.5	0.5	0.1	1.0	1.0	0.3	0.5	−0.542529	−0.631216	0.254711	0.369260	0.852258	1.215730
0.5	0.5	0.1	1.0	1.0	0.4	0.5	−0.540049	−0.62732	0.250116	0.362760	0.926315	1.319476
0.5	0.5	0.1	1.0	1.0	0.1	1.0	−0.653087	−0.752674	0.254270	0.369751	0.659161	0.945279
0.5	0.5	0.1	1.0	1.0	0.1	1.5	−0.706175	−0.808502	0.248188	0.361620	0.651395	0.933777
0.5	0.5	0.1	1.0	1.0	0.1	2.0	−0.738809	−0.842470	0.244589	0.356765	0.647166	0.927493
0.5	0.5	0.1	1.0	1.0	0.1	2.5	−0.761005	−0.865425	0.242207	0.353538	0.644503	0.923531

in mass transfer rates. The influence of ($\mathcal{N}$b) is evident in its impact on the skin-friction coefficient, Nusselt number, and Sherwood number; while the skin-friction coefficient and Nusselt number decrease, the Sherwood number increases with higher (Nb) values. Moreover, elevated (Nt) values relate to reduced skin-friction coefficient, Nusselt number, and Sherwood number in both linear and non-linear cases. The study emphasizes the role of the Cr, as higher Cr values correlate with a decrease in skin-friction coefficient and Nusselt number, and an increase in Sherwood number. Finally, the study underscores the influence of the β, where higher β values correspond to a decreased Nusselt number and Sherwood number, and an increased skin-friction coefficient.

5 Conclusion

The investigation focused on the behavior of MHD Casson nanofluid across both linear and non-linear surfaces, highlighting the roles of nanoparticles, thermophoresis, and Brownian motion. It meticulously examined effects on flow velocity, thermal and solutal fields, alongside evaluations of skin friction, heat transfer, and mass transfer efficiency through Nusselt and Sherwood numbers, across diverse conditions. The results of the present investigation provide valuable insights into the behavior of the Casson nanofluid in the presence of magnetic fields and offer significant contributions to the better understanding of boundary layer heat and mass transfer phenomena. Some significant findings are as follows.

- The increase in the magnetic field parameter ($\mathcal{M}$) leads to significant improvements in the profiles of temperature and concentration.
- An escalation in the magnetic field parameter ($\mathcal{M}$) results in a reduction of the velocity profile.
- As the non-dimensional parameter ($\mathcal{N}r$) is upgraded, temperature and concentration profiles also are elevated.
- The velocity profiles contract as the non-dimensional parameter ($\mathcal{N}r$) escalates.
- As the Brownian motion parameter ($\mathcal{N}b$) in the nanofluid's boundary layer flow rises, both the velocity and heat transfer rates experience a decline.
- An increase in the Brownian motion parameter ($\mathcal{N}$b) results in higher mass rates.
- Elevating the thermophoresis parameter ($\mathcal{N}t$) thickens the thermal and concentration boundary layers in nanofluid flow, with a more marked effect in non-linear compared to linear cases.
- The thermophoresis parameter significantly influences the boundary layer characteristics of nanofluid, particularly in the non-linear scenario.
- As the values of $\mathcal{R}$ increase, both the skin-friction coefficient and the Nusselt number exhibit a decrease.
- The Sherwood number shows an increase with higher values of the $\mathcal{R}$.

References

1. Crane LJ (1970) Flow past a stretching plate. J Appl Math Phys (ZAMP) 21
2. Gupta PS, Gupta AS (1977) Heat and mass transfer on a stretching sheet with suction or blowing. Can J Chem Eng 55:744–746
3. Banks WHH (1983) Similarity solutions of the boundary-layer equations for a stretching wall. J Mec Theor Appl 2:375–392
4. Grubka LJ, Bobba KM (1985) Heat transfer characteristics of a continuous, stretching surface with variable temperature. J Heat Transf 107:248–250
5. Dutta BK, Ray P, Gupta AS (1985) Temperature field in flow over a stretching sheet with uniform heat flux. Intern Comm Heat Mass Transf 12:89–94
6. Dutta BK, Gupta AS (1987) Cooling of a stretching sheet in viscous flow. Ind Eng Chem Res 26:333–336

7. Dutta BK (1988) Heat transfer from a stretching sheet in viscous flow with suction of blowing. ZAMM?J Appl Math Mech/Zeitschrift für Angewandte Mathematik und Mechanik 68(6):231–236
8. Chen C-K, Char M-I (1988) Heat transfer of a continuous, stretching surface with suction or blowing. J Math Anal Appl 135:568–580
9. Soewono E, Vajravelu K, Mohapatra RN (1992) Existence of solutions of a nonlinear boundary value problem, arising in flow and heat transfer over a stretching sheet. Nonlinear Anal 18:93–98
10. Karahalios GT (1992) A note on a time-dependent flow from a stretching plane surface. Int J Non-Linear Mech 27(2):293–298
11. Vajravelu K (1994) Convection heat transfer at a stretching sheet with suction or blowing. J Math Anal Appl 188(3):1002–1011
12. Kumaran V, Ramanaiah G (1996) A note on the flow over a stretching sheet. Acta Mech 116(1–4):229–233
13. Magyari E, Keller B (1999) Heat and mass transfer in the boundary layers on an exponentially stretching continuous surface. J Phys D Appl Phys 32(5):577
14. Magyari E, Keller B (2000) Exact solutions for self-similar boundary-layer flows induced by permeable stretching walls. Eur J Mech-B/Fluids 19(1):109–122
15. Magyari E, Keller B (2001) The wall jet as limiting case of a boundary-layer flow induced by a permeable stretching surface. Zeitschrift für angewandte Mathematik und Physik ZAMP 52(4):696–703
16. Magyari E, Pop I, Keller B (2002) Mixed convection boundary-layer flow past a horizontal permeable flat plate. Fluid Dyn Res 31(3):215
17. Magyari E, Keller B (2003) The opposing effect of viscous dissipation allows for a parallel free convection boundary-layer flow along a cold vertical flat plate. Transp Porous Media 51:227–230
18. Mahapatra TR, Gupta AS (2003) Stagnation?point flow towards a stretching surface. Can J Chem Eng 81(2):258–263
19. Partha MK, Murthy PVSN, Rajasekhar GP (2005) Effect of viscous dissipation on the mixed convection heat transfer from an exponentially stretching surface. Heat Mass Transf 41:360–366
20. Liao S-J, Pop I (2003) Explicit analytic solution for similarity boundary layer equations. Int J Heat Mass Transf 47(1):75–85
21. Liao S-J (2005) A challenging nonlinear problem for numerical techniques. J Comput Appl Math 181(2):467–472
22. Liao S-J, Pop I (2004) Explicit analytic solution for similarity boundary layer equations. Int J Heat Mass Transf 47(1):75–85
23. Xu H (2005) An explicit analytic solution for convective heat transfer in an electrically conducting fluid at a stretching surface with uniform free stream. Int J Eng Sci 43(10):859–874
24. Pavlov KB (1974) Magnetohydrodynamic flow of an incompressible viscous fluid caused by deformation of a plane surface. Magnitnaya Gidrodinamika 4(1):146–147
25. Chakrabarti A, Gupta AS (1979) Hydromagnetic flow and heat transfer over a stretching sheet. Q Appl Math 37(1):73–78
26. Kumari M, Nath G, Takhar HS (1990) MHD flow and heat transfer over a stretching surface with prescribed wall temperature or heat flux. Waerme-und Stoffuebertragung; (Germany, FR) 25(6)
27. Andersson HI (1992) MHD flow of a viscoelastic fluid past a stretching surface. Acta Mech 95(1–4):227–230
28. Vajravelu K, Nayfeh J (1993) Convective heat transfer at a stretching sheet. Acta Mech 96(1–4):47–54
29. Char M-I (1994) Heat and mass transfer in a hydromagnetic flow of the viscoelastic fluid over a stretching sheet. J Math Anal Appl 186(3):674–689
30. Dandapat BS, Ray PC (1994) The effect of thermocapillarity on the flow of a thin liquid film on a rotating disc. J Phys D Appl Phys 27(10):2041

31. Mahapatra, Roy T, Gupta AS (2001) Magnetohydrodynamic stagnation-point flow towards a stretching sheet. Acta Mechanica 152(1–4):191–196
32. Liao S-J (2003) On the analytic solution of magnetohydrodynamic flows of non-Newtonian fluids over a stretching sheet. J Fluid Mech 488:189–212
33. Datti PS, Prasad KV, Subhas Abel M, Joshi A (2004) MHD visco-elastic fluid flow over a non-isothermal stretching sheet. Int J Eng Sci 42(8–9):935–946
34. Afify AA (2004) MHD free convective flow and mass transfer over a stretching sheet with chemical reaction. Heat Mass Transf 40(6–7):495–500
35. Liu I-C (2005) Flow and heat transfer of an electrically conducting fluid of second grade in a porous medium over a stretching sheet subject to a transverse magnetic field. Int J Non-Linear Mech 40(4):465–474
36. Cortell R (2006) Effects of viscous dissipation and work done by deformation on the MHD flow and heat transfer of a viscoelastic fluid over a stretching sheet. Phys Lett A 357(4–5):298–305
37. Massoudi M, Maneschy CE (2004) Numerical solution to the flow of a second grade fluid over a stretching sheet using the method of quasi-linearization. Appl Math Comput 149(1):165–173
38. Layek GC, Mukhopadhyay S, Samad SkA (2007) Heat and mass transfer analysis for boundary layer stagnation-point flow towards a heated porous stretching sheet with heat absorption/generation and suction/blowing. Int Commun Heat Mass Transf 34(3): 347–356
39. Ishak A, Jafar K, Nazar R, Pop I (2009) MHD stagnation point flow towards a stretching sheet. Physica A 388(17):3377–3383
40. Mahapatra TR, Nandy SK, Gupta AS (2009) Analytical solution of magnetohydrodynamic stagnation-point flow of a power-law fluid towards a stretching surface. Appl Math Comput 215(5):1696–1710
41. Mahapatra TR, Nandy SK, Gupta AS (2009) Analytical solution of magnetohydrodynamic stagnation-point flow of a power-law fluid towards a stretching surface. Appl Math Comput 215(5):1696–1710
42. Hayat T, Qasim M, Abbas Z (2010) Radiation and mass transfer effects on the magnetohydrodynamic unsteady flow induced by a stretching sheet. Zeitschrift für Naturforschung A 65(3):231–239
43. Abbas Z, Javed T, Sajid M, Ali N (2010) Unsteady MHD flow and heat transfer on a stretching sheet in a rotating fluid. J Taiwan Inst Chem Eng 41(6):644–650
44. Mustafa M, Hayat T, Pop I, Asghar S, Obaidat S (2011) Stagnation-point flow of a nanofluid towards a stretching sheet. Int J Heat Mass Transf 54(25–26):5588–5594
45. Makinde OD, Aziz A (2011) Boundary layer flow of a nanofluid past a stretching sheet with a convective boundary condition. Int J Thermal Sci 50(7):1326–1332
46. Hassani M, Tabar MM, Nemati H, Domairry G, Noori F (2011) An analytical solution for boundary layer flow of a nanofluid past a stretching sheet. Int J Thermal Sci 50(11):2256–2263
47. Kandasamy R, Hayat T, Obaidat S (2011) Group theory transformation for Soret and Dufour effects on free convective heat and mass transfer with thermophoresis and chemical reaction over a porous stretching surface in the presence of heat source/sink. Nucl Eng Des 241(6):2155–2161
48. Bhattacharyya K, Uddin MS, Layek GC, and Ali PkW (2011) Analysis of boundary layer flow and heat transfer for two classes of viscoelastic fluid over a stretching sheet with heat generation or absorption. Bangladesh J Sci Ind Res 46(4):451–456
49. Noghrehabadi A, Pourrajab R, Ghalambaz M (2012) Effect of partial slip boundary condition on the flow and heat transfer of nanofluids past stretching sheet prescribed constant wall temperature. Int J Therm Sci 54:253–261
50. Bhattacharyya K (2013) Boundary layer stagnation-point flow of Casson fluid and heat transfer towards a shrinking/stretching sheet. Front Heat Mass Transf (FHMT) 4(2)
51. Turkyilmazoglu M, Pop I (2013) Exact analytical solutions for the flow and heat transfer near the stagnation point on a stretching/shrinking sheet in a Jeffrey fluid. Int J Heat Mass Transf 57(1):82–88
52. Ibrahim W, Shankar B (2013) MHD boundary layer flow and heat transfer of a nanofluid past a permeable stretching sheet with velocity, thermal and solutal slip boundary conditions. Comput Fluids 75:1–10

53. Choi SUS (1995) Enhancing thermal conductivity of fluids with nanoparticles. In: Development and applications of non newtonian flow. FED-vol. 231/MD-vol. 66: ASME; pp 99–105
54. Xuan Y, Li Q (2000) Heat transfer enhancement of nanofluids. Int J Heat Fluid Flow 21(1):58–64
55. Eastman JA, Choi SUS, Li S, Yu W, Thompson LJ (2001) Anomalously increased effective thermal conductivities of ethylene glycol-based nanofluids containing copper nanoparticles. Appl Phys Lett 78(6):718–720
56. Das SK, Putra N, Thiesen P, Roetzel W (2003) Temperature dependence of thermal conductivity enhancement for nanofluids. J Heat Transf 125(4):567–574
57. Buongiorno J (2006) Convective transport in nanofluids. J Heat Transf 128:240–250
58. Das SK, Stephen US (2009) A review of heat transfer in nanofluids. Adv Heat Transf 41:81–197
59. Rana P, Bhargava R (2012) Flow and heat transfer of a nanofluid over a nonlinearly stretching sheet: a numerical study. Commun Nonlinear Sci Numer Simul 17:212–226
60. Noghrehabadi A, Pourrajab R, Ghalambaz M (2013) Effect of partial slip boundary condition on the flow and heat transfer of nanofluids past stretching sheet prescribed constant wall temperature. Int J Therm Sci 54:253–261
61. Rahman MM, Eltayeb IA (2013) Radiative heat transfer in a hydromagnetic nanofluid past a non-linear stretching surface with convective boundary condition. Meccanica 48:601–615
62. Rashidi MM, Ganesh NV, Hakeem AKA, Ganga B (2014) Buoyancy effect on MHD flow of nanofluid over a stretching sheet in the presence of thermal radiation. J Mol Liq 198:234–238
63. Sakiadis BC (1961) Boundary-layer behavior on continuous solid surfaces: I. Boundary layer equations for two-dimensional and axisymmetric flow. AIChE J 7:26–28
64. Sakiadis BC (1961) Boundary-layer behavior on continuous solid surfaces: II. The boundary layer on a continuous flat surface. AIChE J 7:221–225
65. Tsou F, Sparrow E, Goldstein R (1967) Flow and heat transfer in the boundary layer on a continuous moving surface. Int J Heat Mass Transf 10:219–235
66. Crane LJ (1970) Flow past a stretching plate. Zeitschrift fr angewandte Mathematik und Physik ZAMP 21:645–647
67. Gupta PS, Gupta AS (1977) Heat and mass transfer on a stretching sheet with suction or blowing. Can J Chem Eng 55:744–746
68. Ali ME (1994) Heat transfer characteristics of a continuous stretching surface. Wrmeund Stoffbertragung 29:227–234
69. Elbashbeshy EMA, Bazid MAA (2004) Heat transfer in a porous medium over a stretching surface with internal heat generation and suction or injection. Appl Math Comput 158:799–807
70. Mahapatra TR, Gupta AS (2003) Stagnation-point flow towards a stretching surface. Can J Chem Eng 81:258–263
71. Lok YY, Amin N, Pop I (2006) Non-orthogonal stagnation point flow towards a stretching sheet. Int J Non-Linear Mech 41:622–627
72. Attia HA (2007) Stagnation point flow towards a stretching surface through a porous medium with heat generation. Turk J Eng Environ Sci 30:299–306
73. Rosali H, Ishak A, Pop I (2011) Stagnation point flow and heat transfer over a stretching/shrinking sheet in a porous medium. Int Commun Heat Mass Transfer 38:1029–1032
74. Andersson HI (1992) MHD flow of a viscoelastic fluid past a stretching surface. Acta Mech 95:227–230
75. Attia HA (2003) Hydromagnetic stagnation point flow with heat transfer over a permeable surface. Arab J Sci Eng 28(1):107–112
76. Sharma PR, Singh G (2009) Effects of ohmic heating and viscous dissipation on steady MHD flow near a stagnation point on an isothermal stretching sheet. Therm Sci 13(1):5–12
77. Singh P, Tomer NS, Kumar S, Sinha D (2010) MHD oblique stagnation-point flow towards a stretching sheet with heat transfer. Int J Appl Math Mech 6(13):94–111
78. Al-Harbi SM (2007) Numerical study of heat transfer over permeable stretching surface with variable viscosity and thermal diffusivity in uniform magnetic field. Soochow J Math 33(2):229–240

79. Pal D, Mondal H (2014) Effects of temperature-dependent viscosity and variable thermal conductivity on MHD non-Darcy mixed convective diffusion of species over a stretching sheet. J Egyptian Math Soc 22:123–133
80. Chen C-K, Char M-I (1988) Heat transfer of a continuous, stretching surface with suction or blowing. J Math Anal Appl 135:568–580
81. Elbashbeshy EMA (1998) Heat transfer over a stretching surface with variable surface heat flux. J Phys D Appl Phys 31(16):1951–1955
82. Devi SA, Raj JWS (2014) Numerical simulation of magnetohydrodynamic forced convective boundary layer flow past a stretching/shrinking sheet prescribed with variable heat flux in the presence of heat source and constant suction. J Appl Fluid Mech 7(3):415–423
83. Pop SR, Grosan T, Pop I (2004) Radiation effects on the flow near the stagnation point of a stretching sheet. Tech Mech 25(2):100–106
84. Singh P, Sinha D, Tomer NS (2012) Oblique stagnation-point darcy flow towards a stretching sheet. J Appl Fluid Mech 5(3):29–37
85. Arthur EM, Seini IY (2014) Hydromagnetic stagnation point flow over a porous stretching surface in the presence of radiation and viscous dissipation. Appl Comput Math 3:191–196
86. Seddeek MA, Abdelmeguid MS (2006) Effects of radiation and thermal diffusivity on heat transfer over a stretching surface with variable heat flux. Phys Lett A 348:172–179
87. Kumar H (2009) Radiative heat transfer with hydromagnetic flow and viscous dissipation over a stretching surface in the presence of variable heat flux. Therm Sci 13(2):163–169
88. Singh P, Singh TN, Kumar S, Sinha D (2011) Effect of radiation and porosity parameter on magnetohydrodynamic flow due to stretching sheet in porous media. Therm Sci 15(2):517–526
89. Cortell R (2014) Fluid flow and radiative nonlinear heat transfer over a stretching sheet. J King Saud Univ - Sci 26:161–167
90. Chiam T (1995) Hydromagnetic flow over a surface stretching with a power-law velocity. Int J Eng Sci 33:429–435
91. Mukhopadhyay S (2013) MHD boundary layer flow and heat transfer over an exponentially stretching sheet embedded in a thermally stratified medium. Alex Eng J 52:259–265
92. Chaudhary S, Singh S, Chaudhary S (2015) Thermal radiation effects on MHD boundary layer flow over an exponentially stretching surface. Appl Math 06(02):295–303
93. Abel MS, Khan SK, Prasad KV (2002) Study of visco-elastic fluid flow and heat transfer over a stretching sheet with variable viscosity. Int J Non-Linear Mech 37:81–88
94. Prasad KV, Vajravelu K, Datti PS (2010) The effects of variable fluid propertiescon the hydromagnetic flow and heat transfer over a non-linearly stretching sheet. Int J Thermal Sci 49:603–610
95. Devi SA, Gururaj ADM (2012) Effects of variable viscosity and nonlinear radiation on MHD flow with heat transfer over a surface stretching with a power-law velocity. Adv Appl Sci Res 3(1):319–334
96. Sreedevi P, Sudarsana Reddy P, Chamkha AJ (2017) Heat and mass transfer analysis of nanofluid over linear and non-linear stretching surfaces with thermal radiation and chemical reaction. Powder Technol 315: 194–204
97. Hayat T, Imtiaz M, Alsaedi A (2015) Partial slip effects in flow over nonlinear stretching surface. Appl Math Mech -Engl Ed 36:1513–1526

Chapter 6
Non-linear Dynamical Functioning of a Time-Independent Uncertain System with Optimization

R. Harini, K. Indhira, and N. Thillaigovindan

Abstract This work investigates a bulk arrival retrial queueing model that offers multi-phases of heterogeneous service. Moreover, as soon as the orbit becomes empty, the server goes on a Bernoulli working vacation (BWV), where it works at a lower speed. At certain times, consumers are also permitted to balk and renege. Further, the busy server may experience a failure at any time. Here the supplementary variable approach has been incorporated in order to derive the probability generating function (PGF) of the number of clients in the system as well as in the orbit. The impact of particular parameters on the system's overall performance has been demonstrated with the aid of a few numerical examples. Finally, this research is accelerated in order to bring about the best possible (optimal) cost for the system by adopting a range of optimization approaches namely particle swarm optimization (PSO), artificial bee colony (ABC), grey wolf optimizer (GWO) and differential evolution (DE).

1 Introduction

Waiting in a queue is a typical phenomenon in real life, owing to its numerous significant purposes as a process. Due to network complexity and an expanding customer base, consumer behaviour and the retry phenomenon may have had a substantial impact on how well computer networks perform. A customer who accesses the service area and notices the server is active might opt to depart for a short while and come back

R. Harini · K. Indhira (✉)
Department of Mathematics, School of Advanced Sciences, Vellore Institute of Technology, Vellore 632 014, Tamil Nadu, India
e-mail: kindhira@vit.ac.in

R. Harini
e-mail: harini.r2020@vitstudent.ac.in

N. Thillaigovindan
Department of Mathematics, College of Natural Sciences, Arba Minch University, Arba Minch, Ethiopia

G. Arulprakash et al. (eds.), *Mathematical Modelling of Complex Patterns Through Fractals and Dynamical Systems*, Studies in Infrastructure and Control,
https://doi.org/10.1007/978-981-97-2343-0_6

at a later time. This is the primary goal of the retrial queue (RQ). Before retrying to occupy a server, such consumers are supposed to be in orbit. Queues of that kind have various implications in a variety of fields, like call centers, re-transmission-enabled communications networks, etc. Moreover, a substantial amount of research has been done recently in which consumers approaching in batches are served with multi-phases of heterogeneous service that are provided by a single server. For instance, the process of withdrawing money from a bank involves multiple stages, the first of which entails getting a withdrawal receipt, the second of which involves filling out that form, and the third of which involves actually withdrawing the money from the bank. Many studies on queueing systems (QS) with server vacations have been conducted in the past. Numerous principles for the working vacation (WV) methodology have been put forth in the literature as well. Also, various RQ with impatient consumers have prompted significant notice as well. Further, in RQ literature, the server is typically believed to be permanently accessible. But given the significant impact of server failures, this seems questionable. Therefore, to implement RQ models, it is essential to examine RQ with server disruption. In addition, when a model is built and ready to use, the designer's or analyst's primary worry typically involves whether the model is cost-effective or not. As a result, "cost analysis"plays a key role in determining how effective a model is. It shrinks the risk aspect and aids system designers in making better choices.

2 Literature Survey

To comprehend the research efforts conducted and the research gaps present, a thorough and methodical review of the prior studies has been presented as follows:

Consumers' repeated efforts when the server is down generate RQ. In recent years, RQ has become a vital topic of research. There is a plethora of research done on RQ. Most often, the works of Artalejo [2] and Yang [35] have been prioritized much. Customers' arrival in clusters displays intriguing behaviour and can be accomplished in a wide range of practical queuing scenarios. Murugan and Vijaykrishnaraj [20] studied a bulk RQ with exponentially distributed multiple WV. Recently, Abdollahi and Salehi Rad [1] analysed a dual phases of heterogeneous services with dependent admission by utilizing a $M^{[X]}/G/1$ retrial QS.

Services provided to consumers can be delivered in a single, comprehensive phase or in a sequence of smaller but equally significant phases. Based on their preferences, customers can select a phase service pattern that consists entirely of required phases or a hybrid model that includes both required and optional phases. This kind of queuing is prevalent nowadays. Radha et al. [24] scrutinized a multi-phase a unreliable RQ with Bernoulli vacation. Multi-stages of heterogeneous service with variant queueing behaviours was explored in detail by Kirupa and Udaya Chandrika [12–14]. Lately, Abdollahi et al. [1] and Sangeetha and Udaya Chandrika [29] addressed multi-discretionary services.

In queueing models, the concept of impatience is a common occurrence. Balking clients opt not to line up if they discover the server is inaccessible upon advent, while reneging consumers quit the line if they have spent too long waiting for service. RQ with non-persistent consumers have gained greater attention recently, for instance, Liu and Song [16] addressed a retrial QS with non-persistent consumers and WV. Nila and Sumitha [21] have investigated a bulk arrival RQ with dual essential services, inconsistent consumers, random breakdown, feedback and orbital search. Lately, a discrete-time bulk arrival $Geo^X/G/1$ RQ with impatient consumers and three varieties of vacations namely single, multiple and J vacations has been analysed by Rajasudha et al. [28]. Moreover, prompted by the study of cellular network performance Peng and Wu [23] analysed a $M^X/G/1$ RQ with impatient customers

Vacation queues in QS allude to the temporary absence of servers, whereas WV periods have servers providing service to consumers at a reduced rate while fully refraining to do so during regular vacation periods. A single server pre-emptive priority RQ with immediate Bernoulli feedback under WV and vacation interruption was initially scrutinized by Rajadurai et al. [26]. Recently, Rajadurai [27] scrutinized a sole server pre-emptive priority RQ with BWVs and vacation interruption. Recently, cost optimization on batch arrival retrial G queue was investigated by Upadhyaya et al. [32].

Recently, Bharathidass et al. [3] analysed a batch service queue with server breakdowns and repairs. A finite Markovian QS with vacation subjected to breakdown was also recently analysed by Dasa et al. [5]. Thus, while developing any system, preventative or remedial steps should be incorporated. Therefore, one of the most crucial topics in this study is server breakdowns and their repair. Many of the systems that have been explored have thought about this issue, including Rajadurai et al. [25]. But occasionally, these repairs are delayed because of certain restrictions. This issue has been taken into account in the framework created by Jain and Bhagat [8] and Choudhury and Ke [4].

There is not much of work on RQ that addresses cost optimization and evaluates appropriate control parameters for the queueing model using many optimization techniques. Lately, Vaishnawi et al. [33] analysed the accuracy of a $Geo^{[X]}/Geo/1$ recurrent model in discrete time using a variety of optimization approach to induce the lowest possible cost for the system. Kumar and Jain [15] studied a Markovian QS and in order to assess the best options for the service system's decision variables, they formulated a cost function and the ideal service rates at the lowest possible cost are calculated using a combination of two distinct optimization techniques. Moreover, refer to articles [6, 9, 18, 34] for a review of published works on cost optimization.

The rest of the work is outlined as follows: Sect. 3 presents a thorough analysis of the framework under consideration and how it can be applied realistically. The system's stability criterion, associated equations and steady state (SS) outcomes are determined in Sect. 4. Several essential system metrics are covered in Sect. 5. The statistical analysis in Sect. 6 investigates the consequences of different attributes on the system's framework. The analysis of cost optimization by using different methods is addressed in Sect. 7. A summary of the study is listed in Sect. 8.

3 Description of the Model

An illustration of the model under consideration is presented below:

Arrival process: A compound "Poisson process" with rate ϵ enables clients to congregate in bulk. Presuming that "$\mathcal{V}_k, k = 1, 2, 3, \ldots$" follows a standard dist., let $\mathcal{V}_k$ be the no. of clients affiliated with the kth batch arrival. "$Pr[\mathcal{V}_k = n] = \mathcal{A}_n, n = 1, 2, 3, \ldots$" and $\mathcal{V}(\breve{\varsigma})$ indicate the PGF of $\mathcal{V}$.

Retrial process: If an approaching priority consumer realizes the server is vacant, he/she starts their service right away. When a priority consumer arrives, the server provides service for another priority consumer or under WV; the newly arrived priority client leaves the system immediately without receiving service. The arrival of a priority consumer will pause the service provided to an ordinary consumer by the regularly busy server, and the server will then start providing service to the priority consumer. In the case that a priority client took precedence over an ordinary client, we predicted that ordinary client, had just been served prior to the priority consumer's service commencement, would remain in the service station until the remaining service was accomplished.

As long as there isn't a defined waiting area, clients are allowed to use the server right away if they arrive notice it vacant. Suppose they show up and the server is full or on WV, they are required to either exit the system with prob. $(1 - b)$ or enrol in a group of obstructed consumers known as an orbit with prob. b in compliance with $FCFS$ discipline. As soon as the server stops processing requests, the client at the top of the RQ engages in a battle of wits with other possible primary consumers to decide who will be given access to the service next. If a regular or priority client enters before the retrial client, the latter may cancel its request for service and either return to its previous position in the RQ with prob. r or exit the system with prob. $(1 - r)$. In this case, an arbitrary dist. $\mathcal{B}(\grave{\aleph})$ with a subsequent Laplace Stieltjes Transform (LST) $\mathcal{B}^*(\beta)$ is used to standardize the retrial duration for all orbiting clients.

Bernoulli Working Vacation process: The server initiates a WV each time the orbit turns empty and it follows an exponential distribution with a parameter δ. If any new clients come in on vacation, the server keeps running but at a slower speed. The WV period is a slower-moving operational period. If any orbiting clients with significantly slower completion periods complete their service during the vacation period, the server will terminate the vacation and resume the usual busy period. This results in a vacation interruption. Otherwise, if there are no clients in the system at the end of the vacation, the server either enters the system and waits passively for a new client with prob. ζ (single WV) to serve the clients as they arrive in regular mode, or it departs for another WV with prob. $1 - \zeta$ (WV). Following a vacation, the server restores regular functions provided there remain clients within the orbit. During the WV period, the service time is represented by the random variable $\mathcal{L}_v$ with the distribution function $\mathcal{L}_v(\hat{\varepsilon})$ having LST $\mathcal{L}_v^*(\beta)$ and the first moment is given by $\mathcal{L}_v^{*'}(\delta) = \int_0^\infty \hat{\varepsilon} e^{-\delta\hat{\varepsilon}} d\mathcal{L}_v(\hat{\varepsilon})$.

Regular Service process: Different C-type services are offered by the server. With prob. p_i $(1 \leq i \leq C)$, clients prefer the ith type of service. One of the clients receives regular service while additional clients join the orbit if the arriving batch of positive clients discovers the server idle. If not, every client in the batch becomes a member of the orbit. It is expected that generic principles with a probability dist. function (PDF) $\mathcal{Q}_i(\hat{\varepsilon})$ regulate the ith (i = 1,2 ...C) stage service periods with a corresponding LST $\mathcal{Q}_i^*(\beta)$.

Breakdown process: The server may malfunction at any point while it is operating in any phase of the service, failing the service stream for a brief period of time and causing the server to go down. The breakdowns are induced by exogenous Poisson processes having rates η_i $(i = 1, 2 \ldots C, \ldots C)$. Repair process:

Repair process: The server that malfunctions is sent out right away for repair. It ceases servicing customers who stroll by over this period and awaits on the repair to start. This period is referred to as the server's "delay period." The delay time $\mathcal{P}_i$ of the server abides a PDF $\mathcal{P}_i(\widetilde{\vartheta})$ with corresponding LST $\mathcal{P}_i^*(\beta)$ $(i = 1, 2 \ldots C)$. The client, who had just begun receiving service before the server went down, is awaiting for the remaining portion of the service to be delivered. The server's repair time (indicated as $\mathcal{G}_i$ $(i = 1, 2 \ldots C)$) distributions for all service phases are taken to be randomly distributed with PDF $\mathcal{G}_i(\widetilde{\vartheta})$ and LST $C_i^*(\beta)$ where $i = 1, 2 \ldots C$.

All stochastic processes in the system are considered to be unrelated of one another.

4 Evaluation of the Steady State Probabilities

When formulating the SS differential equations for the RQ system in this section, we take the elapsed retrial times, elapsed regular service times, elapsed BWV time, elapsed delay times and elapsed repair times as supplementary variables.

The Steady State Equations

In SS, we take into consideration that $\mathcal{B}(0) = 0, \mathcal{B}(\infty) = 1, \mathcal{Q}_{i,b}(0) = 0, \mathcal{Q}_{i,b}(\infty) = 1$, $(i = 1, 2 \ldots C)$ and $\mathcal{L}(0)_v = 0, \mathcal{L}_v(\infty) = 1$, are continuous at $\hat{\varepsilon} = 0$ and $\mathcal{P}_i(0) = 0$, $\mathcal{P}_i(\infty) = 1$, $\mathcal{G}_i(0) = 0$, $\mathcal{G}_i(\infty) = 1$, $(i = 1, 2 \ldots C)$, are continuous at $\widetilde{\vartheta} = 0$. Hence, the fn. The retrial, service on FPS and SPS, vacation, deferred repair and repair on all the C phases have the corresponding "conditional completion rates (hazard rates)" as $\omega(\hat{\varepsilon})$, $\upsilon_{i,b}(\hat{\varepsilon})$, $\psi_v(\hat{\varepsilon})$, $\xi_i(\widetilde{\vartheta})$ and $\theta_i(\widetilde{\vartheta})$

$$\omega(\hat{\varepsilon})d\hat{\varepsilon} = \frac{d\mathcal{B}(\hat{\varepsilon})}{1 - \mathcal{B}(\hat{\varepsilon})}; \quad \upsilon_{i,b}(\hat{\varepsilon})d\hat{\varepsilon} = \frac{d\mathcal{Q}_{i,b}(\hat{\varepsilon})}{1 - \mathcal{Q}_{i,b}(\hat{\varepsilon})}; \quad \psi_v(\hat{\varepsilon})d\hat{\varepsilon} = \frac{d\mathcal{L}_v(\hat{\varepsilon})}{1 - \mathcal{L}_v(\hat{\varepsilon})};$$

$$\xi_i(\widetilde{\vartheta})d\widetilde{\vartheta} = \frac{d\mathcal{P}_i(\widetilde{\vartheta})}{1 - \mathcal{P}_i(\widetilde{\vartheta})}; \quad \theta_i(\widetilde{\vartheta})d\widetilde{\vartheta} = \frac{d\mathcal{G}_i(\widetilde{\vartheta})}{1 - \mathcal{G}_i(\widetilde{\vartheta})}.$$

At time $\check{t}$, the elapsed retrial times, service times, vacation times, delay times and repair times are denotes as $\mathcal{B}^0(\check{t})$, $\mathcal{Q}_{i,b}^0(\check{t})$, $\mathcal{L}_v^0(\check{t})$ and $\mathcal{P}_i^0(\check{t})$ and $\mathcal{G}_i^0(\check{t})$ respectively. Additionally, construct the random variable,

$$\Sigma(\check{t}) = \begin{cases} 0, & \text{the server is inactive on regular busy and BWV period at time } \check{t} \\ 1, & \text{the server is engaged with } i\text{th phases of service at time } \check{t} \\ 2, & \text{the server is on BWV at time } \check{t} \\ 3, & \text{the server is on delaying repair under } i\text{th phase of service at time } \check{t} \\ 4, & \text{the server is under repair with } i\text{th phase of service at time } \check{t}. \end{cases}$$

Moreover, we emphasize the utility of the bivariate Markov process in characterizing the state of the system at a specific time $\check{t}$ $\{\Sigma(\check{t}), \Xi(\check{t}); \check{t} \geq 0\}$, where $\Sigma(\check{t})$ indicates the server state $(0, 1, 2, 3, 4)$ based on whether the server is inactive on regular busy and BWV, under regular busy period, on BWV, under delaying repair or repair. $\Xi(\check{t})$ denotes the no. of clients within orbit at time $\check{t}$. The elapsed retrial time corresponds to $\mathcal{B}^0(\check{t})$ for $\Sigma(\check{t}) = 0$ and $\Xi(\check{t}) > 0$. Suppose $\Sigma(\check{t}) = 1$ and $\Xi(\check{t}) \geq 0$, thereby the client's elapsed time of being served normally on ith phase $(i = 1, 2 \ldots C)$ corresponds to $\mathcal{Q}^0_{i,b}(\check{t})$. If $\Sigma(\check{t}) = 2$ and $\Xi(\check{t}) \geq 0$, then $\mathcal{L}^0_v(\check{t})$ denotes the client's elapsed time under BWV. If $\Sigma(\check{t}) = 3$ and $\Xi(\check{t}) \geq 0$, then $\mathcal{P}^0_i(\check{t})$ refers to the server's elapsed time of being delayed repair on ith phase $(i = 1, 2 \ldots C)$. If $\Sigma(\check{t}) = 4$ and $\Xi(\check{t}) \geq 0$, then $\mathcal{G}^0_i(\check{t})$ is proportional to the server's elapsed time to be fixed ith phase $(i = 1, 2 \ldots C)$.

Theorem 1 *The embedded Markov chain (MC)* $\{D_n; n \epsilon N\}$ *is ergodic iff* $\rho < 1$, *where* $\rho = (1 - \mathcal{B}^*(\epsilon))r\mathcal{I}_1 + \sum_{i=1}^{C} p_i \tau^{(1)}$ *and,*

$$\tau^{(1)} = \epsilon \mathcal{I}_1 \sum_{i=1}^{C} \mathcal{Q}_i^{(1)}(1 + \eta_i[\mathcal{P}_i^{(1)} + \mathcal{G}_i^{(1)}]).$$

Proof It is straightforward to verify that ergodicity is a sufficient condition by applying Foster's criteria [22], which claims that perhaps the chain $\{D_n; n \in N\}$ is irreducible and aperiodic. If a non-negative fn. $g(h)$, $h \in N$ and $\epsilon > 0$, exists that guarantees that the mean drift $\varsigma_h = E[g(m_{n+1}) - g(m_n)/m_n = h]$ is limited for all $h \in N$ and $h \in N$ and $\varsigma_h \leq - \epsilon$, excluding possibly for a finite no. of h's, then the MC is ergodic. When considering the function $g(h) = h$ in our scenario, we get

$$\varsigma_h = \begin{cases} \sum_{i=1}^{C} p_i \tau^{(1)}, & \text{if } h = 0 \\ (1 - \mathcal{B}^*(\epsilon))r\mathcal{I}_1 + \sum_{i=1}^{C} p_i \tau^{(1)} - 1, & \text{if } h = 1, 2, \ldots \end{cases}$$

It is clear that ergodicity must exist in order for the inequality below to exist.

$$(1 - \mathcal{B}^*(\epsilon))r\mathcal{I}_1 + \sum_{i=1}^{C} p_i \tau^{(1)} < 1.$$

If the MC $\{D_n; n \epsilon N\}$ meets Kaplan's criterion, we may easily ensure non-ergodicity in accordance with Sennott et al. [7] particularly $\varsigma_h < \infty$ for all $h \geq 0$ and $\exists\, h_0 \in N$ s.t $\varsigma_h \geq 0$ for $h \geq h_0$. The fact that "Kaplan's condition" is met in our scenario is underscored by the existence of a w such that $(f_{lh}) = 0$ for $h < l - w$ and $l > 0$, where (f_{lh}) is the one-step transition matrix of $\{D_n; n \in N\}$. Consequently, it is indicated that the MC is non-ergodic by

$$(1 - \mathcal{B}^*(\epsilon))r\mathcal{I}_1 + \sum_{i=1}^{C} p_i \tau^{(1)} \geq 1.$$

Let the collection of epochs whereby a regular service period, re-service period, vacation period or repair period terminates be $\{\check{t}_n; n = 1, 2, \ldots\}$. Thus, using a sequence of random vectors $J_n = \{\Sigma(\check{t}_n+), \Xi(\check{t}_n+)\}$, an MC is created and incorporated in the retry QS. Our system must thus be stable according to Theorem 1, which states that $D_n; n \in N$ is ergodic iff $(1 - \mathcal{B}^*(\epsilon))r\mathcal{I}_1 + \sum_{i=1}^{C} p_i \tau^{(1)} < 1$. The probs. $\Omega_0(\check{t}) = P\{\Sigma(\check{t}) = 0, \Xi(\check{t}) = 0\}$ and $\aleph_0(\check{t}) = P\{\Sigma(\check{t}) = 1, \Xi(\check{t}) = 0\}$ and prob. densities for the method $\{\Xi(\check{t}), \check{t} \geq 0\}$, are stated below,

$$\begin{aligned}
\Omega_n(\hat{\varepsilon}, \check{t})d\hat{\varepsilon} &= P\{\Sigma(\check{t}) = 0, \Xi(\check{t}) = n, \hat{\varepsilon} \leq \mathcal{B}^0(\check{t}) < \hat{\varepsilon} + d\hat{\varepsilon}\}, \\
&\quad \text{for } \check{t}, \hat{\varepsilon} \geq 0 \text{ and } n \geq 1 \\
\Psi_{i,b,n}(\hat{\varepsilon}, \check{t})d\hat{\varepsilon} &= P\{\Sigma(\check{t}) = 1, \Xi(\check{t}) = n, \hat{\varepsilon} \leq \mathcal{Q}^0_{i,b}(\check{t}) < \hat{\varepsilon} + d\hat{\varepsilon}\}, \\
&\quad \text{for } \check{t}, \hat{\varepsilon} \geq 0, n \geq 0, i = 1, 2 \ldots C \\
\aleph_{v,n}(\hat{\varepsilon}, \check{t})d\hat{\varepsilon} &= P\{\Sigma(\check{t}) = 2, \Xi(\check{t}) = n, \hat{\varepsilon} \leq \mathcal{L}^0_v(\check{t}) < \hat{\varepsilon} + d\hat{\varepsilon}\}, \\
&\quad \text{for } \check{t}, \hat{\varepsilon} \geq 0, n \geq 0, i = 1, 2 \ldots C \\
\Theta_{i,n}(\hat{\varepsilon}, \widetilde{\vartheta}, \check{t})d\hat{\varepsilon} &= P\{\Sigma(\check{t}) = 3, \Xi(\check{t}) = n, \widetilde{\vartheta} \leq \mathcal{P}^0_i(\check{t}) < \widetilde{\vartheta} + d\widetilde{\vartheta}/\mathcal{Q}^0_i(\check{t}) = \hat{\varepsilon}\}, \\
&\quad \text{for } \check{t} \geq 0, (\hat{\varepsilon}, \widetilde{\vartheta}) \geq 0, n \geq 0, i = 1, 2 \ldots C \\
\Upsilon_{i,n}(\hat{\varepsilon}, \widetilde{\vartheta}, \check{t})d\hat{\varepsilon} &= P\{\Sigma(\check{t}) = 4, \Xi(\check{t}) = n, \widetilde{\vartheta} \leq \mathcal{G}^0_i(\check{t}) < \widetilde{\vartheta} + d\widetilde{\vartheta}/\mathcal{Q}^0_i(\check{t}) = \hat{\varepsilon}\}, \\
&\quad \text{for } \check{t} \geq 0, (\hat{\varepsilon}, \widetilde{\vartheta}) \geq 0, n \geq 0, i = 1, 2 \ldots C
\end{aligned}$$

We presume that the sequel complies with the stability requirements, thus we can assign limiting probs. for $\hat{\varepsilon}, \widetilde{\vartheta} > 0$, $n \geq 0$ and $(i = 1, 2 \ldots C)$

$$\begin{aligned}
&\Omega_0 = lim_{\check{t}\to\infty}\Omega_0(\check{t});\ \ \aleph_0 = lim_{\check{t}\to\infty}\aleph_0(\check{t}); \\
&\Omega_n(\hat{\varepsilon}) = lim_{\check{t}\to\infty}\Omega_n(\hat{\varepsilon}, \check{t});\ \ \Psi_{i,b,n}(\hat{\varepsilon}) = lim_{\check{t}\to\infty}\Psi_{i,b,n}(\hat{\varepsilon}, \check{t});\ \ \aleph_{v,n}(\hat{\varepsilon}) = lim_{\check{t}\to\infty}\aleph_{v,n}(\hat{\varepsilon}, \check{t}); \\
&\Theta_{i,n}(\hat{\varepsilon}, \widetilde{\vartheta}) = lim_{\check{t}\to\infty}\Theta_{i,n}(\hat{\varepsilon}, \widetilde{\vartheta}, \check{t});\ \ \Upsilon_{i,n}(\hat{\varepsilon}, \widetilde{\vartheta}) = lim_{\check{t}\to\infty}\Upsilon_{i,n}(\hat{\varepsilon}, \widetilde{\vartheta}, \check{t}).
\end{aligned}$$

The equations below, which govern the dynamics of the system's behaviour, are generated by using a supplementary variable approach for $(i = 1, 2 \ldots C)$.

$$\epsilon\Omega_0 = \delta\zeta\aleph_0 \tag{1}$$

$$(\epsilon+\delta)\aleph_0=\delta(1-\zeta)\aleph_0+\sum_{i=1}^{C}\left(\int_0^\infty \Psi_{i,b,0}(\hat{\varepsilon})\upsilon_{i,b}(\hat{\varepsilon})d\hat{\varepsilon}+\int_0^\infty \Upsilon_{i,0}(\hat{\varepsilon})\theta_i(\hat{\varepsilon})d\hat{\varepsilon}\right)$$
$$+\int_0^\infty \aleph_{v,0}(\tilde{\varsigma})\psi_v(\hat{\varepsilon})d\hat{\varepsilon}+\int_0^\infty \Lambda_{v,0}(\grave{\upsilon})\chi_v(\grave{\upsilon})d\grave{\upsilon} \tag{2}$$

$$\frac{d\Omega_n(\hat{\varepsilon})}{d\hat{\varepsilon}}+(\epsilon+\omega(\hat{\varepsilon}))\Omega_n(\hat{\varepsilon})=0;\, n\geq 0 \tag{3}$$

$$\frac{d\Psi_{i,b,n}(\hat{\varepsilon})}{d\hat{\varepsilon}}+(\epsilon+\eta_i+\upsilon_{i,b}(\hat{\varepsilon}))\Psi_{i,b,n}(\hat{\varepsilon})=\epsilon(1-b)\Psi_{i,b,n}(\hat{\varepsilon})+\epsilon b\sum_{k=1}^{n}\chi_k\Psi_{i,b,n-k}(\hat{\varepsilon})$$
$$+\int_0^\infty \Upsilon_{i,n}(\hat{\varepsilon},\widetilde{\vartheta})\theta_i(\widetilde{\vartheta})d\widetilde{\vartheta};\; n\geq 1 \tag{4}$$

$$\frac{d\aleph_{v,n}(\hat{\varepsilon})}{d\hat{\varepsilon}}+(\epsilon+\delta+\psi_v(\hat{\varepsilon}))\aleph_{v,n}(\hat{\varepsilon})=\epsilon(1-b)\aleph_{v,n}(\hat{\varepsilon})+\epsilon b\sum_{k=1}^{n}\chi_k\aleph_{v,n-k}(\hat{\varepsilon});\; n\geq 1 \tag{5}$$

$$\frac{d\Theta_{i,n}(\hat{\varepsilon},\widetilde{\vartheta})}{d\widetilde{\vartheta}}+(\epsilon+\xi_i(\widetilde{\vartheta}))\Theta_{i,n}(\hat{\varepsilon},\widetilde{\vartheta})=\epsilon(1-b)\Theta_{i,n}(\hat{\varepsilon},\widetilde{\vartheta})+\epsilon b\sum_{k=1}^{n}\chi_k\Theta_{i,n-k}(\hat{\varepsilon},\widetilde{\vartheta});\; n\geq 1 \tag{6}$$

$$\frac{d\Upsilon_{i,n}(\hat{\varepsilon},\widetilde{\vartheta})}{d\widetilde{\vartheta}}+(\epsilon+\theta_i(\widetilde{\vartheta}))\Upsilon_{i,n}(\hat{\varepsilon},\widetilde{\vartheta})=\epsilon(1-b)\Upsilon_{i,n}(\hat{\varepsilon},\widetilde{\vartheta})+\epsilon b\sum_{k=1}^{n}\chi_k\Upsilon_{i,n-k}(\hat{\varepsilon},\widetilde{\vartheta});\; n\geq 1 \tag{7}$$

At $\hat{\varepsilon}=0,\ \widetilde{\vartheta}=0$ and for $(i=1,2\ldots C)$ the SS boundary conditions are as follows:

$$\Omega_n(0)=\sum_{i=1}^{C}\int_0^\infty \Psi_{i,b,n}(\hat{\varepsilon})\upsilon_{i,b}(\hat{\varepsilon})d\hat{\varepsilon}+\int_0^\infty \aleph_{v,n}(\hat{\varepsilon})\psi_v(\hat{\varepsilon})d\hat{\varepsilon};\; n\geq 1 \tag{8}$$

$$\Psi_{i,b,0}(0)=p_i[\int_0^\infty \Omega_1(\hat{\varepsilon})\omega(\hat{\varepsilon})d\hat{\varepsilon}+\epsilon(1-r)\int_0^\infty \Omega_1(\hat{\varepsilon})d\hat{\varepsilon}]+\delta\int_0^\infty \aleph_{v,0}(\hat{\varepsilon})d\hat{\varepsilon};\; n=0 \tag{9}$$

$$\Psi_{i,b,n}(0)=p_i[\int_0^\infty \Omega_{n+1}(\hat{\varepsilon})\omega(\hat{\varepsilon})d\hat{\varepsilon}+\epsilon r\sum_{k=1}^{n}\chi_k\int_0^\infty \Omega_{n-k+1}(\hat{\varepsilon})d\hat{\varepsilon}$$
$$+\epsilon(1-r)\int_0^\infty \Omega_{n+1}(\hat{\varepsilon})d\hat{\varepsilon}]+\delta\int_0^\infty \aleph_{v,n}(\hat{\varepsilon})d\hat{\varepsilon};\; n\geq 0 \tag{10}$$

$$\aleph_{v,n}(0)=\begin{cases}\epsilon\aleph_0;\; n=0\\ 0;\; n\geq 1\end{cases} \tag{11}$$

$$\Theta_{i,n}(\hat{\varepsilon},0)=\eta_i\Psi_{i,b,n}(\hat{\varepsilon});\; n\geq 1; \tag{12}$$

$$\Upsilon_{i,n}(\hat{\varepsilon},0)=\int_0^\infty \Theta_{1,n}(\hat{\varepsilon},\widetilde{\vartheta})\xi_i(\widetilde{\vartheta})d\widetilde{\vartheta};\, n\geq 1 \tag{13}$$

The normalizing condition is

$$\Omega_0 + \aleph_0 + \sum_{n=1}^{\infty}\int_0^{\infty}\Omega_n(\hat{\varepsilon})d\hat{\varepsilon} + \sum_{n=0}^{\infty}\Bigg[\sum_{i=1}^{C}\Bigg(\int_0^{\infty}\Psi_{i,b,n}(\hat{\varepsilon})d\hat{\varepsilon} + \int_0^{\infty}\int_0^{\infty}\Theta_{i,n}(\hat{\varepsilon},\widetilde{\vartheta})d\widetilde{\vartheta}$$
$$+ \int_0^{\infty}\int_0^{\infty}\Upsilon_{i,n}(\hat{\varepsilon},\widetilde{\vartheta})d\widetilde{\vartheta}\Bigg) + \int_0^{\infty}\aleph_{v,n}(\hat{\varepsilon})d\hat{\varepsilon}\Bigg] = 1 \qquad (14)$$

The Steady State Solution

The retrial queueing framework's SS solution is generated with the aid of generating function approach. Furthermore, the GFs for $|\breve{\varphi}| < 1$ are constructed underneath in order to solve the aforementioned equations:

$$\Omega(\hat{\varepsilon},\breve{\varphi}) = \sum_{n=1}^{\infty}\Omega_n(\hat{\varepsilon})\breve{\varphi}^n;\ \Omega(0,\breve{\varphi}) = \sum_{n=1}^{\infty}\Omega_n(0)\breve{\varphi}^n;\ \chi(\breve{\varphi}) = \sum_{n=1}^{\infty}\chi_n\breve{\varphi}^n;$$

$$\Psi_{i,b}(\hat{\varepsilon},\breve{\varphi}) = \sum_{n=0}^{\infty}\Psi_{i,b,n}(\hat{\varepsilon})\breve{\varphi}^n;\ \Psi_{i,b}(0,\breve{\varphi}) = \sum_{n=0}^{\infty}\Psi_{i,b,0}(0)\breve{\varphi}^n;\ i = 1,2\ldots C$$

$$\aleph_v(\hat{\varepsilon},\breve{\varphi}) = \sum_{n=0}^{\infty}\aleph_{v,n}(\hat{\varepsilon})\breve{\varphi}^n;\ \aleph_v(0,\breve{\varphi}) = \sum_{n=0}^{\infty}\aleph_{v,0}(0)\breve{\varphi}^n$$

$$\Theta_i(\hat{\varepsilon},\widetilde{\vartheta},\breve{\varphi}) = \sum_{n=0}^{\infty}\Theta_{i,n}(\hat{\varepsilon},\widetilde{\vartheta})\breve{\varphi}^n;\ \Theta_i(\hat{\varepsilon},0,\breve{\varphi}) = \sum_{n=0}^{\infty}\Theta_{i,0}(\hat{\varepsilon},0)\breve{\varphi}^n;\ i = 1,2\ldots C$$

$$\Upsilon_i(\hat{\varepsilon},\widetilde{\vartheta},\breve{\varphi}) = \sum_{n=0}^{\infty}\Upsilon_{i,n}(\hat{\varepsilon},\widetilde{\vartheta})\breve{\varphi}^n;\ \Upsilon_i(\hat{\varepsilon},0,\breve{\varphi}) = \sum_{n=0}^{\infty}\Upsilon_{i,0}(\hat{\varepsilon},0)\breve{\varphi}^n;\ i = 1,2\ldots C$$

From (3) to (13), multiply the SS equations and boundary conditions by $\breve{\varphi}^n$ and summing on n, and $(n = 0, 1, 2, \ldots)$.

$$\frac{\partial}{\partial\hat{\varepsilon}}\Omega(\hat{\varepsilon},\breve{\varphi}) + (\epsilon + \omega(\hat{\varepsilon}))\Omega(\hat{\varepsilon},\breve{\varphi}) = 0 \qquad (15)$$

$$\frac{\partial}{\partial\hat{\varepsilon}}\Psi_{i,b}(\hat{\varepsilon},\breve{\varphi}) + [\epsilon b(1-\chi(\breve{\varphi})) + \eta_i + \upsilon_{i,b}(\hat{\varepsilon})]\Psi_{i,b}(\hat{\varepsilon},\breve{\varphi}) = \int_0^{\infty}\Upsilon_i(\hat{\varepsilon},\widetilde{\vartheta},\breve{\varphi})\theta_i(\widetilde{\vartheta})d\widetilde{\vartheta} \qquad (16)$$

$$\frac{\partial}{\partial\hat{\varepsilon}}\aleph_v(\hat{\varepsilon},\breve{\varphi}) + [\epsilon b(1-\chi(\breve{\varphi})) + \delta + \psi_v(\hat{\varepsilon})]\aleph_v(\hat{\varepsilon},\breve{\varphi}) = 0 \qquad (17)$$

$$\frac{\partial}{\partial\widetilde{\vartheta}}\Theta_i(\hat{\varepsilon},\widetilde{\vartheta},\breve{\varphi}) + [\epsilon b(1-\chi(\breve{\varphi})) + \xi_i(\widetilde{\vartheta})]\Theta_i(\hat{\varepsilon},\widetilde{\vartheta},\breve{\varphi}) = 0 \qquad (18)$$

$$\frac{\partial}{\partial\widetilde{\vartheta}}\Upsilon_i(\hat{\varepsilon},\widetilde{\vartheta},\breve{\varphi}) + [\epsilon b(1-\chi(\breve{\varphi})) + \theta_i(\widetilde{\vartheta})]\Upsilon_i(\hat{\varepsilon},\widetilde{\vartheta},\breve{\varphi}) = 0 \qquad (19)$$

$$\Omega_n(0,\breve{\varphi}) = \sum_{i=1}^{C}\int_0^{\infty}\Psi_{i,b}(\hat{\varepsilon},\breve{\varphi})\upsilon_{i,b}(\hat{\varepsilon})d\hat{\varepsilon} + \int_0^{\infty}\aleph_v(\hat{\varepsilon},\breve{\varphi})\psi_v(\hat{\varepsilon})d\hat{\varepsilon} \qquad (20)$$

$$\Psi_{i,b}(0,\breve{\varphi}) = p_i[\frac{1}{\breve{\varphi}}\int_0^{\infty}\Omega(\hat{\varepsilon},\breve{\varphi})\omega(\breve{\varphi})d\hat{\varepsilon} + \frac{\epsilon r}{\breve{\varphi}}\chi(\breve{\varphi})\int_0^{\infty}\Omega(\hat{\varepsilon},\breve{\varphi})d\hat{\varepsilon}$$
$$+ \frac{\epsilon(1-r)}{\breve{\varphi}}\int_0^{\infty}\Omega(\hat{\varepsilon},\breve{\varphi})d\hat{\varepsilon}] + \delta\int_0^{\infty}\aleph_v(\hat{\varepsilon},\breve{\varphi})d\hat{\varepsilon} \tag{21}$$

$$\aleph_v(0,\breve{\varphi}) = \epsilon\aleph_0 \tag{22}$$

$$\Theta_i(\hat{\varepsilon},0,\breve{\varphi}) = \eta_i\Psi_{i,b}(\hat{\varepsilon},\breve{\varphi}) \tag{23}$$

$$\Upsilon_i(\hat{\varepsilon},0,\breve{\varphi}) = \int_0^{\infty}\Theta_i(\hat{\varepsilon},\widetilde{\vartheta},\breve{\varphi})\xi_i(\widetilde{\vartheta})d\widetilde{\vartheta} \tag{24}$$

Solving the partial differential equations (15) to (19), we get

$$\Omega(\hat{\varepsilon},\breve{\varphi}) = \Omega(0,\breve{\varphi})[1-\mathcal{B}(\hat{\varepsilon})]e^{-\epsilon\hat{\varepsilon}} \tag{25}$$

$$\Psi_{i,b}(\hat{\varepsilon},\breve{\varphi}) = \Psi_{i,b}(0,\breve{\varphi})[1-\mathcal{Q}_{i,b}(\hat{\varepsilon})]e^{-\mathcal{A}_i(\breve{\varphi})\hat{\varepsilon}};\ (i=1,2\ldots C) \tag{26}$$

$$\aleph_v(\hat{\varepsilon},\breve{\varphi}) = \aleph_v(0,\breve{\varphi})[1-\mathcal{L}_v(\hat{\varepsilon})]e^{-\mathcal{M}(\breve{\varphi})\hat{\varepsilon}} \tag{27}$$

$$\Theta_i(\hat{\varepsilon},\widetilde{\vartheta},\breve{\varphi}) = \Theta_i(\hat{\varepsilon},0,\breve{\varphi})[1-\mathcal{P}_i(\widetilde{\vartheta})]e^{-\mathcal{D}(\breve{\varphi})\widetilde{\vartheta}};\ (i=1,2\ldots C) \tag{28}$$

$$\Upsilon_i(\hat{\varepsilon},\widetilde{\vartheta},\breve{\varphi}) = \Upsilon_i(\hat{\varepsilon},0,\breve{\varphi})[1-\mathcal{G}_i(\widetilde{\vartheta})]e^{-\mathcal{D}(\breve{\varphi})\widetilde{\vartheta}};\ (i=1,2\ldots C). \tag{29}$$

where

$\mathcal{D}(\breve{\varphi}) = \epsilon b(1-\chi(\breve{\varphi}))$; $\mathcal{A}_i(\breve{\varphi}) = \mathcal{D}(\breve{\varphi}) + \eta_i[1-\mathcal{P}_i^*(\mathcal{D}(\breve{\varphi}))\mathcal{G}_i^*(\mathcal{D}(\breve{\varphi}))]$; and $\mathcal{M}(\breve{\varphi}) = \delta + \epsilon b(1-\chi(\breve{\varphi}))$

By applying (26) and (27) in (20) and performing some calculations, we likely get

$$\Omega(0,\breve{\varphi}) = \breve{\varphi}\aleph_0 \times \frac{\left\{\left[\epsilon\mathcal{H}(\breve{\varphi})+\delta\zeta\right]\sum_{i=1}^{C}\mathcal{Q}_{i,b}^*(\mathcal{A}_i(\breve{\varphi})) + \left[\epsilon\left(\mathcal{L}_v^*(\mathcal{M}(\breve{\varphi}))-1\right)-\delta\zeta\right]\right\}}{\breve{\varphi}-\sum_{i=1}^{C}\mathcal{Q}_{i,b}^*(\mathcal{A}_i(\breve{\varphi}))p_i\left[\mathcal{B}^*(\epsilon)+(1-\mathcal{B}^*(\epsilon))[r\chi(\breve{\varphi})+(1-r)]\right]} \tag{30}$$

where,

$$\mathcal{H}(\breve{\varphi}) = \frac{\delta[1-\mathcal{L}_v^*(\mathcal{M}(\breve{\varphi}))]}{\delta+\epsilon b(1-\chi(\breve{\varphi}))}$$

By putting (25) and (27) in (21) and after performing some calculations, we have for $i = 1,2\ldots C$

$$\Psi_{i,b}(0,\breve{\varphi}) = \aleph_0 \times \frac{\left\{\left[\epsilon(\mathcal{L}_v^*(\mathcal{M}(\breve{\varphi}))-1)-\delta\zeta\right]\sum_{i=1}^{C}p_i\left[\mathcal{B}^*(\epsilon)+(1-\mathcal{B}^*(\epsilon))[r\chi(\breve{\varphi})+(1-r)]\right] + \breve{\varphi}(\epsilon\mathcal{H}(\breve{\varphi})+\delta\zeta)\right\}}{\breve{\varphi}-\sum_{i=1}^{C}\mathcal{Q}_{i,b}^*(\mathcal{A}_i(\breve{\varphi}))p_i\left[\mathcal{B}^*(\epsilon)+(1-\mathcal{B}^*(\epsilon))[r\chi(\breve{\varphi})+(1-r)]\right]} \tag{31}$$

Inserting (26) in (23), we obtain for $i = 1, 2 \ldots C$,

$$\Theta_i(\hat{\varepsilon}, 0, \breve{\varphi}) = \aleph_0 \eta_i \times \frac{\left\{\left[\epsilon(\mathcal{L}_v^*(\mathcal{M}(\breve{\varphi})) - 1) - \delta\zeta\right] \sum_{i=1}^{C} p_i\left[\mathcal{B}^*(\epsilon) + (1 - \mathcal{B}^*(\epsilon))[r\chi(\breve{\varphi}) + (1-r)]\right] + \breve{\varphi}(\epsilon\mathcal{H}(\breve{\varphi}) + \delta\zeta)\right\}[1 - Q_{i,b}(\hat{\varepsilon})]e^{-\mathcal{A}_i(\breve{\varphi})\hat{\varepsilon}}}{\breve{\varphi} - \sum_{i=1}^{C} Q_{i,b}^*(\mathcal{A}_i(\breve{\varphi}))p_i\left[\mathcal{B}^*(\epsilon) + (1 - \mathcal{B}^*(\epsilon))[r\chi(\breve{\varphi}) + (1-r)]\right]} \quad (32)$$

Inserting (28) in (24), we obtain for $i = 1, 2 \ldots C$,

$$\Upsilon_i(\hat{\varepsilon}, 0, \breve{\varphi}) = \aleph_0 \eta_i \times \frac{\left\{\left[\epsilon(\mathcal{L}_v^*(\mathcal{M}(\breve{\varphi})) - 1) - \delta\zeta\right] \sum_{i=1}^{C} p_i\left[\mathcal{B}^*(\epsilon) + (1 - \mathcal{B}^*(\epsilon))[r\chi(\breve{\varphi}) + (1-r)]\right] + \breve{\varphi}(\epsilon\mathcal{H}(\breve{\varphi}) + \delta\zeta)\right\}\mathcal{P}_i^*(\mathcal{D}(\breve{\varphi}))[1 - Q_{i,b}(\hat{\varepsilon})]e^{-\mathcal{A}_i(\breve{\varphi})\hat{\varepsilon}}}{\breve{\varphi} - \sum_{i=1}^{C} Q_{i,b}^*(\mathcal{A}_i(\breve{\varphi}))p_i\left[\mathcal{B}^*(\epsilon) + (1 - \mathcal{B}^*(\epsilon))[r\chi(\breve{\varphi}) + (1-r)]\right]} \quad (33)$$

Theorem 2 *Underneath the stability criterion $\rho < 1$, the stationary dist. of the system's client population while the server is vacant, is engaged on ith phase, on BWV, during deferring repair time on ith phase and during repair on ith phase are provided by,*

$$\Omega(\breve{\varphi}) = \frac{\breve{\varphi}\aleph_0\left\{\left[\epsilon\mathcal{H}(\breve{\varphi}) + \delta\zeta\right]\sum_{i=1}^{C} Q_{i,b}^*(\mathcal{A}_i(\breve{\varphi})) + \left[\epsilon\left(\mathcal{L}_v^*(\mathcal{M}(\breve{\varphi})) - 1\right) - \delta\zeta\right]\right\}(1 - \mathcal{B}^*(\epsilon))}{\epsilon\left\{\breve{\varphi} - \sum_{i=1}^{C} Q_{i,b}^*(\mathcal{A}_i(\breve{\varphi}))p_i\left[\mathcal{B}^*(\epsilon) + (1 - \mathcal{B}^*(\epsilon))[r\chi(\breve{\varphi}) + (1-r)]\right]\right\}} \quad (34)$$

$$\Psi_{i,b}(\breve{\varphi}) = \aleph_0 \times \frac{\left\{\left[\epsilon(\mathcal{L}_v^*(\mathcal{M}(\breve{\varphi})) - 1) - \delta\zeta\right] \sum_{i=1}^{C} p_i\left[\mathcal{B}^*(\epsilon) + (1 - \mathcal{B}^*(\epsilon))[r\chi(\breve{\varphi}) + (1-r)]\right] + \breve{\varphi}(\epsilon\mathcal{H}(\breve{\varphi}) + \delta\zeta)\right\}[1 - Q_{i,b}^*(\mathcal{A}_i(\breve{\varphi}))]}{\mathcal{A}_i(\breve{\varphi})\left\{\breve{\varphi} - \sum_{i=1}^{C} Q_{i,b}^*(\mathcal{A}_i(\breve{\varphi}))p_i\left[\mathcal{B}^*(\epsilon) + (1 - \mathcal{B}^*(\epsilon))[r\chi(\breve{\varphi}) + (1-r)]\right]\right\}} \quad (35)$$

$$\aleph_v(\breve{\varphi}) = \frac{\epsilon\aleph_0\mathcal{H}(\breve{\varphi})}{\delta} \quad (36)$$

$$\Theta_i(\breve{\varphi}) = \aleph_0 \eta_i \times \frac{\left\{\left[\epsilon(\mathcal{L}_v^*(\mathcal{M}(\breve{\varphi})) - 1) - \delta\zeta\right] \sum_{i=1}^{C} p_i\left[\mathcal{B}^*(\epsilon) + (1 - \mathcal{B}^*(\epsilon))[r\chi(\breve{\varphi}) + (1-r)]\right] + \breve{\varphi}(\epsilon\mathcal{H}(\breve{\varphi}) + \delta\zeta)\right\}[1 - Q_{i,b}^*(\mathcal{A}_i(\breve{\varphi}))][1 - \mathcal{P}_i^*(\mathcal{D}_i(\breve{\varphi}))]}{\mathcal{A}_i(\breve{\varphi})\mathcal{D}_i(\breve{\varphi})\left\{\breve{\varphi} - \sum_{i=1}^{C} Q_{i,b}^*(\mathcal{A}_i(\breve{\varphi}))p_i\left[\mathcal{B}^*(\epsilon) + (1 - \mathcal{B}^*(\epsilon))[r\chi(\breve{\varphi}) + (1-r)]\right]\right\}} \quad (37)$$

$$\Upsilon_i(\breve{\varphi}) = \aleph_0 \eta_i \times \frac{\left\{ \left[\epsilon(\mathcal{L}_v^*(\mathcal{M}(\breve{\varphi})) - 1) - \delta\zeta\right] \sum_{i=1}^{C} p_i \left[\mathcal{B}^*(\epsilon) + (1 - \mathcal{B}^*(\epsilon))[r\chi(\breve{\varphi}) + (1-r)]\right] + \breve{\varphi}(\epsilon \mathcal{H}(\breve{\varphi}) + \delta\zeta) \right\} \mathcal{P}_i^*(\mathcal{D}(\breve{\varphi}))[1 - \mathcal{Q}_{i,b}^*(\mathcal{A}_i(\breve{\varphi}))][1 - \mathcal{G}_i^*(\mathcal{D}_i(\breve{\varphi}))]}{\mathcal{A}_i(\breve{\varphi})\mathcal{D}_i(\breve{\varphi}) \left\{ \breve{\varphi} - \sum_{i=1}^{C} \mathcal{Q}_{i,b}^*(\mathcal{A}_i(\breve{\varphi})) p_i \left[\mathcal{B}^*(\epsilon) + (1 - \mathcal{B}^*(\epsilon))[r\chi(\breve{\varphi}) + (1-r)]\right] \right\}} \tag{38}$$

where,

$$\aleph_0 = \frac{1 - (1 - \mathcal{B}^*(\epsilon)) r \mathcal{I}_1 + \sum_{i=1}^{C} p_i \tau^{(1)}}{[\frac{\delta\zeta}{\epsilon} + 1 + \frac{\epsilon(1 - \mathcal{L}_v^*(\delta))}{\delta}]\left[1 - (1 - \mathcal{B}^*(\epsilon)) r \mathcal{I}_1 + p_i \tau^{(1)}\right] + \mathcal{Q}_i^{(1)} \left\{ [\epsilon(1 - \mathcal{L}_v^*(\delta)) + \delta\zeta][1 - (1 - \mathcal{B}^*(\epsilon)) r \mathcal{I}_1] - (\frac{\epsilon^2 b \mathcal{I}_1 (1 - \mathcal{L}_v^*(\delta))}{\delta}) \right\} \left[1 + \eta_i \mathcal{P}_i^{(1)} [1 + \epsilon b \mathcal{I}_1 \mathcal{G}_i^{(1)}]\right] + (1 - \mathcal{B}^*(\epsilon)) \left[\frac{\epsilon b (\mathcal{L}_v^*(\delta))}{\delta} \mathcal{I}_1 + [\epsilon(1 - \mathcal{L}_v^*(\delta)) + \delta\zeta] \tau^{(1)}\right]} \tag{39}$$

$$\Omega_0 = \frac{\delta\zeta}{\epsilon} \times \aleph_0 \tag{40}$$

and

$$\tau^{(1)} = \epsilon \mathcal{I}_1 \sum_{i=1}^{C} \mathcal{Q}_i^{(1)} (1 + \eta_i [\mathcal{P}_i^{(1)} + \mathcal{G}_i^{(1)}]).$$

Proof By utilizing the Eqs. (30) to (33) and (22) in (25) to (29) and further by integrating them with respect to $\hat{\varepsilon}$ and $\widetilde{\vartheta}$ and finally by formulating the PGF as $\Omega(\breve{\varphi}) = \int_0^\infty \Omega(\hat{\varepsilon}, \breve{\varphi}) d\hat{\varepsilon}$, $\Psi_{i,b}(\breve{\varphi}) = \int_0^\infty \Psi_{i,b}(\hat{\varepsilon}, \breve{\varphi}) d\hat{\varepsilon}$ $(i = 1, 2 \ldots C)$, $\aleph_v(\breve{\varphi}) = \int_0^\infty \aleph_v(\hat{\varepsilon}, \breve{\varphi}) d\hat{\varepsilon}$, $\Theta_i(\hat{\varepsilon}, \breve{\varphi}) = \int_0^\infty \Theta_i(\hat{\varepsilon}, \widetilde{\vartheta}, \breve{\varphi}) d\widetilde{\vartheta}$, $\Theta_i(\breve{\varphi}) = \int_0^\infty \Theta_i(\hat{\varepsilon}, \breve{\varphi}) d\hat{\varepsilon}$, $\Upsilon_i(\hat{\varepsilon}, \breve{\varphi}) = \int_0^\infty \Upsilon_i(\hat{\varepsilon}, \widetilde{\vartheta}, \breve{\varphi}) d\widetilde{\vartheta}$, $\Upsilon_i(\breve{\varphi}) = \int_0^\infty \Upsilon_i(\hat{\varepsilon}, \breve{\varphi}) d\hat{\varepsilon}$ $(i = 1, 2 \ldots C)$ we get the above mentioned equations. Since the normalizing condition can be used to compute the single unknown, Ω_0, the prob. in which there aren't any clients in the orbit, the server becomes idle. As a result, by using the rule of L-Hospitals whenever necessary and by utilizing $\breve{\varphi} = 1$ in the Eqs. (34)–(38), we get,

$$\Omega_0 + \aleph_0 + \Omega(1) + \aleph_v(1) + \sum_{i=1}^{C} (\Psi_{i,b}(1) + \Theta_i(1) + \Upsilon_i(1)) = 1.$$

Theorem 3 *Given the stability requirement $\rho < 1$. the PGF of the no. of clients in the system and the orbit size dist. at a stationary point of period are estimated as follows,*

$$K_e(\breve{\varphi}) = \frac{N r_e(\breve{\varphi})}{D r_e(\breve{\varphi})} \tag{41}$$

$$Nr_e(\breve{\varphi}) = \aleph_0\Bigg\{b(1-\chi(\breve{\varphi}))\Bigg\{\Bigg[\frac{\epsilon}{\delta}(\delta+\breve{\varphi}\epsilon\mathcal{H}(\breve{\varphi}))+\delta\zeta\Bigg]Dr(\breve{\varphi})+\breve{\varphi}(1-\mathcal{B}^*(\epsilon))$$
$$\times\Bigg[[\epsilon\mathcal{H}(\breve{\varphi})+\delta\zeta]\sum_{i=1}^{C}Q^*_{i,b}(\mathcal{A}_i(\breve{\varphi}))+[\epsilon(\mathcal{L}^*_v(\mathcal{M}(\breve{\varphi}))-1)-\delta\zeta]\Bigg]\Bigg\}$$
$$+\breve{\varphi}\sum_{i=1}^{C}(1-Q^*_{i,b}(\mathcal{A}_i(\breve{\varphi})))\Bigg\{[\epsilon(\mathcal{L}^*_v(\mathcal{M}(\breve{\varphi}))-1)-\delta\zeta]p_i\Bigg[\mathcal{B}^*(\epsilon)$$
$$+(1-\mathcal{B}^*(\epsilon))[r\chi(\breve{\varphi})+(1-r)]\Bigg]+\breve{\varphi}(\epsilon\mathcal{H}(\breve{\varphi})+\delta\zeta)\Bigg\}\Bigg\}$$
$$Dr_e(\breve{\varphi}) = (1-\chi(\breve{\varphi}))Dr(\breve{\varphi})$$

$$R_e(\breve{\varphi}) = \frac{Nr_u(\breve{\varphi})}{Dr_e(\breve{\varphi})} \tag{42}$$

$$Nr_u(\breve{\varphi}) = \aleph_0\Bigg\{b(1-\chi(\breve{\varphi}))\Bigg\{\Bigg[\frac{\epsilon}{\delta}(\delta+\epsilon\mathcal{H}(\breve{\varphi}))+\delta\zeta\Bigg]Dr(\breve{\varphi})+\breve{\varphi}(1-\mathcal{B}^*(\epsilon))$$
$$\times\Bigg[[\epsilon\mathcal{H}(\breve{\varphi})+\delta\zeta]\sum_{i=1}^{C}Q^*_{i,b}(\mathcal{A}_i(\breve{\varphi}))+[\epsilon(\mathcal{L}^*_v(\mathcal{M}(\breve{\varphi}))-1)-\delta\zeta]\Bigg]\Bigg\}$$
$$+\sum_{i=1}^{C}(1-Q^*_{i,b}(\mathcal{A}_i(\breve{\varphi})))\Bigg\{[\epsilon(\mathcal{L}^*_v(\mathcal{M}(\breve{\varphi}))-1)-\delta\zeta]p_i\Bigg[\mathcal{B}^*(\epsilon)$$
$$+(1-\mathcal{B}^*(\epsilon))[r\chi(\breve{\varphi})+(1-r)]\Bigg]+\breve{\varphi}(\epsilon\mathcal{H}(\breve{\varphi})+\delta\zeta)\Bigg\}\Bigg\}$$

where $Dr(\breve{\varphi}) = \Big\{\breve{\varphi}-\sum_{i=1}^{C}Q^*_{i,b}(\mathcal{A}_i(\breve{\varphi}))p_i\big[\mathcal{B}^*(\epsilon)+(1-\mathcal{B}^*(\epsilon))[r\chi(\breve{\varphi})+(1-r)]\big]\Big\}$ *and* Ω_0 *is represented by Eq.* (39).

Proof The following equations are utilized to calculate the PGF for the no. of clients in the system $(K_e(\breve{\varphi}))$ and in the orbit $(R_e(\breve{\varphi}))$. $K_e(\breve{\varphi}) = \Omega_0+\aleph_0+\Omega(\breve{\varphi})+\breve{\varphi}\Big(\aleph_v(\breve{\varphi})+\sum_{i=1}^{C}(\Psi_{i,b}(\breve{\varphi})+\Theta_i(\breve{\varphi})+\Upsilon_i(\breve{\varphi}))\Big)$ and $R_e(\breve{\varphi}) = \Omega_0+\aleph_0+\Omega(\breve{\varphi})+\aleph_v(\breve{\varphi})+\sum_{i=1}^{C}(\Psi_{i,b}(\breve{\varphi})+\Theta_i(\breve{\varphi})+\Upsilon_i(\breve{\varphi}))$. The Eqs. (41) and (42) may be computed directly when the Eqs. (34) to (38) are substituted in the preceding findings. □

5 System Performance Measures

The mean busy time and mean busy cycle of the model, as well as several other pertinent system probability and efficiency metrics, are computed in this section for a variety of system states. Note that the Eqs. (39) and (40) gives the SS prob. when the server is vacant but available in the system. Thus the probabilities of server state are derived from (34) to (38) which is given below

1. Let Ω represent the SS prob. the server remains inactive while on the retrial time,

$$\Omega = lim_{\breve{\varphi}\to 1}\Omega(\breve{\varphi})$$
$$= \aleph_0(1-\mathcal{B}^*(\epsilon))\frac{\left\{\frac{\epsilon b}{\delta}(1-\mathcal{L}_v^*(\delta))\mathcal{I}_1+[\epsilon(1-\mathcal{L}_v^*(\delta))+\delta\zeta]\tau^{(1)}\right\}}{1-(1-\mathcal{B}^*(\epsilon))r\mathcal{I}_1+\sum_{i=1}^{C}p_i\tau^{(1)}}$$

2. Let Ψ_b represent the SS prob. that the server is busy during ith phase service

$$\Psi_b = lim_{\breve{\varphi}\to 1}\sum_{i=1}^{C}\Psi_{i,b}(\breve{\varphi})$$
$$= \aleph_0 Q_i^{(1)}\frac{\left\{[\epsilon(1-\mathcal{L}_v^*(\delta))+\delta\zeta][1-(1-\mathcal{B}^*(\epsilon))r\mathcal{I}_1]-\left(\frac{\epsilon^2 b\mathcal{I}_1(1-\mathcal{L}_v^*(\delta))}{\delta}\right)\right\}}{1-(1-\mathcal{B}^*(\epsilon))r\mathcal{I}_1+\sum_{i=1}^{C}p_i\tau^{(1)}}$$

3. Let $\aleph_v$ represent the SS prob. that the server is under BWV

$$\aleph_v = lim_{\breve{\varphi}\to 1}\aleph_v(\breve{\varphi}) = \frac{\epsilon\aleph_0}{\delta}(1-\mathcal{L}_v^*(\delta))$$

4. Let Θ represent the SS prob. that the server is under delaying repair time during ith phase of service

$$\Theta = lim_{\breve{\varphi}\to 1}\sum_{i=1}^{C}\Theta_i(\breve{\varphi})$$
$$= \aleph_0 Q_i^{(1)}\mathcal{P}_i^{(1)}\frac{\left\{[\epsilon(1-\mathcal{L}_v^*(\delta))+\delta\zeta][1-(1-\mathcal{B}^*(\epsilon))r\mathcal{I}_1]-\left(\frac{\epsilon^2 b\mathcal{I}_1(1-\mathcal{L}_v^*(\delta))}{\delta}\right)\right\}}{1-(1-\mathcal{B}^*(\epsilon))r\mathcal{I}_1+\sum_{i=1}^{C}p_i\tau^{(1)}}$$

5. Let Υ represent the SS prob. that the server is under repair time during ith phase of service

$$\Upsilon = lim_{\breve{\varphi}\to 1}\sum_{i=1}^{C}\Upsilon_i(\breve{\varphi})$$

$$= \aleph_0 \epsilon b\mathcal{I}_1 Q_i^{(1)}\mathcal{G}_i^{(1)}\mathcal{P}_i^{(1)}\frac{\left\{[\epsilon(1-\mathcal{L}_v^*(\delta))+\delta\zeta][1-(1-\mathcal{B}^*(\epsilon))r\mathcal{I}_1]-(\frac{\epsilon^2 b\mathcal{I}_1(1-\mathcal{L}_v^*(\delta))}{\delta})\right\}}{1-(1-\mathcal{B}^*(\epsilon))r\mathcal{I}_1+\sum_{i=1}^{C}p_i\tau^{(1)}}$$

Average System Size and Orbit Size

Upon entering an SS, (i) By differentiating (42) with respect to $\breve{\varphi}$ and providing $\breve{\varphi} = 1$ under stability conditions, the mean no. of clients in the orbit (L_q) is computed.

$$L_q = R_e^{'}(1) = \lim_{\breve{\varphi}\to 1}\frac{d}{d\breve{\varphi}}R_e(\breve{\varphi}) = \aleph_0\left[\frac{Nr_q^{'''}(1)Dr_q^{''}(1) - Dr_q^{'''}(1)Nr_q^{''}(1)}{3(Dr_q^{''}(1))^2}\right] \tag{43}$$

$$\begin{aligned} Nr_q^{''}(1) = & -2\left\{b\mathcal{I}_1\left[\frac{\epsilon}{\delta}(1-\mathcal{L}_v^*(\delta))[\epsilon\nu + b\mathcal{I}_1(1-\mathcal{B}^*(\epsilon)) + \delta\tau^{(1)}] + \nu[\epsilon+\delta\zeta]\right]\right.\\ & + \tau^{(1)}\left[[\epsilon(1-\mathcal{L}_v^*(\delta))+\delta\zeta][1-\sum_{i=1}^{C}p_i(1-\mathcal{B}^*(\epsilon))r\mathcal{I}_1]\right.\\ & \left.\left. +\epsilon\sum_{i=1}^{C}p_i\mathcal{L}_v^{*'}(\delta)+\epsilon\mathcal{H}^{'}(\breve{\varphi})\right]\right\}\\ Nr_q^{'''}(1) = & -3\left\{b\mathcal{I}_2\left[\left(\epsilon+\frac{\epsilon^2}{\delta}(1-\mathcal{L}_v^*(\delta))+\delta\zeta\right)\nu + (1-\mathcal{B}^*(\epsilon))\left[\frac{\epsilon b\mathcal{I}_1(1-\mathcal{L}_v^*(\delta))}{\delta}\right.\right.\right.\\ & \left.\left. +\epsilon(1-\mathcal{L}_v^*(\delta))\tau^{(1)}\right]\right] + b\mathcal{I}_1\left[\left(\epsilon+\frac{\epsilon^2}{\delta}(1-\mathcal{L}_v^*(\delta))+\delta\zeta\right)\nu^1 + 2\mathcal{H}^{'}(\breve{\varphi})\right.\\ & \left.\times\left[\frac{\epsilon^2}{\delta}\nu + (1-\mathcal{B}^*(\epsilon))[\epsilon+\tau^{(1)}]\right] + 2\epsilon(1-\mathcal{B}^*(\epsilon))(1-\mathcal{L}_v^*(\delta))\tau^{(1)}\right]\\ & + \left(\epsilon(1-\mathcal{L}_v^*(\delta))+\delta\zeta\right)\left[1+(1-\mathcal{B}^*(\epsilon))\left[\mathcal{I}_1\tau^{(2)}\left[b-\sum_{i=1}^{C}p_i r\right]\right.\right.\\ & \left.\left. - r\mathcal{I}_2\sum_{i=1}^{C}p_i\tau^{(1)}\right]\right] + \epsilon\left[(1-\mathcal{B}^*(\epsilon))b\mathcal{I}_1+\tau^{(1)}\right]\left[\mathcal{H}^{''}(\breve{\varphi})+\mathcal{L}_v^{*''}(\delta)\right]\\ & - \tau^{(2)}\sum_{i=1}^{C}p_i\mathcal{L}_v^*(\delta) + \epsilon\mathcal{H}^{'}(\breve{\varphi})[2\tau^{(1)}+\tau^{(2)}] + 2\mathcal{I}_1 - (1-\mathcal{B}^*(\epsilon))\\ & \left.\times\mathcal{L}_v^{*'}(\delta)[\epsilon+\sum_{i=1}^{C}p_i\tau^{(1)}r]\right\} \end{aligned}$$

$$Dr_q^{''}(1) = -2b\mathcal{I}_1\big[1-(1-\mathcal{B}^*(\epsilon))r\mathcal{I}_1 - \sum_{i=1}^{C} p_i\tau^{(1)}\big]$$

$$Dr_q^{'''}(1) = -3b\Big[2\mathcal{I}_1(1-\mathcal{B}^*(\epsilon))r\mathcal{I}_2 + \sum_{i=1}^{C} p_i\tau^{(1)}[2(1-\mathcal{B}^*(\epsilon))r(\mathcal{I}_1)^2 + \mathcal{I}_2] - \mathcal{I}_2 + \sum_{i=1}^{C} p_i\tau^{(2)}\mathcal{I}_1\Big]$$

(ii) By differentiating (41) with regard to $\breve{\varphi}$ and providing $\breve{\varphi} = 1$ under stability conditions, the mean no. of clients in the orbit (L_s) is computed.

$$L_s = K_e^{'}(1) = \lim_{\breve{\varphi}\to 1}\frac{d}{d\breve{\varphi}}K_e(\breve{\varphi}) = \aleph_0\left[\frac{Nr_s^{'''}(1)Dr_q^{''}(1) - Dr_q^{'''}(1)Nr_q^{''}(1)}{3(Dr_q^{''}(1))^2}\right] \tag{44}$$

$$Nr_s^{'''}(1) = Nr_q^{'''}(1) - 6\frac{\epsilon^2}{\delta}\nu(1-\mathcal{L}_v^*(\delta)) - 6\tau^{(1)}\Big\{\big[\epsilon(1-\mathcal{L}_v^*(\delta)) + \delta\zeta\big]\big[1+\sum_{i=1}^{C} p_i(1-\mathcal{B}^*(\epsilon))r\mathcal{I}_1\big] + \sum_{i=1}^{C} p_i(1-\mathcal{L}_v^{*'}(\delta)) + \epsilon\mathcal{H}^{'}(\breve{\varphi})\Big\} \tag{45}$$

(iii) With the aid of Little's approach, it is possible to predict the amount of time a user can expect to spend in the system (W_s) and the amount of time a user can expect to spend in the queue (W_q), (i.e.) $W_s = \frac{L_s}{\epsilon\mathcal{I}_1}$ and $W_q = \frac{L_q}{\epsilon\mathcal{I}_1}$.

where,

$$\tau^{(2)} = \sum_{i=1}^{C}\Big\{-\tau^{(1)}\Big[\epsilon\mathcal{I}_2 + \eta_i[\epsilon\mathcal{I}_2[\mathcal{P}_i^{(1)} + \mathcal{G}_i^{(1)}] - \eta_i(\epsilon\mathcal{I}_1)^2][\mathcal{P}_i^{(2)} + \mathcal{G}_i^{(2)}]\Big] + (\epsilon\mathcal{I}_1\mathcal{Q}_i^{(1)})^2\mathcal{Q}_i^{(2)}(1+\eta_i[\mathcal{P}_i^{(1)} + \mathcal{G}_i^{(1)}])\Big\}$$

$$\nu = 1-(1-\mathcal{B}^*(\epsilon))r\mathcal{I}_1 + \sum_{i=1}^{C} p_i\tau^{(1)};$$

$$\nu^{(1)} = \sum_{i=1}^{C} p_i\tau^{(2)} - 2\sum_{i=1}^{C} p_i\tau^{(1)}(1-\mathcal{B}^*(\epsilon))r\mathcal{I}_1 - (1-\mathcal{B}^*(\epsilon))r\mathcal{I}_2$$

$$\mathcal{H}^{'}(\breve{\varphi}) = \frac{\epsilon}{\delta}\Big(1-\mathcal{L}_v^*(\delta) + \delta\mathcal{L}_v^{*'}(\delta)\Big);\ \mathcal{H}^{''}(\breve{\varphi}) = \Big(\frac{\epsilon}{\delta}\Big)^2\Big[2\Big(1-\mathcal{L}_v^*(\delta) + \delta\mathcal{L}_v^{*'}(\delta)\Big) - (\delta)^2\mathcal{L}_v^{*''}(\delta)\Big]$$

6 Numerical Examples

To demonstrate the many alternatives for the system's dynamic responsiveness, we'll employ MATLAB in this part. In addition, the exponentially distributed retrial times, dual-stage and re-service, vacation and repair times haven been analysed significantly. Random selection is made in order to determine whether numerical measures satisfy the stability requirements. The proximal values of distinct parameters of several frameworks, including the prob. that the server is inactive Ω_0, the mean queue size (L_q) and the mean queue waiting time (W_q), are presented in Tables 1, 2, 3 and 4.

Table 1 depicts that as arrival rate (ϵ) rises, L_q and W_q rises, whereas Ψ_b and $\aleph_v$ declines for the value of $\aleph_v = 0.6$, $\zeta = 0.5$ $b = 0.7$, $r = 0.8$, $\eta = 4$, $\theta = 2.1$.

Table 2 displays that as the service rate ($\upsilon_{i,b}$) rises, L_s, W_q and Θ reduces, whereas Ω_0 mounts for the value of $\epsilon = 1.2$, $\aleph_v = 0.6$, $\zeta = 0.5$ $b = 0.7$, $r = 0.8$, $\eta = 4$, $\theta = 2.1$.

Table 3 displays that as the balking prob. (b) mounts, $\aleph_0$ and $\aleph_v$ also mounts but L_q and Ψ_b decreases for the value of $\epsilon = 1.2$, $\zeta = 0.5$ $b = 0.7$, $r = 0.8$, $\eta = 4$, $\theta = 2.1$, $\omega = 1$.

Table 4 displays that as BWV rate (ψ_v) mounts, Ω_0 and $\aleph_0$ declines whereas Θ and Υ declines for the value of $\epsilon = 1.8$, $\zeta = 0.5$ $b = 0.7$, $r = 0.8$, $\eta = 4$, $\zeta = 2.1$, $\omega = 1$.

Here, Fig. 1a depicts that as ϵ elevates, W_q also elevates whereas Ψ_b diminish. Figure 1b depicts that as $\upsilon_{i,b}$ mounts, Ω_0 mounts whereas L_s reduces. Figure 1c

Table 1 The effect of arrival rate (ϵ) on L_q, W_q, Ψ_b, $\aleph_v$

Arrival rate (ϵ)	L_q	W_q	Ψ_b	$\aleph_v$
3.01	0.2886	8.9134	0.0651	1.3433
3.03	0.3307	10.4698	0.0650	1.3363
3.05	0.5593	12.0600	0.0650	1.3294
3.07	0.9709	13.6840	0.0649	1.3227
3.09	1.4946	15.3422	0.0648	1.3162

Table 2 The effect of service rate ($\upsilon_{i,b}$) on Ω_0, L_s, W_q, Θ

Service rate ($\upsilon_{i,b}$)	Ω_0	L_s	W_q	Θ
5.1	0.1878	0.0524	13.1545	0.0201
5.2	0.1881	0.0499	12.3412	0.0199
5.3	0.1885	0.0480	11.3952	0.0196
5.4	0.1888	0.0466	10.2929	0.0194
5.5	0.1891	0.0454	9.0064	0.0192

Table 3 The effect of balking prob. (b) on $\aleph_0$, L_q, $\aleph_v$, Ψ_b

Balking prob. (b)	$\aleph_0$	L_q	$\aleph_v$	Ψ_b
0.60	0.2600	0.0186	0.5211	0.0520
0.65	0.2633	0.0135	0.5277	0.0486
0.70	0.2668	0.0109	0.5346	0.0452
0.75	0.2705	0.0093	0.5420	0.0416
0.80	0.2745	0.0082	0.5498	0.0380

Table 4 The effect of BWV rate (ψ_v) on Ω_0, $\aleph_0$, Θ, Υ

BWV rate (ψ_v)	Ω_0	$\aleph_0$	Θ	Υ
0.6	0.0361	0.2898	0.0305	0.0073
0.7	0.0424	0.3403	0.0303	0.0072
0.8	0.0514	0.4122	0.0300	0.0072
0.9	0.0651	0.5224	0.0296	0.0071
1.0	0.0889	0.7133	0.0289	0.0069

depicts that as b rises $\aleph_0$ and $\aleph_v$ also rises. In Fig. 1d as $\aleph_v$ mounts, Ω_0 rises whereas Θ diminish.

The influence of diverse characteristics on the system's performance standards can perhaps be ascertained using the numerical findings shown above.

7 Cost Analysis

The cost optimization approach is employed to estimate the optimal parameters for the service rates (τ_1, τ_v). A linear cost structure in terms of cost elements tied to diverse system activities is presumed for the expected cost function.

For per unit of time (PUT), the succeeding cost components are included in the expected total cost function TC (τ_1, τ_v,):

$\mathcal{N}_h$ – Each consumer's holding cost per unit time spent within the system
$\mathcal{N}_b$ – Cost per unit time when the server is normally active (both phases)
$\mathcal{N}_v$ – Cost per unit time when the server is in vacation
$\mathcal{N}_f$ – Cost per unit time when the server is under repair (both phases)
$\mathcal{N}_1$ – Cost per consumer served during server's service mode
$\mathcal{N}_2$ – Cost per consumer served during server's vacation mode

$$TC = \mathcal{N}_h L_q + \mathcal{N}_b \Psi_b + \mathcal{N}_v \aleph_v + \mathcal{N}_f \Upsilon + \mathcal{N}_1 \tau_1 + \mathcal{N}_2 \tau_v \tag{46}$$

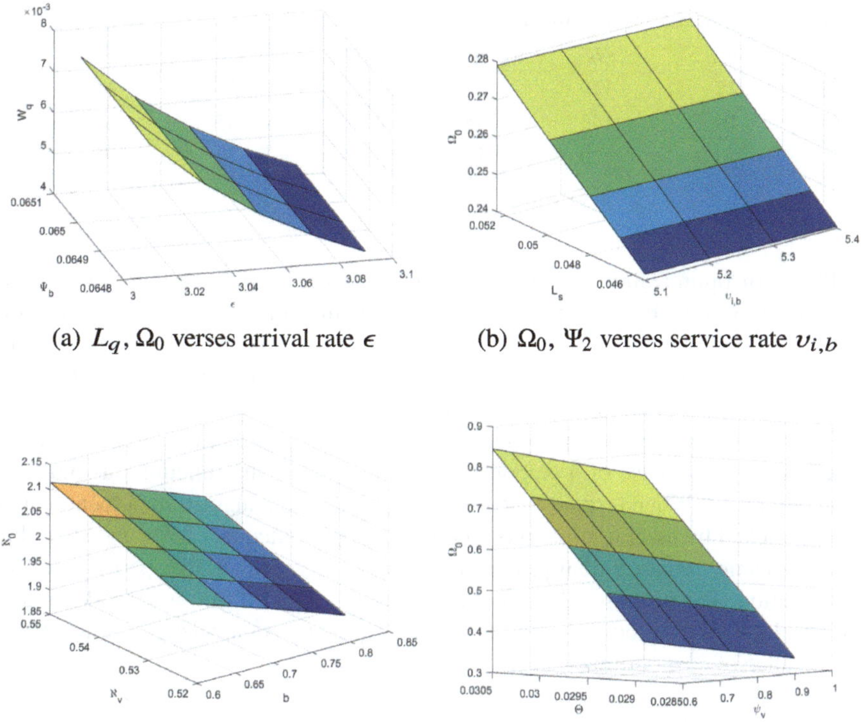

(a) L_q, Ω_0 verses arrival rate ϵ

(b) Ω_0, Ψ_2 verses service rate $\upsilon_{i,b}$

(c) L_s, $\aleph_2$ verses balking prob. b

(d) W_s, Υ_2 verses BWV rate ψ_v

Fig. 1 The impact of various parameters depicted via three dimensional graphs

The difficulty in analytically optimizing the cost function described by (46) stems from its considerable non-linearity. Thus, we infer that total cost is a function of service rates (τ_1, τ_v) and employ a heuristic method to optimize it.

Cost Optimization

For any given objective function, "optimization" refers to the process of discovering its "optimal" or "minimal" collection of inputs. Cost optimization is a business-centric, iterative process that seeks to lessen outlays and boost profitability. Our primary goal is to minimize the total cost function in order to determine the optimal service rates τ_1^* and τ_v^* for the server's busy mode and its vacation mode respectively. The cost-minimization problem can be expressed mathematically as:

$$TC(\tau_1^*, \tau_v^*) = \underset{\tau_1^*, \tau_v^*}{Minimize} TC(\tau_1, \tau_v)$$

Table 5 contains the values of the cost elements used to conduct a graphical analysis of the cost function's sensitivity.

Table 5 Different cost sets for cost analysis

Cost sets	$\mathcal{N}_h$	$\mathcal{N}_b$	$\mathcal{N}_v$	$\mathcal{N}_f$	$\mathcal{N}_1$	$\mathcal{N}_2$
Set 1	25	110	25	50	55	60
Set 2	10	145	35	45	35	75
Set 3	20	120	15	75	20	40

Plenty of optimization strategies have been constructed since the 1960s. Each of these algorithms has proven its mettle in implementing on various optimization problems. We ought to employ a local optimization strategy when we have an inkling that we are close to the global optimum, or when our objective function has a single optimum, as in the case of a unimodal distribution; we should use a global optimization method when we have a poor idea of the structure of the objective function (as in the case of a response surface), or when the function has multiple local optimums. When a situation calls for a global search algorithm, a local search will be tricked by local optima into producing subpar outcomes. We conducted this research using certain global search optimization approaches like particle swarm optimization (PSO), artificial bee colony (ABC), grey wolf optimizer (GWO) and differential evolution optimization (DE) given that we acknowledge the need and significance of cost optimization.

Particle Swarm Optimization (PSO)

In 1995, Kennedy and Eberhart [11] introduced the world to particle swarm optimization (PSO), a technique for optimizing continuous non-linear functions that was driven by initial research on the behaviour of bird flocks as simulated in computer programmes. Personal bests (pbest) and global bests (gbest) are obtained by assigning a fitness value to each particle. After then, the 'pbest' value of a particle is utilized to replace the 'gbest' value if it is more precise. Until the optimum amount of iterations has been achieved, this procedure is repeated. Several modifications to PSO have been examined by researchers in an effort to address the local minimum and boost PSO's productivity as a multimodal functional approach. Upadhyaya [31] elaborated on this method to talk about cost optimization in a discrete-time retrial queue with Bernoulli feedback and starting failure. For a single-server recurrent system with state-dependent service, Zhang et al. [36] generated cost and computational solutions using a PSO algorithmic programme. For further information on how PSO works, the study by Malik et al. [17] has been referred to. Algorithm 1 lays out the PSO algorithm's pseudo coded sequence of operations. In addition, Table 6 details the effect of ϵ, δ, η on total minimal cost TC using PSO optimization approaches, on the optimal service rate pairs (τ_1^*, τ_v^*).

Artificial Bee Colony Optimization (ABC)

The artificial bee colony (ABC) technique was proposed by Karaboga and Basturk [10], and it is a swarm-based solution for optimization issues that takes its cues

Table 6 Effect of ϵ, δ, η on (TC, τ_1^*, τ_v^*) using PSO

Parameters	TC			
		Cost set 1	Cost set 2	Cost set 3
ϵ	3	(119.8539)	(121.2948)	(98.6297)
	3.5	(123.1613)	(124.7386)	(101.8343)
	4	(154.9853)	(158.9167)	(131.8393)
δ	1	(115.2614)	(116.3027)	(94.3839)
	1.5	(115.7542)	(116.8894)	(94.7904)
	2	(116.7078)	(117.9254)	(95.6726)
η	4	(120.9331)	(122.4926)	(114.2839)
	4.5	(118.6514)	(120.0240)	(109.6129)
	5	(116.8275)	(118.0507)	(106.0928)

Table 7 Effect of ϵ, δ, η on TC using ABC

Parameters	TC			
		Cost set 1	Cost set 2	Cost set 3
ϵ	3	(71.6183)	(73.0862)	(60.3846)
	3.5	(78.3832)	(80.1298)	(66.9379)
	4	(110.4571)	(131.4128)	(107.2688)
δ	1	(65.3723)	(66.3220)	(54.5859)
	1.5	(66.1337)	(67.2112)	(55.2303)
	2	(67.5286)	(68.7181)	(56.5290)
η	4	(71.4954)	(72.9802)	(60.2436)
	4.5	(69.1300)	(70.4352)	(58.0319)
	5	(67.2428)	(68.4046)	(56.2673)

from the smart conduct of honey bees during foraging. There are two main parts to the algorithm: "foraging" and "food source." Foraging bees might be classified as workers, observers or scouts, depending on their roles in the current situation. Both working and jobless bees use foraging to find good food sources. In this algorithm (a good answer), a colony of robotic forager bees searches for plentiful synthetic food sources. In order to optimize the objective function with this method, the correct parameter vector must be used. Then, the robotic bees find a random distribution of starting solution vectors. The closest neighbour search method is employed during the iterative process of reducing the error. The probes are put through their paces in a variety of benchmarking situations. The findings indicate that ABC is either superior to, equivalent to or offers the advantage of fewer control factors than some of the other population approaches. The phases of the ABC method are described in pseudocode in Algorithm 2. Moreover, the impacts of ϵ, δ, η on optimal service rate pairs (τ_1^*, τ_v^*), respectively, as well as total minimal cost TC via ABC optimization methods are laid out in Table 7.

Table 8 Effect of ϵ, δ, η on TC using GWO

Parameters	TC			
		Cost set 1	Cost set 2	Cost set 3
ϵ	3	(120.5566)	(122.0099)	(99.3263)
	3.5	(125.0565)	(126.7030)	(103.6788)
	4	(146.9448)	(144.2378)	(131.9470)
δ	1	(116.2781)	(117.3307)	(95.3969)
	1.5	(116.6181)	(117.7799)	(95.6345)
	2	(117.3189)	(118.5620)	(96.2631)
η	4	(124.9759)	(126.8383)	(106.1517)
	4.5	(122.1496)	(123.7789)	(103.1067)
	5	(119.9104)	(121.3550)	(100.7158)

Grey Wolf Optimizer (GWO)

One of the more prominent recent swarm intelligence metaheuristics is called the Grey Wolf Optimizer (GWO). Due to its superior qualities over other swarm intelligence approaches, such as its small number of parameters and the absence of a need for derivation information in the first search, it has been extensively adapted for a wide range of optimization issues. Furthermore, it is straightforward, user-friendly, adaptable and scalable, with a unique capacity to find a balance between exploration and exploitation during the search, resulting in favorable convergence. As a result, the GWO has quickly attracted a massive readership from a variety of study fields. Mirjalili et al. [19] was the pioneer of GWO. The GWO algorithm was designed after observing grey wolves in the wild as they looked for the most efficient means of catching their meal. The GWO algorithm uses the same method found in nature, which is based on the wolf pack's natural hierarchy. The phases of the GWO method are described in pseudocode in Algorithm 3. Moreover, the impacts of ϵ, δ, η on optimal service rate pairs (τ_1^*, τ_v^*), respectively, as well as total minimal cost TC via GWO optimization methods are laid out in Table 8.

Differential Evolution (DE)

Differential Evolution (DE) is an optimization algorithm for tackling non-linear optimization problems using a stochastic population-based approach. It optimizes a given issue by repeatedly searching for better solutions until one is identified that meets certain specified criteria. Storn and Price [30] introduced the algorithm to the world. The conceptual simplicity and ease of use of DE are its primary assets, with its good convergence properties and adaptability for parallelization. In order to find the optimal solution to a problem, DE first generates a group of applicants, which are then combined and crossed across to produce additional applicants, from which the best is ultimately chosen. Moreover, DE's control variables are held constant throughout the optimization process, making it a breeze to implement. The phases of the DE method are described in pseudocode in Algorithm 4. Moreover, the impacts

Table 9 Effect of ϵ, δ, η on TC using DE

Parameters	TC			
		Cost set 1	Cost set 2	Cost set 3
ϵ	3	(116.9243)	(118.1953)	(95.8039)
	3.5	(118.5462)	(119.8744)	(97.3824)
	4	(129.1828)	(131.0012)	(107.6163)
δ	1	(114.7149)	(115.7588)	(93.7990)
	1.5	(114.9086)	(116.0201)	(93.9284)
	2	(115.2943)	(116.4543)	(94.2700)
η	4	(122.8509)	(124.6108)	(101.2972)
	4.5	(120.2189)	(121.7617)	(98.8576)
	5	(118.1387)	(119.5099)	(96.9295)

of ϵ, δ, η on optimal service rate pairs (τ_1^*, τ_v^*), respectively, as well as total minimal cost TC via DE optimization methods are laid out in Table 9.

Comparative Analysis within PSO, ABC, GWO and DE

Here we examine five methods, namely PSO, ABC, GWO and DE to estimate the lowest possible cost using their respective MATLAB programmes. In this case, we analyse four distinct sets of costs, depicted in Table 5. The MATLAB programmes for each of these algorithms are then executed in turn. As a result of carrying this approach forward, Tables 6, 7, 8 and 9 have been created.

There was considerable similarity between the outcomes of the four programmes. Thus, the optimal solutions of those four strategies are adjacent to one another. This proves that the aforementioned heuristics provide trustworthy, optimal solutions. Based on the data shown in Tables 6, 7, 8 and 9, ABC has the lowest ideal cost value. Thus, we can use any strategy to determine the ideal cost; nevertheless, as we juxtapose for the proposed framework, DE is a highly profitable approach for determining the best feasible cost. It is simple to configure, performs admirably in global queries and is insensitive to scaling changes in design variables. ABC tends to lead to a slow and quick convergence in mid-optimal locations, as well as a ponderous convergence in a widened search domain.

At the outset of any optimization method- PSO, ABC, GWO or DE -the particles are not stable, and it is vital to know whether or not they return to normal and whether or not they roam around in search of a better solution. That's why convergence is so important to cost analysis. Here Fig. 2 depicts the convergence of the cost function via PSO, ABC, GWO and DE techniques. In addition, by employing the aforementioned techniques, we have been able to arrive at combined optimum values, which offer the lowest predicted costs.

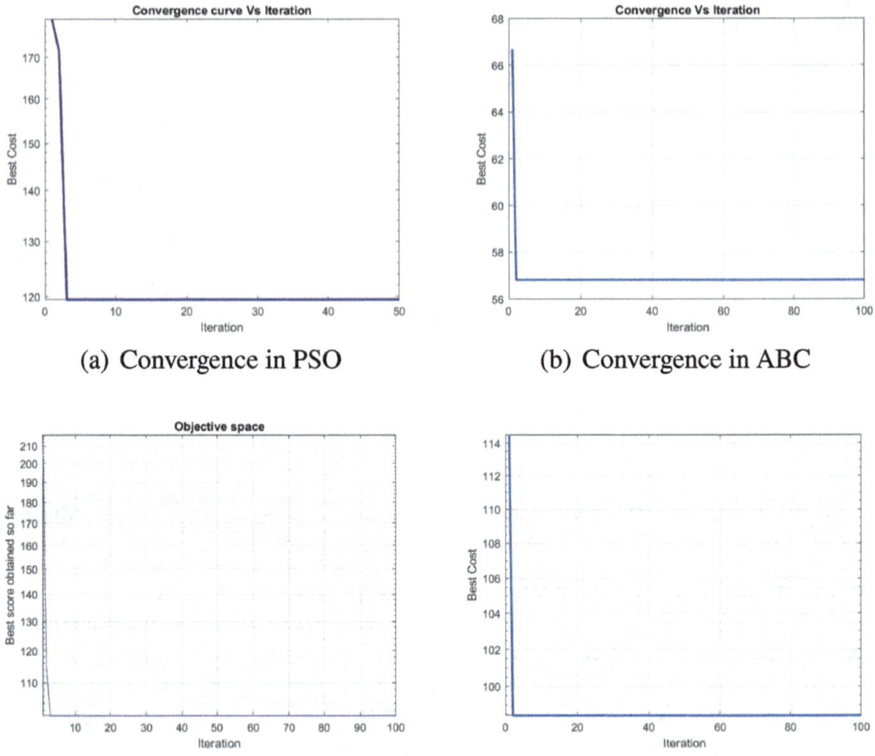

(a) Convergence in PSO (b) Convergence in ABC

(c) Convergence in GWO (d) Convergence in DE

Fig. 2 Convergence of the cost function

Convergence in PSO, ABC, GWO, DE and BA

Knowing whether or not a particle lays off and how it will wander over in search of a more effective outcome is actually crucial when applying optimization techniques like PSO, ABC, GWO or DE, because the constituent parts (people, groups or signals) are inherently out of equilibrium. Hence, cost analysis convergence is crucial here since it eliminates all of these challenges. Tables 6, 7, 8 and 9 display the typical outcomes of applying the methods PSO, ABC, GWO and DE to every one of the cost sets listed in Table 5. Further, as shown in Fig. 2, particles converge to the optimal solution following certain trials (generations) in PSO, ABC, GWO and DE. After a brief amount of instances (generations), the particles in all four methods converge on the minimum possible cost. Overall, the cost study above suggests that ABC converges faster. We acknowledge that the overall estimated cost is sensitive to the parameters chosen for priority, voluntary service and vacation, and that admins should exercise great caution when making these choices. The analysts can afford this system in its entirety, which will help alleviate some of their financial strain. Besides demonstrating the reliability of our model, the cost analysis's close resemblance to real-world examples which helps system designers and analysts to minimize problems.

8 Conclusion

In this work, a bulk arrival retrial QS with multi-phase of heterogeneous service phases under BWV has been examined. Additionally, it is presumed that the consumers are allowed to balk and renege. Further, the server may undergo delaying repair as well. It is determined what conditions must be met for the system to remain stable. Some of the existing findings are drawn from the paradigm under study as special cases. When developing such models in a wide range of real-world circumstances, the analytical conclusions that have been verified by numerical examples are beneficial. Therefore, our model is more flexible to deal with the real-time systems of many different business units in multiple actual queueing circumstances.

Declarations

Conflicts of Interest: The authors report no conflict of interest.

Authors Contribution: All the authors made substantial contributions to the conception or design of the work.

Acknowledgments: Not applicable.

Availability of Data: Not applicable.

Funding: Not applicable.

Appendix

See Algorithms 1, 2, 3 and 4.

Algorithm 1 Pseudo Code for PSO Algorithm

INPUT: Objective function $=TC(\tau_b, \tau_v)$,
OUTPUT: The cost function's value $TC(\tau_b^*, \tau_v^*)$
Initialization
for each particle j=1,2,...N, **do**
Start with a uniform distribution of particle positions as $M_j(0) \sim (LB, UB)$, in which LB and UB are the lower and upper limits of the search area.
Setup abest to its starting position abest(j, 0)= M_j(0.)
Initialize cbest to the minimal value of the swarm:$cbest(0) = argmin\ f[M_j(0).]$
end for
Initiate the velocity: $V_j \sim U(-|UB - LB|, |UB - LB|)$.
Iterate until the endpoint is reached
for each particle j=1,2,...N **do**
Choose arbitrary numbers:$s_1, s_2 \sim U(0, 1)$.
Enhance the particle's velocity.
Enhance the particle's position.
Suppose $[M_j(0).] \leq [abest(j, t)(0)]$
Update the particle's current predicted position. $j : abest(j, t) = M_j(t)$.
Suppose $[M_j(0)] \leq [cbest(t)(0)]$, enhance the swarm's best known position: $j : abest(t) = M_j(t)$. t ←(t+1;)
end for
$cbest(t)$ holds the optimal found solution.

Algorithm 2 Pseudo Code for ABC Algorithm

INPUT: Objective function $=TC(\tau_b, \tau_v)$,
OUTPUT: The cost function's value $TC(\tau_b^*, \tau_v^*)$
Generate an assembly of outcome $M_i, i = 1$
Check out the assembly, period 1, $h = 0$
Choose the optimal outcome, M best and fix M best1 = M best
Redo
Obtain a new way out Mnew = M_i for the worker bees and to get them.
Use the greedy selection method for the worker bees.
Give each result a grade, and then choose the best one.
Find the probability P_i of the solution M_i.
With P_1 as a starting point, generate a fresh outcome M_i for the onlookers.
Use the greedy selection method to the onlookers.
If the scout's result has been cancelled, then proceed with a newly created result M_i.
Keep in mind the M new conclusion you've reached thus far.
Put $h = h + 1$ period = period + 1.
Until (the end condition has been met, i.e., period = MCN)

Algorithm 3 Pseudo Code for GWO Algorithm

INPUT: Objective function $=TC(\tau_1, \tau_v), a, B, D$
OUTPUT: The cost function's value $TC(\tau_1^*, \tau_v^*)$
Initiate the grey wolf population $\mathcal{Y}_i$, i= 1,n
Initiate a, B, D
Estimate the fitness of every search factor
$\mathcal{Y}_\alpha$ = the best search factor
$\mathcal{Y}_\beta$ = the second best search factor
$\mathcal{Y}_\delta$ = the third best search factor
while t < maximum no. of iteration **do**
  for every search factor **do**
    Arbitrarily initiate c_1 and c_2
    Update the place of the present search factor
    Update a, B and D
  end for
  Estimate the fitness of every search factors
  Update $\mathcal{Y}_\alpha$, $\mathcal{Y}_\beta$ and $\mathcal{Y}_\delta$
  t=t+1
end while
return $\mathcal{Y}_\alpha$

Algorithm 4 Pseudo Code for DE Algorithm

INPUT: Objective function = $TC(\tau_1, \tau_v)$, no. of iterations, crossover, mutation
OUTPUT: The cost function's value $TC(\tau_1^*, \tau_v^*)$
Initiate population $\mathcal{W}$
for each member i in the population $\mathcal{W}$ **do**
  Select three candidates b_1, b_2 and b_3 such that, $1 \geq b_1, b_2, b_3 \leq$ N
  Generate a arbitrary integer i $\in$ (1,N)
  while iteration < MaxIteration **do**
    for each factor i **do**
      Estimate the fitness of every member
      Generate mutant vectors utilizing mutation approach
      Generate trial vectors using recombining noisy vectors with parent vectors
      Estimate trial vectors with their fitness values
    end for
    Choose winning vectors as members in the new generation
    iteration ++
  end while
end for
Return the best values

References

1. Abdollahi S, Salehi Rad MR (2022) Analysis of a batch arrival retrial queue with two-phase services, feedback and admission. Bull Iran Math Soc 1–14 (2022)
2. Artalejo JR (2010) Accessible bibliography on retrial queues: progress in 2000–2009. Math Comput Modell 51:1071–81
3. Bharathidass S, Arivukkarasu V, Ganesan V (2018) Bulk service queue with server breakdown and repairs. Int J Stat Appl 3:136–142
4. Choudhury G, Ke JC (2012) A batch arrival retrial queue with general retrial times under Bernoulli vacation schedule for unreliable server and delaying repair. Appl Math Model 36:255–269
5. Das RR, Devi V, Rathore A, Chandan K (2022) Analysis of Markovian queueing system with server failures, N-policy and second optional service. Int J Non-linear Anal Appl 13:3073–3083
6. Deora P, Kumari U, Sharma DC (2021) Cost analysis and optimization of machine repair model with working vacation and feedback-policy. Int J Appl Comput Math 7:1–14
7. Humblet Sennott (1983) Tweedi: average drifts and the non-Ergodicity of Markov chains. Oper Res 31:783–789
8. Jain M, Bhagat A (2014) Unreliable bulk retrial queues with delayed repairs and modified vacation policy. J Ind Eng Int 10:1–19
9. Jain M, Kumar A (2022) Unreliable Server $M^{[X]}/G/1$ queue with working vacation and multi-phase repair with delay in verification. Int J Appl Comput Math 8(4):210
10. Karaboga D, Basturk B (2007) A powerful and efficient algorithm for numerical function optimization: artificial bee colony (ABC) algorithm. J Global Optim 39:459–471
11. Kennedy J, Eberhart R (1995) Particle swarm optimization. In: Proceedings of ICNN95 - international conference on neural networks, Perth, WA, Australia, vol 4, pp 1942–1948
12. Kirupa K, Chandrika KU (2015) Unreliable batch arrival retrial queue with negative customers, multi-types of heterogeneous service, setup time and reserved time. Int J Comput Appl 975:0975–8887
13. Kirupa K (2015) Udaya Chandrika K (2015) Analysis of batch arrival retrial G-queue with multi-types of heterogeneous service, feedback, randomized J vacations and orbital search. Appl Math 86:34870–34876
14. Kirupa K, Chandrika K (2015) Unreliable batch arrival retrial queue with negative arrivals, multi-types of heterogeneous service, feedback and randomized J vacations. Int J Comput Appl 127:38–42
15. Kumar A, Jain M (2023) Cost Optimization of an Unreliable server queue with two stage service process under hybrid vacation policy. Math Comput Simul 204:259–281
16. Liu Z, Song Y (2013) $Geo/Geo/1$ retrial queue with non-persistent consumers and working vacations. J Appl Math Comput 42:103–115
17. Malik G, Upadhyaya S, Sharma R (2021) Cost inspection of a Geo/G/1 retrial model using particle swarm optimization and Genetic algorithm. Ain Shams Eng 12:2241–2254
18. Malik G, Upadhyaya S, Sharma R (2021) Particle swarm optimization and maximum entropy results for $M^{[X]}/G/1$ retrial G queue with delayed repair. Int J Math Eng Manag Sci 6:541
19. Mirjalili S, Mirjalili SM, Lewis A (2014) Grey wolf optimizer. Adv Eng Softw 69:46–61
20. Murugan SPB, Vijaykrishnaraj R (2019) A bulk arrival retrial queue with orbital search and exponentially distributed multiple working vacation. In: AIP conference proceedings, vol 2177(1), AIP Publishing LLC
21. Nila M, Sumitha D (2021) A feedback $M^X/G/1$ retrial queue with impatient consumers, active breakdown and orbital search. Adv Math: Sci J 10:1397–1405
22. Pakes AG (1969) Some conditions for Ergodicity and recurrence of Markov chains. Oper Res 17:1058–1061
23. Peng Y, Wu J (2021) Analysis of a batch arrival retrial queue with impatient customers subject to the server disasters. J Ind Manag Opt 17(4):2243–2264

24. Radha J, Indhira K, Chandrasekaran VM (2017) Multi stage unreliable retrial Queueing system with Bernoulli vacation. In: IOP conference series. Materials science and engineering, vol 263, no 4. IOP Publishing
25. Rajadurai P, Varalakshmi M, Saravanarajan MC, Chandrasekaran VM (2015) Analysis of $M^{[x]}/G/1$ retrial queue with two phase service under Bernoulli vacation schedule and random breakdown. Int J Math Oper Res 7:19–41
26. Rajadurai R, Yuvarani S, Saravanarajan MC (2016) Performance analysis of preemptive priority retrial queue with immediate Bernoulli feedback under working vacations and vacation interruption. Songklanakarin J Sci & Technol 38:507–520
27. Rajadurai P (2019) A study on $M/G/1$ preemptive priority retrial queue with Bernoulli working vacations and vacation interruption. Int J Process Manag Benchmarking 9:193–215
28. Rajasudha R, Arumuganathan R, Dharmaraja S (2022) Performance analysis of discrete-time $Geo^{(X)}/G/1$ retrial queue with various vacation policies and impatient consumers. RAIRO Oper Res 56:1089–1117
29. Sangeetha N, Chandrika KU (2020) Multi stage and multi optional retrial G-queue with feedback and starting failure. In: AIP conference proceedings, vol 2261, no 1, AIP publishing
30. Storn R, Price K (1996) Minimizing the real functions of the ICEC'96 contest by differential evolution. In: Proceedings of IEEE international conference on evolutionary computation, pp 842–844
31. Upadhyaya S (2020) Cost optimization of a discrete-time retrial queue with Bernoulli feedback and starting failure. Int J Ind Syst Eng 36:165–196
32. Upadhyaya S, Ghosh S, Malik G (2022) Cost investigation of a batch arrival retrial G-Queue with working malfunction and working vacation using particle swarm optimization. Nonlinear Studies 29(3)
33. Vaishnawi M, Upadhyaya S, Kulshrestha R (2022) Optimal cost analysis for discrete-time recurrent queue with Bernoulli feedback and emergency vacation. Int J Appl Comput Math 8(5):254
34. Vijaya Laxmi P, Jyohsna K (2022) Cost and revenue analysis of an impatient customer queue with second optional service and working vscations. Commun Stat - Simul Comput 51(8):4799–4814
35. Yang T, Templeton JGC (1987) A survey on retrial queue. Que Sys 201–233
36. Zhang X, Wang J, Ma Q (2017) Optimal design for a retrial queueing system with state dependent service rate. J Syst Sci Complex 30:883–900

Chapter 7
Topological Indices on Fractal Patterns

A. Divya, A. Manimaran, Intan Muchtadi Alamsyah, and Ahmad Erfanian

Abstract Sierpinski and Koch Snowflake are the two most studied topics in fractal geometry. Sierpinski Rhombus (SR_n) is formed by a pattern of n sequences of a graph that results in a planar fractal. Koch Snowflake (KS_n) is also formed by some patterns similar to the Sierpinski Rhombus. In this study, topological indices are used to study fractal structures. This chapter is the first to derive a closely related formula of M-polynomials and entropy measures for the fractal structures (SR_n) and (KS_n). The calculated topological value is usually correlated with the physical properties of the structure. The scope of this work is to relate the derivation of topological indices with the fractal dimension for the graph sequences (SR_n) and (KS_n).

1 An Overview of Topological Indices

Fractals

In 1975, Mandelbrot presented fractal geometry as a new type of natural geometry, and it has since evolved into a modern discipline of pure and applied mathematics.

A. Divya
Assistant Professor, Department of Mathematics, Madanapalle Institute of Technology & Science, Madanapalle 517325, Andhra Pradesh, India
e-mail: divyaa@mits.ac.in

A. Manimaran (✉)
Department of Mathematics, School of Advanced Sciences, Vellore Institute of Technology, Vellore 632014, Tamil Nadu, India
e-mail: manimaran.a@vit.ac.in

I. M. Alamsyah
Algebra Research Group, Faculty of Mathematics and Natural Sciences, Institut Teknologi Bandung, Bandung, Indonesia
e-mail: ntan@math.itb.ac.d

A. Erfanian
Department of Pure Mathematics and the Center of Excellence in Analysis on Algebraic Structures, Ferdowsi University of Mashhad, Mashhad, Iran
e-mail: erfanian@um.ac.ir

G. Arulprakash et al. (eds.), *Mathematical Modelling of Complex Patterns Through Fractals and Dynamical Systems*, Studies in Infrastructure and Control,
https://doi.org/10.1007/978-981-97-2343-0_7

Fractal geometry is a powerful tool for the structural evaluation of natural and idealized phenomena used in many scientific fields. The use of fractal geometry can be found in many areas; refer [12, 16, 22, 23]. In [13], according to Mandelbrot, the fractal is a set with a Hausdorff dimension that is strictly greater than its topological dimension. In fractals, there is always some form of repeating pattern that is rigorously self-similar. According to [24], iteration is one of the most common origins of self-similarity. The Cantor set, Sierpiński gasket, and Koch curve are the three well-known self-similar fractals. In early 1915, Sierpinski introduced the Sierpinski gasket, a classic fractal. Sierpinski gasket is used in many areas, such as graph theory, topology, and probability. The curve Koch Snowflake was created in 1904 by Swedish mathematician Helge von Koch. Refer to [11] for more information on the Koch curve, a continuous curve that is not differentiable everywhere.

Sierpiński Rhombus

In this part, we will study the sketch of the Sierpiński rhombus graph in detail below. Combining two Sierpiński triangle graphs [denoted by S_n] form the n-level Sierpiński rhombus (SR_n) by identifying the edges along with one of their sides.

In Fig. 1, we can see that the graph SR_n contains $SR_{n,T,T}$, $SR_{n,R,R}$, $SR_{n,B,B}$ and $SR_{n,L,L}$ has four corner vertices. One can see that the division of SR_n consists of four partitions, particularly the Sierpiński rhombus of an upper and lower triangle is considered as the $n-1$ level, and it is denoted by (S_n, T) and (S_n, B). Likewise, the left and right Sierpiński rhombus contain an $n-1$ level named (S_n, L) and (S_n, R). The fractal dimension of the Sierpiński Rhombus is $\frac{log(3)}{log(2)} \approx 1.585$.

Fig. 1 SR_n

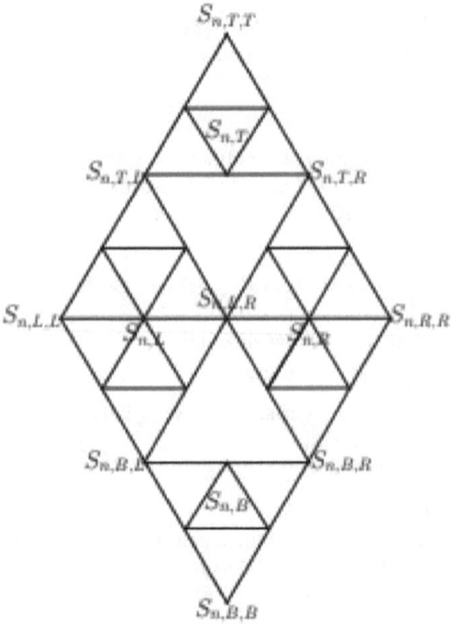

Fig. 2 KS_n

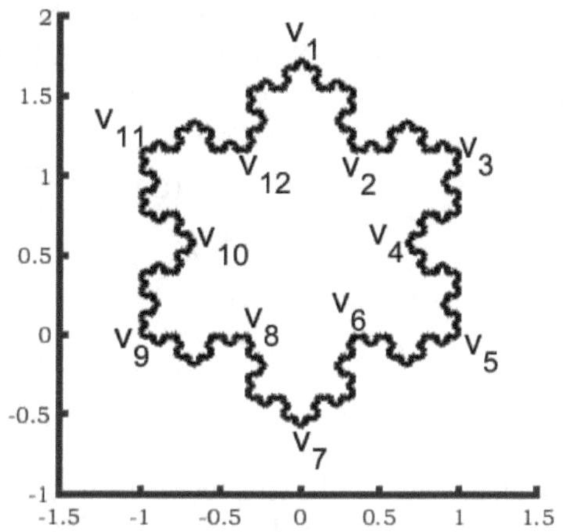

Koch Snowflake
Combining the three externally pointed Koch curves together makes a hexagonal symmetrical Koch snowflake see Fig. 2. For more details, refer [5]. The fractal dimension of the Koch Snowflake (KS_n) is $\frac{log(4)}{log(3)} \approx 1.26186$.

Topological Indices
Topological Index (TI) is a numerical parameter obtained mathematically for a molecular graph. For this concept, the term graph-theoretical index is more accurate than the topological index. In QSAR/QSPR studies, topological indices are an easy way to convert chemical properties into numerical values in a molecular graph of a compound. It correlates chemical structure with biological reactions or with various physical properties. Generally, any index has identical values for isomorphic graphs; hence, it is called a graph variant. Moreover, topological indices frequently represent the size and as well as the shape of molecules.

Molecular descriptors were developed to identify numbers to molecular graphs, and we can describe the molecule using that numbers. The QSAR models use physicochemical properties, experimentally, theoretically and mechanically computed quantum parameters as descriptors.

Let G be the pair of $(V(G), E(G))$ in a graph, with $V(G)$ being the vertex set and $E(G)$ being the edge set. Then the symbol $\mid V(G) \mid$ and $\mid E(G) \mid$ suggests the number of vertices and the edges in the graph G. Now let us have a look at the primitive definition of the indices. In [19], the Milan Randić is the first who introduce the degree-dependent index called Randić index and which is defined as

$$R_{-1/2}(G) = \sum_{uv \in E(G)} \frac{1}{\sqrt{d_u \times d_v}},$$

where d_s is the degree of a vertex s, which means the number of edges connecting the vertex s. In 1988, Bollobás et al. [3] and Amić et al. [1] proposed the general Randić index, and it is provided by

$$R_\alpha(G) = \sum_{uv \in E(G)} \frac{1}{(d_u \times d_v)^\alpha},$$

in which α is a real number and the inverse Randić index is given by

$$RR_\alpha(G) = \sum_{uv \in E(G)} (d_u \times d_v)^\alpha.$$

Of all the indices, the vital topological index was the first Zagreb index, discovered in 1972 refer [9]. In [8], the discovery of the second Zagreb index is obtained following the first Zagreb index, and it is known as

$$M_1(G) = \sum_{uv \in E(G)} d_u + d_v.$$

$$M_2(G) = \sum_{uv \in E(G)} d_u \times d_v.$$

In [18], the modified second Zagreb index is defined as

$$^m M_2(G) = \sum_{uv \in E(G)} \frac{1}{d_u \times d_v}.$$

The harmonic index [29] is another variant of the Randić index, and it is studied as

$$H(G) = \sum_{uv \in E(G)} \frac{2}{d_u + d_v}.$$

In [27], among 148 discrete Adriatic indices, the symmetric division deg index is the one. This one estimates the polychlorobiphenyl total surface area better than all the indices. It is defined by

$$SDD(G) = \sum_{uv \in E(G)} \left(\frac{d_u}{d_v} + \frac{d_u}{d_v} \right).$$

Balaban [2] introduced the inverse sum index in 1982; the prediction for the total surface area of octane isomers is significant and invariant. The definition of this index is

$$I(G) = \sum_{uv \in E(G)} \left(\frac{d_u \times d_v}{d_u + d_v} \right).$$

In [7], Furtula et al. proposed the augmented Zagreb index of a graph G by

$$AZI(G) = \sum_{uv \in E(G)} \left(\frac{d_u \times d_v}{d_u + d_v - 2} \right)^3.$$

The atom bond connectivity index is proposed by Estrada et al. [6]; it is presented as follows:

$$ABC(G) = \sum_{uv \in E(G)} \sqrt{\frac{d_u + d_v - 2}{d_u \times d_v}}.$$

In [28], the geometric arithmetic index is developed by Vukičević et al., which is considered as

$$GA(G) = \sum_{uv \in E(G)} \frac{2\sqrt{d_u \times d_v}}{d_u + d_v}.$$

In 2013, Ranjini et al. [15, 20] recently initiated the modified Zagreb indices, specifically the redefined first, second and third Zagreb indices and are stated below:

$$ReZG_1(G) = \sum_{uv \in E(G)} \frac{d_u + d_v}{d_u d_v}.$$

$$ReZG_2(G) = \sum_{uv \in E(G)} \frac{d_u \times d_v}{d_u + d_v}.$$

$$ReZG_3(G) = \sum_{uv \in E(G)} (d_u \times d_v)(d_u + d_v).$$

In [26], the hyper-Zagreb index is defined by Shirdel et al. in 2013 as

$$HM(G) = \sum_{uv \in E(G)} (d_u + d_v)^2.$$

Following this, we can discuss which indices are well correlated with the physical properties of alkanes. The Randić index and the geometric arithmetic index significantly correlated with all the physical properties of alkanes. The atom bond connectivity index and hyper Zagreb are the most helpful methods for estimating alkane boiling points. The vaporization of heat and critical temperature of an alkane is well communicated, and this can be calculated using the harmonic index. The inverse sum index substantially predicts the total surface area for octane isomers and the symmetric division deg index for polychlorobiphenyls. The augmented Zagreb index well indicates the heat of formation. Finally, the Sanskruti index is one that is well connected with entropy.

The work process is formalized as follows: In Sect. 2, we have found a derivative of M-polynomial and neighbourhood M-polynomial to generate topological indices results. The study of Sect. 3 encloses the discussion based on the derivation of indices for the entropy measure. The discussion of numerical results for the M-polynomial and entropy measures is covered in Sect. 4. And then finally, Sect. 5 ends with the conclusion.

2 M-Polynomial Measure on Topological Indices

The primary goal of this section is to find the general closed form of the M-polynomial that is related to the graph Sierpiński rhombus and Koch snowflake sequences.

Graph-related algebraic objects are known as graph polynomials. Graph polynomials are employed to configure various algebraic techniques to encrypt the information shown on the graph and draw the hidden information. In general, beneath graph isomorphism, it will be invariant. Generally, we obtain the topological indices by definition. Although, a compact and standard way to generate a type of topological index is to calculate a common polynomial. A topological index is obtained by evaluating the derivative or integral at a particular point.

Within graph theory, a large number of algebraic polynomial graphs are found. Wherein Hosoya polynomial, forgotten polynomial, pi polynomial, Schultz polynomial, modified Schultz polynomial, matching polynomial, Tutte polynomial and M-polynomial are the few prominent polynomials.

In 2015 [4], Deutsch and Klavžar introduced the M-polynomial for a graph G, and it is proposed as

$$M(G; p, q) = \sum_{\delta \leq k \leq l \leq \Delta} m_{k,l} p^k q^l. \tag{1}$$

Here, the notation δ and Δ represent the minimum and maximum degree of vertex d_u, $\forall u \in V(G)$. Let $m_{k,l}$ stand for the number of edges $uv \in E(G)$ of G so that $d_u(1 \leq \delta \leq d_u) = k$ and $d_v(d_v \leq \Delta \leq\mid V(G) \mid -1) = l$.

In [10, 14], several graphs are introduced as M-polynomials. Tables 1 and 2 show the relationship between degree-based topological indices and the M-polynomial.

$$D_p = p \times \frac{\partial(h(p,q))}{\partial p},\ D_q = q \times \frac{\partial(h(p,q))}{\partial q},\ S_p = \int_0^p \frac{h(t,q)}{t}\, dt,$$
$$S_q = \int_0^q \frac{h(p,t)}{t}\, dt,\ J(h(p,q)) = h(p,p),\ Q_\alpha(h(p,q)) = p^\alpha h(p,q),$$
$$D_p^{\frac{1}{2}}(h(p,q)) = \sqrt{p\frac{\partial(h(p,q))}{\partial p}}\sqrt{h(p,q)},\ D_q^{\frac{1}{2}}(h(p,q)) = \sqrt{q\frac{\partial(h(p,q))}{\partial q}}\sqrt{h(p,q)},$$
$$S_p^{\frac{1}{2}}(h(p,q)) = \sqrt{\int_0^p \frac{h(t,q)}{t}\, dt}\sqrt{h(p,q)},\ S_q^{\frac{1}{2}}(h(p,q)) = \sqrt{\int_0^q \frac{h(p,t)}{t}\, dt}\sqrt{h(p,q)}$$

Table 1 Derivation of degree-based topological indices from $M(H, p, q)$

TI	$h(p,q)$	$Derivation\ from\ M(H,p,q)$
M_1	$p+q$	$(D_p + D_q)(M(H,p,q))\ \vert_{p=q=1}$
M_2	pq	$(D_p D_q)(M(H,p,q))\ \vert_{p=q=1}$
$^m M_2$	$\frac{1}{pq}$	$(S_p S_q)(M(H,p,q))\ \vert_{p=q=1}$
R_α	$(pq)^\alpha$	$(D_q^\alpha D_p^\alpha)(M(H,p,q))\ \vert_{p=q=1}$
RR_α	$\frac{1}{(pq)^\alpha}$	$(S_p^\alpha S_q^\alpha)(M(H,p,q))\ \vert_{p=q=1}$
SDD	$\frac{p^2+q^2}{pq}$	$(S_q D_p + S_p D_q)(M(H,p,q))\ \vert_{p=q=1}$
H	$\frac{2}{p+q}$	$2(S_p J)(M(H,p,q))\ \vert_{p=q=1}$
I	$\frac{pq}{p+q}$	$(S_p J D_p D_q)(M(H,p,q))\ \vert_{p=q=1}$
A	$\left(\frac{pq}{p+q-3}\right)^3$	$(S_p^3 Q_{-2} J D_x^3 D_q^3)(M(H,p,q))\ \vert_{p=q=1}$
ABC	$\sqrt{\frac{p+q-2}{pq}}$	$(D_p^{\frac{1}{2}} Q_{-2} J S_p^{\frac{1}{2}} S_q^{\frac{1}{2}})(M(H,p,q))\ \vert_{p=q=1}$
GA	$\frac{2\sqrt{pq}}{p+q}$	$(2 S_p J D_p^{\frac{1}{2}} D_q^{\frac{1}{2}})(M(H,p,q))\ \vert_{p=q=1}$

Table 2 Derivation of neighbourhood-based topological indices from $M(H, p, q)$

TI	$h(p,q)$	$Derivation\ from\ NM(H,p,q)$
M_1'	$p+q$	$(D_p + D_q)(NM(H,p,q))\ \vert_{p=q=1}$
M_2'	pq	$(D_p D_q)(NM(H,p,q))\ \vert_{p=q=1}$
F_N^*	p^2+q^2	$(D_p^2 + D_q^2)(NM(H,p,q))\ \vert_{p=q=1}$
M_2^{nm}	$\frac{1}{pq}$	$(S_p S_q)(NM(H,p,q))\ \vert_{p=q=1}$
NR_α	$(pq)^\alpha$	$(D_q^\alpha D_p^\alpha)(NM(H,p,q))\ \vert_{p=q=1}$
ND_3	$pq(p+q)$	$(D_p D_q)(D_p + D_q)(NM(H,p,q))\ \vert_{p=q=1}$
ND_5	$\frac{p^2+q^2}{pq}$	$(S_q D_p + S_p D_q)(NM(H,p,q))\ \vert_{p=q=1}$
NH	$\frac{2}{p+q}$	$2(S_p J)(NM(H,p,q))\ \vert_{p=q=1}$
NI	$\frac{pq}{p+q}$	$(S_p J D_p D_q)(NM(H,p,q))\ \vert_{p=q=1}$
S	$\left(\frac{pq}{p+q-3}\right)^3$	$(S_p^3 Q_{-2} J D_x^3 D_q^3)(NM(H,p,q))\ \vert_{p=q=1}$

Theorem 1 *Let a Sierpiński Rhombus be denoted by SR_n. Then, the M-Polynomial is*

$$M(SR_n, p, q) = 4p^2q^4 + 4p^3q^4 + 2p^3q^6 + ((2 \times 3^n) - (3 \times 2^n) - 4)p^4q^4 + (2^{n+1} - 4)p^4q^6 + (2^{n-1} - 2)p^6q^6.$$

Proof From the graph sequences SR_n, a Sierpiński Rhombus, as we can see

$$\mid V(SR_n) \mid = 3^n - 2^{n-1} + 2.$$
$$\mid E(SR_n) \mid = (2 \times 3^n) - 2^{n-1}.$$

For the graph SR_n, the vertex set is partitioned into four divisions based on the degrees of the vertices.

$$
\begin{aligned}
| V_2 | &= | \{v \in V(SR_n) \mid d_v = 2\} | = 2. \\
| V_3 | &= | \{v \in V(SR_n) \mid d_v = 3\} | = 2. \\
| V_4 | &= | \{v \in V(SR_n) \mid d_v = 4\} | = 3^n - 2^n - 1. \\
| V_6 | &= | \{v \in V(SR_n) \mid d_v = 6\} | = 2^{n-1} - 1.
\end{aligned}
$$

Similarly, we can see that the partition of the edge set of the SR_n is as follows:

$$
\begin{aligned}
| E_1 | &= | \{uv \in E(SR_n) \mid d_u = 2 \, and \, d_v = 4\} | = 4. \\
| E_2 | &= | \{uv \in E(SR_n) \mid d_u = 3 \, and \, d_v = 4\} | = 4. \\
| E_3 | &= | \{uv \in E(SR_n) \mid d_u = 3 \, and \, d_v = 6\} | = 2. \\
| E_4 | &= | \{uv \in E(SR_n) \mid d_u = 4 \, and \, d_v = 4\} | = (2 \times 3^n) - (3 \times 2^n) - 4. \\
| E_5 | &= | \{uv \in E(SR_n) \mid d_u = 4 \, and \, d_v = 6\} | = 2^{n+1} - 4. \\
| E_6 | &= | \{uv \in E(SR_n) \mid d_u = 6 \, and \, d_v = 6\} | = 2^{n-1} - 2.
\end{aligned}
$$

Let $d_u = k$ and $d_v = l$ such that $k \leq l$, for all $k, l \in V(G)$ and $uv \in E(G)$. As a result of the definition, the M-polynomial of SR_n is

$$
\begin{aligned}
M(SR_n, p, q) &= \sum_{k \leq l} m_{kl} p^k q^l \\
&= | E_1 | p^2 q^4 + | E_2 | p^3 q^4 + | E_3 | p^3 q^6 + | E_4 | p^4 q^4 \\
&\quad + | E_5 | p^4 q^6 + | E_6 | p^6 q^6 \\
&= 4p^2q^4 + 4p^3q^4 + 2p^3q^6 + \big((2 \times 3^n) - (3 \times 2^n) - 4\big) p^4 q^4 \\
&\quad + (2^{n+1} - 4) p^4 q^6 + (2^{n-1} - 2) p^6 q^6.
\end{aligned}
$$

□

Theorem 2 *Let a Koch Snowflake be denoted by KS_n. Then, the M-Polynomial is*

$$M(KS_n, p, q) = (3 \times 4^n) p^2 q^2.$$

Proof From the graph sequences KS_n, a Koch Snowflake, as we can see

$$
\begin{aligned}
| V(KS_n) | &= 3 \times 4^n. \\
| E(KS_n) | &= 3 \times 4^n.
\end{aligned}
$$

For the graph KS_n, the vertex set is partitioned into the division based on the degree of the vertices.

$$| V_2 | = | \{v \in V(KS_n) \mid d_v = 2\} | = 3 \times 4^n.$$

Similarly, we can see that the partition of the edge set of the KS_n is as follows:

$$| E_1 | = | \{uv \in E(KS_n) \mid d_u = 2 \, and \, d_v = 2\} | = 3 \times 4^n.$$

Let $d_u = k$ and $d_v = l$ such that $k \leq l$, for all $k, l \in V(G)$ and $uv \in E(G)$. As a result of the definition, the M-polynomial of KS_n is

$$M(KS_n, p, q) = \sum_{k \leq l} m_{kl} p^k q^l = | E_1 | p^2 q^2 = (3 \times 4^n) p^2 q^2.$$

□

Degree-Based M-Polynomial on Topological Indices

In this part, we use Theorems 1 and 2 to evaluate some topological indices for the graph SR_n and KS_n. Moreover, we characterize the calculated indices with the help of Maple.

Theorem 3 *Let a Sierpiński Rhombus be denoted by* SR_n *and*

$$M(SR_n, p, q) = 4p^2q^4 + 4p^3q^4 + 2p^3q^6 + \big((2 \times 3^n) - (3 \times 2^n) - 4\big)p^4q^4 + (2^{n+1} - 4)p^4q^6 + (2^{n-1} - 2)p^6q^6.$$

Then,

1. $M_1(SR_n) = (16 \times 3^n) + (2 \times 2^n) - 26.$
2. $M_2(SR_n) = (32 \times 3^n) + (18 \times 2^n) - 116.$
3. ${}^mM_2(SR_n) = (0.125 \times 3^n) + (0.007 \times 2^n) - 0.090.$
4. $R_\alpha(SR_n) = (0.5 \times 3^n) - (0.258 \times 2^n) + 0.832.$
5. $RR_\alpha(SR_n) = (8 \times 3^n) + (0.798 \times 2^n) - 11.960.$
6. $SDD(SR_n) = (4 \times 3^n) - (0.667 \times 2^n) + 2.667.$
7. $H(SR_n) = (0.5 \times 3^n) - (0.267 \times 2^n) + 0.787.$
8. $I(SR_n) = (4 \times 3^n) + (0.3 \times 2^n) - 7.410.$
9. $A(SR_n) = (37.926 \times 3^n) + (105.11 \times 2^n) + 536.022.$
10. $ABC(SR_n) = (0.5 \times 3^n) - (0.419 \times 2^n) + 0.845.$
11. $GA(SR_n) = (2 \times 3^n) - (0.540 \times 2^n) - 0.303.$

Proof Let the M-polynomial of SR_n be $h(p, q)$. Hence,

$$h(p, q) = 4p^2q^4 + 4p^3q^4 + 2p^3q^6 + \big((2 \times 3^n) - (3 \times 2^n) - 4\big)p^4q^4 + (2^{n+1} - 4)p^4q^6 + (2^{n-1} - 2)p^6q^6.$$

The first Zagreb index is obtained as under:

$$D_p\big(h(p, q)\big) = 8p^2q^4 + 12p^3q^4 + 6p^3q^6 + 4\big((2 \times 3^n) - (3 \times 2^n) - 4\big)p^4q^4 + (2^{n+1} - 4)p^4q^6 + 6(2^{n-1} - 2)p^6q^6.$$

$$D_q\big(h(p, q)\big) = 16p^2q^4 + 16p^3q^4 + 12p^3q^6 + 4\big((2 \times 3^n) - (3 \times 2^n) - 4\big)p^4q^4 + 6(2^{n+1} - 4)p^4q^6 + 6(2^{n-1} - 2)p^6q^6.$$

$$M_1(SR_n) = (D_p + D_q)\big(h(p, q)\big)\big|_{p=q=1} = (16 \times 3^n) + (2 \times 2^n) - 26.$$

The second Zagreb index is evaluated as follows:

$$D_qD_p\big(h(p, q)\big) = 32p^2q^4 + 48p^3q^4 + 36p^3q^6 + 16\big((2 \times 3^n) - (3 \times 2^n) - 4\big)p^4q^4 + 24(2^{n+1} - 4)p^4q^6 + 36(2^{n-1} - 2)p^6q^6.$$

$$M_2(SR_n) = (D_pD_q)\big(h(p, q)\big)\big|_{p=q=1} = (32 \times 3^n) + (18 \times 2^n) - 116.$$

The modified second Zagreb index is estimated as below:

$$\begin{aligned}
S_p\big(h(p,q)\big) &= 2p^2q^4 + 1.333p^3q^4 + 0.667p^3q^6 + 0.25\big((2\times 3^n) - (3\times 2^n)\\
&\quad -4\big)p^4q^4 + 0.25(2^{n+1}-4)p^4q^6 + 0.167(2^{n-1}-2)p^6q^6.\\
S_q\big(h(p,q)\big) &= p^2q^4 + p^3q^4 + 0.333p^3q^6 + 0.25\big((2\times 3^n) - (3\times 2^n) - 4\big)p^4q^4\\
&\quad +0.167(2^{n+1}-4)p^4q^6 + 0.167(2^{n-1}-2)p^6q^6.\\
S_pS_q\big(h(p,q)\big) &= 0.5p^2q^4 + 0.333p^3q^4 + 0.111p^3q^6 + 0.063\big((2\times 3^n) - (3\times 2^n)\\
&\quad -4\big)p^4q^4 + 0.042(2^{n+1}-4)p^4q^6 + 0.028(2^{n-1}-2)p^6q^6.\\
{}^mM_2(SR_n) &= (S_pS_q)\big(h(p,q)\big)\mid_{p=q=1} = (0.125\times 3^n) + (0.007\times 2^n) - 0.090.
\end{aligned}$$

We calculate the Randić index as below:

$$\begin{aligned}
D_p^{\alpha}(h(p,q)) &= (4\times 2^{\alpha})p^2q^4 + (4\times 3^{\alpha})p^3q^4 + (2\times 3^{\alpha})p^3q^6 + 4^{\alpha}\big((2\times 3^n)\\
&\quad -(3\times 2^n) - 4\big)p^4q^4 + 4^{\alpha}(2^{n+1}-4)p^4q^6 + 6^{\alpha}(2^{n-1}-2)p^6q^6.\\
D_q^{\alpha}(h(p,q)) &= (4\times 4^{\alpha})p^2q^4 + (4\times 4^{\alpha})p^3q^4 + (2\times 6^{\alpha})p^3q^6 + \big(4^{\alpha}\times((2\times 3^n)\\
&\quad -(3\times 2^n) - 4)\big)p^4q^4 + 6^{\alpha}(2^{n+1}-4)p^4q^6 + 6^{\alpha}(2^{n-1}-2)p^6q^6.\\
D_q^{\alpha}D_p^{\alpha}(h(p,q)) &= 2^{3\alpha+2}p^2q^4 + (3^{\alpha}\times 2^{2\alpha+2})p^3q^4 + (2\times 3^{\alpha}\times 6^{\alpha})p^3q^6\\
&\quad +2^{4\alpha}((2\times 3^n) - (3\times 2^n) - 4)p^4q^4 + (4^{\alpha}\times 6^{\alpha}(2^{n+1}-4))p^4q^6\\
&\quad +6^{2\alpha}(2^{n-1}-2)p^6q^6.\\
R_{\alpha=-1/2}(SR_n) &= (D_q^{\alpha}D_p^{\alpha})\big(h(p,q)\big)\mid_{p=q=1} = (0.5\times 3^n) - (0.258\times 2^n) + 0.832.
\end{aligned}$$

We assess the inverse Randić index as

$$\begin{aligned}
S_p^{\alpha}\big(h(p,q)\big) &= (4\times 2^{-\alpha})p^2q^4 + (4\times 3^{-\alpha})p^3q^4 + (2\times 3^{-\alpha})p^3q^6\\
&\quad +4^{-\alpha}\big((2\times 3^n) - (3\times 2^n) - 4\big)p^4q^4 + 4^{-\alpha}(2^{n+1}-4)p^4q^6\\
&\quad +6^{-\alpha}(2^{n-1}-2)p^6q^6.\\
S_q^{\alpha}\big(h(p,q)\big) &= (4\times 4^{-\alpha})p^2q^4 + (4\times 4^{-\alpha})p^3q^4 + (2\times 6^{-\alpha})p^3q^6\\
&\quad +4^{-\alpha}\big((2\times 3^n) - (3\times 2^n) - 4\big)p^4q^4 + 6^{-\alpha}(2^{n+1}-4)p^4q^6\\
&\quad +6^{-\alpha}(2^{n-1}-2)p^6q^6.\\
S_q^{\alpha}S_p^{\alpha}\big(h(p,q)\big) &= (4\times 2^{-3\alpha})p^2q^4 + (3^{-\alpha}\times 4^{1-\alpha})p^3q^4 + (2\times 3^{-\alpha}\times 6^{-\alpha})p^3q^6\\
&\quad +4^{-2\alpha}((2\times 3^n) - (3\times 2^n) - 4)p^4q^4 + (4^{-\alpha}\times 6^{-\alpha}(2^{n+1}-4))p^4q^6\\
&\quad +6^{-2\alpha}(2^{n-1}-2)p^6q^6.\\
RR_{\alpha=-1/2}(SR_n) &= (S_p^{\alpha}S_q^{\alpha})\big(h(p,q)\big)\mid_{p=q=1} = (8\times 3^n) + (0.798\times 2^n) - 11.960.
\end{aligned}$$

The symmetric division deg index is determined by

$$\begin{aligned}
S_qD_p\big(h(p,q)\big) &= 2p^2q^4 + 3p^3q^4 + p^3q^6 + \big((2\times 3^n) - (3\times 2^n) - 4\big)p^4q^4\\
&\quad +0.667(2^{n+1}-4)p^4q^6 + (2^{n-1}-2)p^6q^6.\\
S_pD_q\big(h(p,q)\big) &= 8p^2q^4 + 5.333p^3q^4 + 4p^3q^6 + \big((2\times 3^n) - (3\times 2^n) - 4\big)p^4q^4\\
&\quad +1.5(2^{n+1}-4)p^4q^6 + (2^{n-1}-2)p^6q^6.\\
SDD(SR_n) &= (S_qD_p + S_pD_q)\big(h(p,q)\big)\mid_{p=q=1}\\
&= (4\times 3^n) - (0.667\times 2^n) + 2.667.
\end{aligned}$$

The harmonic index is appraised as below:

$$J\big(h(p,q)\big) = 4p^6 + 4p^7 + 2p^9 + \big((2\times 3^n) - (3\times 2^n) - 4\big)p^8 + (2^{n+1} - 4)p^{10} + (2^{n-1} - 2)p^{12}.$$
$$S_pJ\big(h(p,q)\big) = 0.667p^6 + 0.571p^7 + 0.222p^9 + 0.125\big((2\times 3^n) - (3\times 2^n) - 4\big)p^8 + 0.1(2^{n+1} - 4)p^{10} + 0.083(2^{n-1} - 2)p^{12}.$$
$$H(SR_n) = 2(S_pJ)\big(h(p,q)\big)\mid_{p=1} = (0.5\times 3^n) - (0.267\times 2^n) + 0.787.$$

The assessment of the inverse sum index is done as follows:

$$S_pJD_pD_q\big(h(p,q)\big) = 5.333p^6 + 6.857p^7 + 4p^9 + 2\big((2\times 3^n) - (3\times 2^n) - 4\big)p^8 + 2.4(2^{n+1} - 4)p^{10} + 3(2^{n-1} - 2)p^{12}.$$
$$I(SR_n) = (S_pJD_pD_q)\big(h(p,q)\big)\mid_{p=1} = (4\times 3^n) + (0.3\times 2^n) - 7.410.$$

We figure the augmented index as

$$S_p^3Q_{-2}JD_x^3D_q^3\big(h(p,q)\big) = 32p^4 + 13.824p^5 + 34.006p^7 + 18.963\big((2\times 3^n) - (3\times 2^n) - 4\big)p^6 + 27(2^{n+1} - 4)p^8 + 6^3(2^{n-1} - 2)p^{10}.$$
$$A(SR_n) = (S_p^3Q_{-2}JD_x^3D_q^3)\big(h(p,q)\big)\mid_{p=1} = (37.926\times 3^n) + (105.11\times 2^n) + 536.022.$$

The computation of the atom bond connectivity index is implemented as below:

$$D_p^{\frac{1}{2}}Q_{-2}JS_p^{\frac{1}{2}}S_q^{\frac{1}{2}}\big(h(p,q)\big) = 2\sqrt{2}p^4 + 2.582p^5 + 1.247p^7 + 0.612\big((2\times 3^n) - (3\times 2^n) - 4\big)p^6 + 1.155(2^n - 2)p^8 + 0.264(2^n - 4)p^{10}.$$
$$ABC(SR_n) = (D_p^{\frac{1}{2}}Q_{-2}JS_p^{\frac{1}{2}}S_q^{\frac{1}{2}})\big(h(p,q)\big)\mid_{p=1} = (0.5\times 3^n) - (0.419\times 2^n) + 0.845.$$

The result of the geometric arithmetic index is carried out as follows:

$$2S_pJD_p^{\frac{1}{2}}D_q^{\frac{1}{2}}\big(h(p,q)\big) = 3.771p^6 + 3.959p^7 + 1.886p^9 + \big((2\times 3^n) - (3\times 2^n) - 4\big)p^8 + 1.960p^{10} + 0.5(2^n - 4)p^{12}.$$
$$GA(SR_n) = (2S_pJD_p^{\frac{1}{2}}D_q^{\frac{1}{2}})\big(h(p,q)\big)\mid_{p=1} = (2\times 3^n) - (0.540\times 2^n) - 0.303.$$

□

Theorem 4 *Let a Koch Snowflake be denoted by KS_n and*

$$M(KS_n, p, q) = (3\times 4^n)p^2q^2.$$

Then,

1. $M_1(SR_n) = 12\times 4^n$.
2. $M_2(SR_n) = 12\times 4^n$.

3. ${}^{m}M_2(SR_n) = 3 \times 4^{n-1}$.
4. $R_\alpha(SR_n) = 1.5 \times 4^n$.
5. $RR_\alpha(SR_n) = 6 \times 4^n$.
6. $SDD(SR_n) = 6 \times 4^n$.
7. $H(SR_n) = 12 \times 4^n$.
8. $I(SR_n) = 3 \times 4^n$.
9. $A(SR_n) = 24 \times 4^n$.
10. $ABC(SR_n) = 2.121 \times 4^n$.
11. $GA(SR_n) = 3 \times 4^n$.

Proof Let the M-polynomial of KS_n be $h(p, q)$. Hence,

$$h(p, q) = (3 \times 4^n) p^2 q^2.$$

The first Zagreb index is obtained as under:

$$D_p\big(h(p,q)\big) = D_q\big(h(p,q)\big) = (6 \times 4^n)p^2q^2.$$
$$M_1(SR_n) = (D_p + D_q)\big(h(p,q)\big)\,|_{p=q=1} = 12 \times 4^n.$$

The second Zagreb index is evaluated as follows:

$$D_qD_p\big(h(p,q)\big) = (12 \times 4^n)p^2q^2.$$
$$M_2(SR_n) = (D_pD_q)\big(h(p,q)\big)\,|_{p=q=1} = 12 \times 4^n.$$

The modified second Zagreb index is estimated as below:

$$S_p\big(h(p,q)\big) = S_q\big(h(p,q)\big) = (1.5 \times 4^n)p^2q^2.$$
$$S_pS_q\big(h(p,q)\big) = (0.75 \times 4^n)p^2q^2.$$
$${}^{m}M_2(SR_n) = (S_pS_q)\big(h(p,q)\big)\,|_{p=q=1} = 3 \times 4^{n-1}.$$

We calculate the Randić index as below,

$$D_p^\alpha\big(h(p,q)\big) = D_q^\alpha\big(h(p,q)\big) = (3 \times 2^\alpha \times 4^n)p^2q^2.$$
$$D_q^\alpha D_p^\alpha\big(h(p,q)\big) = (3 \times 2^{2\alpha} \times 4^n)p^2q^2.$$
$$R_{\alpha=-1/2}(SR_n) = (D_q^\alpha D_p^\alpha)\big(h(p,q)\big)\,|_{p=q=1} = 1.5 \times 4^n.$$

We assess the inverse Randić index as

$$S_p^\alpha\big(h(p,q)\big) = S_q^\alpha\big(h(p,q)\big) = (3 \times 2^{-\alpha} \times 4^n)p^2q^2.$$
$$S_q^\alpha S_p^\alpha\big(h(p,q)\big) = (3 \times 2^{-2\alpha} \times 4^n)p^2q^2.$$
$$RR_{\alpha=-1/2}(SR_n) = (S_p^\alpha S_q^\alpha)\big(h(p,q)\big)\,|_{p=q=1} = 6 \times 4^n.$$

The symmetric division deg index is determined by

$$S_qD_p\big(h(p,q)\big) = S_pD_q\big(h(p,q)\big) = (3 \times 4^n)p^2q^2.$$
$$SDD(SR_n) = (S_qD_p + S_pD_q)\big(h(p,q)\big)\mid_{p=q=1} = 6 \times 4^n.$$

The harmonic index is appraised as below:

$$J\big(h(p,q)\big) = (3 \times 4^n)p^4.$$
$$S_pJ\big(h(p,q)\big) = (12 \times 4^n)p^4.$$
$$H(SR_n) = 2(S_pJ)\big(h(p,q)\big)\mid_{p=1} = 12 \times 4^n.$$

The assessment of the inverse sum index is done as follows:

$$S_pJD_pD_q\big(h(p,q)\big) = (3 \times 4^n)p^4.$$
$$I(SR_n) = (S_pJD_pD_q)\big(h(p,q)\big)\mid_{p=1} = 3 \times 4^n.$$

We figure the augmented index as

$$S_p^3Q_{-2}JD_x^3D_q^3\big(h(p,q)\big) = (24 \times 4^n)p^2.$$
$$A(SR_n) = (S_p^3Q_{-2}JD_x^3D_q^3)\big(h(p,q)\big)\mid_{p=1} = 24 \times 4^n.$$

The computation of the atom bond connectivity index is implemented as below:

$$D_p^{\frac{1}{2}}Q_{-2}JS_p^{\frac{1}{2}}S_q^{\frac{1}{2}}\big(h(p,q)\big) = (2.121 \times 4^n)p^2.$$
$$ABC(SR_n) = (D_p^{\frac{1}{2}}Q_{-2}JS_p^{\frac{1}{2}}S_q^{\frac{1}{2}})\big(h(p,q)\big)\mid_{p=1} = 2.121 \times 4^n.$$

The result of the geometric arithmetic index is carried out as follows:

$$2S_pJD_p^{\frac{1}{2}}D_q^{\frac{1}{2}}\big(h(p,q)\big) = (3 \times 4^n)p^4.$$
$$GA(SR_n) = (2S_pJD_p^{\frac{1}{2}}D_q^{\frac{1}{2}})\big(h(p,q)\big)\mid_{p=1} = 3 \times 4^n.$$

□

Next, we discuss the neighbourhood M-polynomial of the graph for Sierpiński Rhombus and Koch Snowflake and to derive the general closed-form for the sequences of graphs.

Theorem 5 *Let a Sierpiński Rhombus be denoted by SR_n. Then, the neighbourhood M-polynomial is*

$$\begin{aligned} NM(SR_n, p, q) = {} & 4p^8q^{14} + 4p^{14}q^{17} + 2p^{14}q^{25} + 2p^{16}q^{17} + 4p^{17}q^{20} + ((2 \times 3^n) \\ & -(2^{n+2}) - 8)p^{16}q^{16} + (2^n - 4)p^{16}q^{20} + (2^{n+1} - 2)p^{20}q^{25} \\ & +4p^{17}q^{25} + 2p^{25}q^{28} + (2^{n-1} - 4)p^{28}q^{28}. \end{aligned}$$

Proof From the graph SR_n sequences, we can see the partition of the edge set for neighbourhood is as follows:

$$
\begin{aligned}
| E_1 | &=| \{uv \in E(SR_n) \mid d_u = 8 \; and \; d_v = 14\} |= 4.\\
| E_2 | &=| \{uv \in E(SR_n) \mid d_u = 14 \; and \; d_v = 17\} |= 4.\\
| E_3 | &=| \{uv \in E(SR_n) \mid d_u = 14 \; and \; d_v = 25\} |= 2.\\
| E_4 | &=| \{uv \in E(SR_n) \mid d_u = 16 \; and \; d_v = 16\} |= (2 \times 3^n) - (2^{n+2}) - 8.\\
| E_5 | &=| \{uv \in E(SR_n) \mid d_u = 16 \; and \; d_v = 17\} |= 2.\\
| E_6 | &=| \{uv \in E(SR_n) \mid d_u = 16 \; and \; d_v = 20\} |= 2^n - 4.\\
| E_7 | &=| \{uv \in E(SR_n) \mid d_u = 17 \; and \; d_v = 20\} |= 4.\\
| E_8 | &=| \{uv \in E(SR_n) \mid d_u = 17 \; and \; d_v = 25\} |= 4.\\
| E_9 | &=| \{uv \in E(SR_n) \mid d_u = 20 \; and \; d_v = 25\} |= 2^{n+1} - 2.\\
| E_{10} | &=| \{uv \in E(SR_n) \mid d_u = 25 \; and \; d_v = 28\} |= 2.\\
| E_{11} | &=| \{uv \in E(SR_n) \mid d_u = 28 \; and \; d_v = 28\} |= 2^{n-1} - 4.
\end{aligned}
$$

Let $d_u = k$ and $d_v = l$ such that $k \leq l$, for all $k, l \in V(G)$ and $uv \in E(G)$. As a result of the definition, the neighbourhood M-polynomial of SR_n is

$$
\begin{aligned}
NM(SR_n, p, q) &= \sum_{k \leq l} m_{kl} p^k q^l \\
&= | E_1 | p^8 q^{14} + | E_2 | p^{14} q^{17} + | E_3 | p^{14} q^{25} + | E_4 | p^{16} q^{16} \\
&\quad + | E_5 | p^{16} q^{17} + | E_6 | p^{16} q^{20} + | E_7 | p^{17} q^{20} + | E_8 | p^{17} q^{25} \\
&\quad + | E_9 | p^{20} q^{25} + | E_{10} | p^{25} q^{28} + | E_{11} | p^{28} q^{28}. \\
&= 4p^8 q^{14} + 4p^{14} q^{17} + 2p^{14} q^{25} + 2p^{16} q^{17} + 4p^{17} q^{20} + ((2 \times 3^n) \\
&\quad -(2^{n+2}) - 8)p^{16} q^{16} + (2^n - 4)p^{16} q^{20} + (2^{n+1} - 2)p^{20} q^{25} \\
&\quad +4p^{17} q^{25} + 2p^{25} q^{28} + (2^{n-1} - 4)p^{28} q^{28}.
\end{aligned}
$$

□

Theorem 6 *Let a Koch Snowflake be denoted by KS_n. Then, the neighbourhood M-polynomial is*

$$NM(KS_n, p, q) = (3 \times 4^n) p^4 q^4.$$

Proof From the graph KS_n sequences, we can see the partition of the edge set for the neighbourhood is as follows:

$$| E_1 | =| \{uv \in E(KS_n) \mid d_u = 4 \; and \; d_v = 4\} |= 3 \times 4^n.$$

Let $d_u = k$ and $d_v = l$ such that $k \leq l$, for all $k, l \in V(G)$ and $uv \in E(G)$. As a result of the definition, the neighbourhood M-polynomial of KS_n is

$$NM(KS_n, p, q) = \sum_{k \leq l} m_{kl} p^k q^l =| E_1 | p^4 q^4 = (3 \times 4^n) p^4 q^4.$$

□

Neighbourhood-Based M-polynomial on Topological Indices
This section uses the Theorems 5 and 6 to calculate the topological indices of the graphs SR_n and KS_n. In addition, maple is used to characterize the calculated indices.

Theorem 7 *Let a Sierpiński Rhombus be denoted by SR_n, and*

$$NM(SR_n, p, q) = 4p^8q^{14} + 4p^{14}q^{17} + 2p^{14}q^{25} + 2p^{16}q^{17} + 4p^{17}q^{20} + ((2 \times 3^n) -(2^{n+2}) - 8)p^{16}q^{16} + (2^n - 4)p^{16}q^{20} + (2^{n+1} - 2)p^{20}q^{25} +4p^{17}q^{25}2p^{25}q^{28} + (2^{n-1} - 4)p^{28}q^{28}.$$

Then,

1. $M_1^1(SR_n) = (64 \times 3^n) + (126 \times 2^n) + 64.$
2. $M_2^*(SR_n) = (512 \times 3^n) + (688 \times 2^n) - 360.$
3. $F_N^*(SR_n) = (1024 \times 3^n) - (606 \times 2^n) - 100.$
4. $M_2^{nm}(SR_n) = (0.008 \times 3^n) - (0.008 \times 2^n) + 0.037.$
5. $NR_\alpha(SR_n) = (0.125 \times 3^n) - (0.087 \times 2^n) - 358.119.$
6. $ND_3(SR_n) = (8192 \times 3^n) + (19712 \times 2^n) + 221680.$
7. $ND_5(SR_n) = (4 \times 3^n) - (0.85 \times 2^n) + 10.570.$
8. $NH(SR_n) = (0.125 \times 3^n) - (0.088 \times 2^n) + 0.374.$
9. $NI(SR_n) = (16 \times 3^n) + (6.111 \times 2^n) + 11.377.$
10. $S(SR_n) = (1242.757 \times 3^n) + (3022.730 \times 2^n) - 4097.965.$

Proof Let the neighbourhood M-polynomial of SR_n be $h(p, q)$. Hence,

$$h(p, q) = 4p^8q^{14} + 4p^{14}q^{17} + 2p^{14}q^{25} + 2p^{16}q^{17} + 4p^{17}q^{20} + \big((2 \times 3^n) -(2^{n+2}) - 8\big)p^{16}q^{16} + (2^n - 4)p^{16}q^{20} + (2^{n+1} - 2)p^{20}q^{25} +4p^{17}q^{25}2p^{25}q^{28} + (2^{n-1} - 4)p^{28}q^{28}.$$

The third version of the Zagreb index is obtained as under:

$$D_p\big(h(p, q)\big) = 32p^8q^{14} + 56p^{14}q^{17} + 28p^{14}q^{25} + 32p^{16}q^{17} + 68p^{17}q^{20} + 16\big((2 \times 3^n) -(2^{n+2}) - 8\big)p^{16}q^{16} + 16(2^n - 4)p^{16}q^{20} + 20(2^{n+1} - 2)p^{20}q^{25} +68p^{17}q^{25} + 50p^{25}q^{28} + 28(2^{n-1} - 4)p^{28}q^{28}.$$

$$D_q\big(h(p, q)\big) = 56p^8q^{14} + 68p^{14}q^{17} + 50p^{14}q^{25} + 34p^{16}q^{17} + 80p^{17}q^{20} + 16\big((2 \times 3^n) -(2^{n+2}) - 8\big)p^{16}q^{16} + 20(2^n - 4)p^{16}q^{20} + 25(2^{n+1} - 2)p^{20}q^{25} +100p^{17}q^{25} + 56p^{25}q^{28} + 28(2^{n-1} - 4)p^{28}q^{28}.$$

$$M_1'(SR_n) = (D_p + D_q)\big(h(p, q)\big) |_{p=q=1} = (64 \times 3^n) + (126 \times 2^n) + 64.$$

The neighbourhood second Zagreb index is evaluated as follows:

$$D_pD_q\big(h(p, q)\big) = 448p^8q^{14} + 952p^{14}q^{17} + 700p^{14}q^{25} + 544p^{16}q^{17} + 1360p^{17}q^{20} +256\big((2 \times 3^n) - (2^{n+2}) - 8\big)p^{16}q^{16} + 320(2^n - 4)p^{16}q^{20} +500(2^{n+1} - 2)p^{20}q^{25} + 1700p^{17}q^{25} + 1400p^{25}q^{28} +784(2^{n-1} - 4)p^{28}q^{28}.$$

$$M_2^*(SR_n) = (D_pD_q)\big(h(p, q)\big) |_{p=q=1} = (512 \times 3^n) + (688 \times 2^n) - 360.$$

The neighbourhood forgotten index is estimated as below:

$$\begin{aligned}
D_p^2 + D_q^2\big(h(p,q)\big) &= 1040p^8q^{14} + 1940p^{14}q^{17} + 1642p^{14}q^{25} + 1090p^{16}q^{17}\\
&+2756p^{17}q^{20} + 512\big((2\times 3^n) - (2^{n+2}) - 8\big)p^{16}q^{16}\\
&+656(2^n - 4)p^{16}q^{20} + 1025(2^{n+1} - 2)p^{20}q^{25}\\
&+3656p^{17}q^{25} + 2818p^{25}q^{28} + 1568(2^{n-1} - 4)p^{28}q^{28}.\\
F_N^*(SR_n) &= (D_p^2 + D_q^2)\big(h(p,q)\big)\mid_{p=q=1}\\
&= (1024\times 3^n) - (606\times 2^n) - 100.
\end{aligned}$$

We calculate the second modified Zagreb index as below:

$$\begin{aligned}
S_p\big(h(p,q)\big) &= 0.5p^8q^{14} + 0.286p^{14}q^{17} + 0.143p^{14}q^{25} + 0.125p^{16}q^{17}\\
&+0.235p^{17}q^{20} + 0.063\big((2\times 3^n) - (2^{n+2}) - 8\big)p^{16}q^{16}\\
&+0.063(2^n - 4)p^{16}q^{20} + 0.05(2^{n+1} - 2)p^{20}q^{25} + 0.235p^{17}q^{25}\\
&+0.08p^{25}q^{28} + 0.036(2^{n-1} - 4)p^{28}q^{28}.\\
S_q\big(h(p,q)\big) &= 0.286p^8q^{14} + 0.286p^{14}q^{17} + 0.08p^{14}q^{25} + 0.118p^{16}q^{17}\\
&+0.2p^{17}q^{20} + 0.063\big((2\times 3^n) - (2^{n+2}) - 8\big)p^{16}q^{16}\\
&+0.05(2^n - 4)p^{16}q^{20} + 0.04(2^{n+1} - 2)p^{20}q^{25} + 0.16p^{17}q^{25}\\
&+0.071p^{25}q^{28} + 0.036(2^{n-1} - 4)p^{28}q^{28}.\\
S_pS_q\big(h(p,q)\big) &= 0.036p^8q^{14} + 0.017p^{14}q^{17} + 0.006p^{14}q^{25} + 0.007p^{16}q^{17}\\
&+0.012p^{17}q^{20} + 0.003\big((2\times 3^n) - (2^{n+2}) - 8\big)p^{16}q^{16}\\
&+0.003(2^n - 4)p^{16}q^{20} + 0.002(2^{n+1} - 2)p^{20}q^{25}\\
&+0.009p^{17}q^{25} + 0.003p^{25}q^{28} + 0.001(2^{n-1} - 4)p^{28}q^{28}.\\
M_2^{nm}(SR_n) &= (S_pS_q)\big(h(p,q)\big)\mid_{p=q=1} = (0.008\times 3^n) - (0.008\times 2^n) + 0.037.
\end{aligned}$$

We assess the general Randić index as

$$\begin{aligned}
D_p^\alpha\big(h(p,q)\big) &= (4\times 8^\alpha)p^8q^{14} + (4\times 14^\alpha)p^{14}q^{17} + (2\times 14^\alpha)p^{14}q^{25} + (2\times 16^\alpha)p^{16}q^{17}\\
&+(4\times 17^\alpha)p^{17}q^{20} + 16^\alpha\big((2\times 3^n) - (2^{n+2}) - 8\big)p^{16}q^{16}\\
&+16^\alpha(2^n - 4)p^{16}q^{20} + 20^\alpha(2^{n+1} - 2)p^{20}q^{25} + (4\times 17^\alpha)p^{17}q^{25}\\
&+(2\times 25^\alpha)p^{25}q^{28} + 28^\alpha(2^{n-1} - 4)p^{28}q^{28}.\\
D_q^\alpha\big(h(p,q)\big) &= (4\times 14^\alpha)p^8q^{14} + (4\times 17^\alpha)p^{14}q^{17} + (2\times 25^\alpha)p^{14}q^{25} + (2\times 17^\alpha)p^{16}q^{17}\\
&+(4\times 20^\alpha)p^{17}q^{20} + 16^\alpha\big((2\times 3^n) - (2^{n+2}) - 8\big)p^{16}q^{16}\\
&+20^\alpha(2^n - 4)p^{16}q^{20} + 25^\alpha(2^{n+1} - 2)p^{20}q^{25} + (4\times 25^\alpha)p^{17}q^{25}\\
&+(2\times 28^\alpha)p^{25}q^{28} + 28^\alpha(2^{n-1} - 4)p^{28}q^{28}.\\
D_q^\alpha D_p^\alpha\big(h(p,q)\big) &= (4\times 14^\alpha\times 8^\alpha)p^8q^{14} + (4\times 17^\alpha\times 14^\alpha)p^{14}q^{17} + (2\times 25^\alpha\times 14^\alpha)p^{14}q^{25}\\
&+(2\times 17^\alpha\times 16^\alpha)p^{16}q^{17} + (4\times 20^\alpha\times 17^\alpha)p^{17}q^{20}\\
&+16^{2\alpha}\big((2\times 3^n) - (2^{n+2}) - 8\big)p^{16}q^{16} + (20^\alpha\times 16^\alpha)(2^n - 4)p^{16}q^{20}\\
&+(25^\alpha\times 20^\alpha)(2^{n+1} - 2)p^{20}q^{25} + (4\times 25^\alpha\times 17^\alpha)p^{17}q^{25}\\
&+(2\times 28^\alpha\times 25^\alpha)p^{25}q^{28} + (28^{2\alpha})(2^{n-1} - 4)p^{28}q^{28}.\\
NR_{\alpha=-1/2}(SR_n) &= (D_p^\alpha D_q^\alpha)\big(h(p,q)\big)\mid_{p=q=1} = (0.125\times 3^n) - (0.087\times 2^n) - 358.119.
\end{aligned}$$

The third ND_e index is appraised as below:

$$\begin{aligned}(D_pD_q)(D_p+D_q)\big(h(p,q)\big) &= 14336p^8q^{14}+53312p^{14}q^{17}+19600p^{14}q^{25}\\ &+17408p^{16}q^{17}+92480p^{17}q^{20}+4096\big((2\times 3^n)\\ &-(2^{n+2})-8\big)p^{16}q^{16}+5120(2^n-4)p^{16}q^{20}\\ &+10000(2^{n+1}-2)p^{20}q^{25}+115600p^{17}q^{25}\\ &+70000p^{25}q^{28}+21952(2^{n-1}-4)p^{28}q^{28}.\\ ND_3(SR_n) &= (D_pD_q)(D_p+D_q)\big(h(p,q)\big)\mid_{p=q=1}\\ &= (8192\times 3^n)+(19712\times 2^n)+221680.\end{aligned}$$

The assessment of the fifth ND_e index is done as follows:

$$\begin{aligned}S_qD_p\big(h(p,q)\big) &= 2.286p^8q^{14}+3.294p^{14}q^{17}+1.12p^{14}q^{25}+1.882p^{16}q^{17}\\ &+3.4p^{17}q^{20}+\big((2\times 3^n)-(2^{n+2})-8\big)p^{16}q^{16}+0.8\\ &\times(2^n-4)p^{16}q^{20}+0.8(2^{n+1}-2)p^{20}q^{25}+2.72p^{17}q^{25}\\ &+1.786p^{25}q^{28}+(2^{n-1}-4)p^{28}q^{28}.\\ S_pD_q\big(h(p,q)\big) &= 7p^8q^{14}+4.857p^{14}q^{17}+3.571p^{14}q^{25}+2.125p^{16}q^{17}\\ &+4.706p^{17}q^{20}+\big((2\times 3^n)-(2^{n+2})-8\big)p^{16}q^{16}+1.25\\ &\times(2^n-4)p^{16}q^{20}+1.25(2^{n+1}-2)p^{20}q^{25}+5.882\\ &\times p^{17}q^{25}+2.24p^{25}q^{28}+(2^{n-1}-4)p^{28}q^{28}.\\ S_qD_p+S_pD_q\big(h(p,q)\big) &= 9.286p^8q^{14}+8.151p^{14}q^{17}+4.691p^{14}q^{25}+4.007p^{16}q^{17}\\ &+8.106p^{17}q^{20}+2\big((2\times 3^n)-(2^{n+2})-8\big)p^{16}q^{16}\\ &+2.05(2^n-4)p^{16}q^{20}+2.05(2^{n+1}-2)p^{20}q^{25}\\ &+8.602p^{17}q^{25}+4.026p^{25}q^{28}+2(2^{n-1}-4)p^{28}q^{28}.\\ ND_5(SR_n) &= (D_pS_q+S_pD_q)\big(h(p,q)\big)\mid_{p=q=1}\\ &= (4\times 3^n)-(0.85\times 2^n)+10.570.\end{aligned}$$

We figure the neighbourhood harmonic index as

$$\begin{aligned}J\big(h(p,q)\big) &= 4p^{22}+4p^{31}+2p^{39}+2p^{33}+4p^{37}+\big((2\times 3^n)-2^{n+2}-8\big)p^{32}\\ &+(2^n-4)p^{36}+(2^{n+1}-2)p^{45}+4p^{42}+2p^{53}+(2^{n-1}-4)p^{56}.\\ S_pJ\big(h(p,q)\big) &= 0.181p^{22}+0.129p^{31}+0.051p^{39}+0.061p^{33}+0.108p^{37}+0.031\\ &\times\big((2\times 3^n)-2^{n+2}-8\big)p^{32}+0.028(2^n-4)p^{36}+0.022\\ &\times(2^{n+1}-2)p^{45}+40.095p^{42}+0.038p^{53}+0.018(2^{n-1}-4)p^{56}.\\ NH(SR_n) &= 2(S_pJ)\big(h(p,q)\big)\mid_{p=1}=(0.125\times 3^n)-(0.088\times 2^n)+0.374.\end{aligned}$$

The computation of the neighbourhood inverse sum index is implemented as below:

$$\begin{aligned}S_pJD_pD_q\big(h(p,q)\big) &= 20.364p^{22}+30.710p^{31}+17.949p^{39}+16.485p^{33}+36.757p^{37}\\ &+8\big((2\times 3^n)-2^{n+2}-8\big)p^{32}+8.889(2^n-4)p^{36}+11.111\\ &\times(2^{n+1}-2)p^{45}+40.476p^{42}+26.415p^{53}+14(2^{n-1}-4)p^{56}.\\ NI(SR_n) &= (S_pJD_pD_q)\big(h(p,q)\big)\mid_{p=1}=(16\times 3^n)+(6.111\times 2^n)+11.377.\end{aligned}$$

The result of the Sanskruti index is carried out as follows:

$$\begin{aligned}S_p^3Q_{-2}JD_x^3D_q^3\big(h(p,q)\big) &= 702.464p^{20} + 2211.041p^{29} + 1692.890p^{37}\\ &+1350.988p^{31} + 3666.845p^{35} + 621.378\big((2\times 3^n)\\ &-2^{n+2}-8\big)p^{30} + 833.706(2^n-4)p^{34} + 1572.189\\ &\times(2^{n+1}-2)p^{43} + 4797.852p^{40} + 5171.465p^{51}\\ &+3060.320(2^{n-1}-4)p^{54}.\\ S(SR_n) &= (S_p^3Q_{-2}JD_x^3D_q^3)\big(h(p,q)\big)\mid_{p=1}\\ &= (1242.757\times 3^n) + (3022.730\times 2^n) - 4097.965.\end{aligned}$$

□

Theorem 8 *Let a Koch Snowflake be denoted by KS_n, and*

$$NM(KS_n, p, q) = (3\times 4^n)p^4q^4.$$

Then,

1. $M_1^1(SR_n) = 6\times 4^{n+1}$.
2. $M_2^*(SR_n) = 3\times 4^{n+2}$.
3. $F_N^*(SR_n) = 6\times 4^{n+2}$.
4. $M_2^{nm}(SR_n) = 3\times 4^{n-2}$.
5. $NR_\alpha(SR_n) = 3\times 4^{n-1}$.
6. $ND_3(SR_n) = 18\times 2^{4n+6}$.
7. $ND_5(SR_n) = 6\times 2^{2n}$.
8. $NH(SR_n) = 3\times 4^{n-1}$.
9. $NI(SR_n) = 6\times 4^n$.
10. $S(SR_n) = 2^{2n+11}$.

Proof Let the neighbourhood M-polynomial of KS_n be $h(p,q)$. Hence,

$$h(p,q) = (3\times 4^n)p^4q^4.$$

The third version of Zagreb index is obtained as under:

$$\begin{aligned}D_p\big(h(p,q)\big) = D_q\big(h(p,q)\big) &= (3\times 4^{n+1})p^4q^4.\\ D_p + D_q\big(h(p,q)\big) &= (6\times 4^{n+1})p^4q^4.\\ M_1'(SR_n) &= (D_p+D_q)\big(h(p,q)\big)\mid_{p=q=1} = 6\times 4^{n+1}.\end{aligned}$$

The neighbourhood second Zagreb index is evaluated as follows:

$$\begin{aligned}D_pD_q\big(h(p,q)\big) &= (3\times 4^{n+2})p^4q^4.\\ M_2(SR_n) &= (D_pD_q)\big(h(p,q)\big)\mid_{p=q=1} = 3\times 4^{n+2}.\end{aligned}$$

The neighbourhood forgotten index is estimated as below:

$$D_p^2 + D_q^2\big(h(p,q)\big) = (6\times 4^{n+2})p^4q^4.$$

$$F_N^*(SR_n) = (D_p^2 + D_q^2)\big(h(p,q)\big) \mid_{p=q=1} = 6 \times 4^{n+2}.$$

We calculate the second modified Zagreb index as below:

$$S_p\big(h(p,q)\big) = S_q\big(h(p,q)\big) = (3 \times 4^{n-1})p^4q^4.$$
$$S_pS_q\big(h(p,q)\big) = (3 \times 4^{n-2})p^4q^4.$$
$$M_2^{nm}(SR_n) = (S_pS_q)\big(h(p,q)\big) \mid_{p=q=1} = 3 \times 4^{n-2}.$$

We assess the general Randić index as

$$D_p^\alpha\big(h(p,q)\big) = D_q^\alpha\big(h(p,q)\big) = (3 \times 4^{n+\alpha})p^4q^4.$$
$$D_q^\alpha D_p^\alpha\big(h(p,q)\big) = (3 \times 4^{n+2\alpha})p^4q^4.$$
$$NR_{\alpha=-1/2}(SR_n) = (D_p^\alpha D_q^\alpha)\big(h(p,q)\big) \mid_{p=q=1} = 3 \times 4^{n-1}.$$

The third ND_e index is appraised as below:

$$(D_pD_q)(D_p + D_q)\big(h(p,q)\big) = (18 \times 2^{4n+6})p^4q^4.$$
$$ND_3(SR_n) = (D_pD_q)(D_p + D_q)\big(h(p,q)\big) \mid_{p=q=1} = 18 \times 2^{4n+6}.$$

The assessment of the fifth ND_e index is done as follows:

$$S_qD_p\big(h(p,q)\big) = S_pD_q\big(h(p,q)\big) = (3 \times 4^n)p^4q^4.$$
$$S_qD_p + S_pD_q\big(h(p,q)\big) = (6 \times 2^{2n})p^4q^4.$$
$$ND_5(SR_n) = (D_pS_q + S_pD_q)\big(h(p,q)\big) \mid_{p=q=1} = 6 \times 2^{2n}.$$

We figure the neighbourhood harmonic index as

$$J\big(h(p,q)\big) = (3 \times 4^n)p^8.$$
$$S_pJ\big(h(p,q)\big) = (3 \times 2^{2n-3})p^8.$$
$$NH(SR_n) = 2(S_pJ)\big(h(p,q)\big) \mid_{p=1} = 3 \times 4^{n-1}.$$

The computation of the neighbourhood inverse sum index is implemented as below:

$$S_pJD_pD_q\big(h(p,q)\big) = (6 \times 4^n)p^8.$$
$$NI(SR_n) = (S_pJD_pD_q)\big(h(p,q)\big) \mid_{p=1} = 6 \times 4^n.$$

The result of the Sanskruti index is carried out as follows:

$$S_p^3Q_{-2}JD_x^3D_q^3\big(h(p,q)\big) = 2^{2n+11} \times p^6.$$
$$S(SR_n) = (S_p^3Q_{-2}JD_x^3D_q^3)\big(h(p,q)\big)\mid_{p=1} = 2^{2n+11}. \qquad \square$$

3 Entropy Measure on Topological Indices

In [25], Shannon is the one who initiated the development of the entropy concept in 1948, which is a famous paper by Shannon. Afterwards, the idea of entropy is implemented into the graphs. The graph entropy concept proposed by Rashevsky is based on the vertex orbit classification [21].

The graph entropy is based on the probability distribution on the vertex set of the graph as well as the graph itself. In mathematical chemistry, graph entropy is availed on a broad scale to describe graph-based structure systems. There are different kinds of measures in graph entropy which can be seen in [17]. Intrinsic and extrinsic are the two measures of categories in graph entropy classification. Probability distribution with few structural aspects is prompted for intrinsic graph entropy measures. To the graph elements, a probability distribution is spontaneously assigned for extrinsic graph entropy measures. In classical graph entropy, the majority of the entropies are intrinsic.

The purpose of entropy is to study the measure of information present in the structure of graphs and acts as a complexity measure. The measures of graph entropy combine the probability distributions with the graph elements, i.e. vertices, edges, etc. An equivalence relation defined on a finite graph is related to such a measure. Allows to define the probability distribution of the induced partition by an equivalence relation. Now, let us learn about the Shannon entropy and the definition of the entropy of graph G.

The probability vector of Shannon's entropy be $q = (q_1, q_2, \ldots, q_n)$ and $\sum_{i=1}^{n} q_i = 1$ is defined as follows:

$$I(Q) = -\sum_{i=1}^{n} q_i \log(q_i), \tag{2}$$

where $0 < q_i \leq 1$.

Cao et al. provided a probability value to every vertex $v_i \in V(G)$ to define the entropy of a graph G with the set $V(G) = \{v_1, v_2, \ldots, v_n\}$ of vertices as

$$q(v_i) = \frac{\mathrm{f}(v_i)}{\sum_{j=1}^{n} \mathrm{f}(v_j)}, \tag{3}$$

in which f is arbitrary information functional. Through this, the entropy of a graph G is known to be below as

$$I_{\mathrm{f}}(G) = -\sum_{i=1}^{n} \frac{\mathrm{f}(v_i)}{\sum_{j=1}^{n}\mathrm{f}(v_j)} \log\Big(\frac{\mathrm{f}(v_i)}{\sum_{j=1}^{n}\mathrm{f}(v_j)}\Big) = \log\Big(\sum_{j=1}^{n}\mathrm{f}(v_j)\Big) - \sum_{i=1}^{n} \frac{\mathrm{f}(v_i)\log(\mathrm{f}(v_i))}{\sum_{j=1}^{n}\mathrm{f}(v_j)}. \quad (4)$$

Degree-Based Entropy on Topological Indices

Some degree-based topological entropy indices are considered, and discussions are provided below for SR_n and KS_n.

Theorem 9 *Let ENT represent the entropy. Then the entropy of degree-based topological indices for SR_n is calculated as follows:*

1. $$ENT_{M_1}(SR_n) = log\big((16 \times 3^n) + (2 \times 2^n) - 26\big) - \frac{log}{(16\times3^n)+(2\times2^n)-26}\Big[3^n(2 \times 8^8) + 2^n\Big((2 \times 10^{10}) - (3 \times 8^8) + \frac{12^{12}}{2}\Big) + (4 \times 6^6) + (4 \times 7^7) + (2 \times 9^9) - (4 \times 10^{10}) - (2 \times 12^{12})\Big].$$

Proof By definition, the first Zagreb index is derived using Theorem 1 edge partition, and we obtain the result using it

$$\begin{aligned}
M_1(SR_n) &= (16 \times 3^n) + (2 \times 2^n) - 26.\\
ENT_{M_1}(SR_n) &= log\big(M_1(SR_n)\big) - \frac{log}{M_1(SR_n)}\Big[\sum_{st\in\Xi} \big(\Xi(s) + \Xi(t)\big)^{\Xi(s)+\Xi(t)}\Big]\\
&= log\big((16 \times 3^n) + (2 \times 2^n) - 26\big) - \frac{log}{(16\times3^n)+(2\times2^n)-26}\\
&\quad \times\Big[4(2+4)^{(2+4)} + 4(3+4)^{(3+4)} + 2(3+6)^{(3+6)} + \big((2+3^n)\\
&\quad -(3+2^n) - 4\big)(4+4)^{(4+4)} + (2^{n+1} - 4)(4+6)^{(4+6)}\\
&\quad +(2^{n-1} - 2)(6+6)^{(6+6)}\Big]\\
&= log\big((16 \times 3^n) + (2 \times 2^n) - 26\big) - \frac{log}{(16\times3^n)+(2\times2^n)-26}\\
&\quad \times\Big[3^n(2 \times 8^8) + 2^n\Big((2 \times 10^{10}) - (3 \times 8^8) + \frac{12^{12}}{2}\Big)\\
&\quad +(4 \times 6^6) + (4 \times 7^7) + (2 \times 9^9) - (4 \times 10^{10}) - (2 \times 12^{12})\Big].
\end{aligned}$$

□

2. $$ENT_{M_2}(SR_n) = log\big((32 \times 3^n) + (18 \times 2^n) - 116\big) - \frac{log}{(32\times3^n)+(18\times2^n)-116} \times\Big[3^n(2 \times 16^{16}) + 2^n\Big((2 \times 24^{24}) - (3 \times 16^{16}) + \frac{36^{36}}{2}\Big) + (4 \times 12^{12}) + (2 \times 18^{18}) - (4 \times 16^{16}) - (4 \times 24^{24}) - (2 \times 36^{36}) + (4 \times 8^8)\Big].$$

Proof Applying Theorem 1 and by definition, we attain the second Zagreb index. Hence, the result follows:

$$
\begin{aligned}
M_2(SR_n) &= (32 \times 3^n) + (18 \times 2^n) - 116.\\
ENT_{M_2}(SR_n) &= log\big(M_2(SR_n)\big) - \tfrac{log}{M_2(SR_n)}\Big[\sum_{st\in\Xi} \big(\Xi(s) * \Xi(t)\big)^{\Xi(s)*\Xi(t)}\Big]\\
&= log\big((32 \times 3^n) + (18 \times 2^n) - 116\big) - \tfrac{log}{(32\times 3^n)+(18\times 2^n)-116}\\
&\quad \times\Big[4(2 \times 4)^{2\times 4} + 4(3 \times 4)^{(3\times 4)} + 2(3 \times 6)^{(3\times 6)} + \big((2+3^n)\\
&\quad -(3+2^n) - 4\big)(4 \times 4)^{(4\times 4)} + (2^{n+1} - 4)(4 \times 6)^{(4\times 6)}\\
&\quad +(2^{n-1} - 2)(6 \times 6)^{(6\times 6)}\Big]\\
&= log\big((32 \times 3^n) + (18 \times 2^n) - 116\big) - \tfrac{log}{(32\times 3^n)+(18\times 2^n)-116}\\
&\quad \times\Big[3^n(2 \times 16^{16}) + 2^n\Big((2 \times 24^{24}) - (3 \times 16^{16}) + \tfrac{36^{36}}{2}\Big) + (4 \times 12^{12})\\
&\quad +(2 \times 18^{18}) - (4 \times 16^{16}) - (4 \times 24^{24}) - (2 \times 36^{36}) + (4 \times 8^8)\Big].
\end{aligned}
$$

□

3. $$
\begin{aligned}
ENT_{HM}(SR_n) &= log\big((128 \times 3^n) + (26 \times 2^n) - 226\big) - \tfrac{log}{(128\times 3^n)+(26\times 2^n)-226}\\
&\quad \times\Big[3^n(2 \times 64^{64}) + 2^n\big((2 \times 100^{100}) - (3 \times 64^{64}) + \tfrac{36^{36}}{2}\big)\\
&\quad +(4 \times 36^{36}) + (4 \times 49^{49}) + (2 \times 81^{81}) - (4 \times 64^{64})\\
&\quad -(4 \times 100^{100}) - (2 \times 36^{36})\Big].
\end{aligned}
$$

Proof By definition, the hyper Zagreb index is derived using Theorem 1 edge partition, and we obtain the result using it

$$
\begin{aligned}
HM(SR_n) &= (128 \times 3^n) + (26 \times 2^n) - 226.\\
ENT_{HM}(SR_n) &= log\big(HM(SR_n)\big) - \tfrac{log}{HM(SR_n)}\Big[\sum_{st\in\Xi} \big(\Xi(s) + \Xi(t)\big)^{2\left(\Xi(s)+\Xi(t)\right)^2}\Big]\\
&= log\big((128 \times 3^n) + (26 \times 2^n) - 226\big) - \tfrac{log}{(128\times 3^n)+(26\times 2^n)-226}\\
&\quad \times\Big[4(2+4)^{2(2+4)^2} + 4(3+4)^{2(3+4)^2} + 2(3+6)^{2(3+6)^2}\\
&\quad +\big((2+3^n) - (3+2^n) - 4\big)(4+4)^{2(4+4)^2} + (2^{n+1} - 4)\\
&\quad \times(4+6)^{2(4+6)^2} + (2^{n-1} - 2)(6+6)^{2(6+6)^2}\Big]\\
&= log\big((128 \times 3^n) + (26 \times 2^n) - 226\big) - \tfrac{log}{(128\times 3^n)+(26\times 2^n)-226}\\
&\quad \times\Big[3^n(2 \times 64^{64}) + 2^n\big((2 \times 100^{100}) - (3 \times 64^{64}) + \tfrac{36^{36}}{2}\big)\\
&\quad +(4 \times 36^{36}) + (4 \times 49^{49}) + (2 \times 81^{81}) - (4 \times 64^{64})\\
&\quad -(4 \times 100^{100}) - (2 \times 36^{36})\Big].
\end{aligned}
$$

□

4. $ENT_F(SR_n) = log\big((64 \times 3^n) + (44 \times 2^n) - 210\big) - \frac{log}{(64\times3^n)+(44\times2^n)-210}$
$\times\Big[3^n(2 \times 32^{32}) + 2^n\Big((2 \times 52^{52}) - (3 \times 32^{32}) + \frac{72^{72}}{2}\Big)$
$+(4 \times 20^{20}) + (4 \times 25^{25}) + (2 \times 45^{45}) - (4 \times 32^{32})$
$-(4 \times 52^{52}) - (2 \times 72^{72})\Big].$

Proof Applying Theorem 1 and by definition, we attain the forgotten index. Hence, the result follows:

$$F(SR_n) = (64 \times 3^n) + (44 \times 2^n) - 210.$$

$$\begin{aligned}ENT_F(SR_n) &= log\big(F(SR_n)\big) - \frac{log}{F(SR_n)}\Big[\sum_{st\in\Xi}\big(\Xi(s)^2 + \Xi(t)^2\big)^{\Xi(s)^2+\Xi(t)^2}\Big]\\ &= log\big((64 \times 3^n) + (44 \times 2^n) - 210\big) - \frac{log}{(64\times3^n)+(44\times2^n)-210}\\ &\times\Big[4(2^2+4^2)^{(2^2+4^2)} + 4(3^2+4^2)^{(3^2+4^2)} + 2(3^2+6^2)^{(3^2+6^2)}\\ &+\big((2+3^n) - (3+2^n) - 4\big)(4^2+4^2)^{(4^2+4^2)} + (2^{n+1}-4)\\ &(4^2+6^2)^{(4^2+6^2)} + (2^{n-1}-2)(6^2+6^2)^{(6^2+6^2)}\Big]\\ &= log\big((64 \times 3^n) + (44 \times 2^n) - 210\big) - \frac{log}{(64\times3^n)+(44\times2^n)-210}\\ &\times\Big[3^n(2 \times 32^{32}) + 2^n\Big((2 \times 52^{52}) - (3 \times 32^{32}) + \frac{72^{72}}{2}\Big)\\ &+(4 \times 20^{20}) + (4 \times 25^{25}) + (2 \times 45^{45}) - (4 \times 32^{32})\\ &-(4 \times 52^{52}) - (2 \times 72^{72})\Big].\end{aligned}$$

□

5. $ENT_{AZI}(SR_n) = log\big((37.926 \times 3^n) + (20.439 \times 2^n) - 155.862\big)$
$-\frac{log}{(37.926\times3^n)+(20.439\times2^n)-155.862}\Bigg[3^n\Bigg(\Bigg(2\times\Big(\frac{16}{6}\Big)^{3\left(\frac{16}{6}\right)^3}\Bigg)$
$+2^n\Bigg(2\times\Big(\frac{24}{8}\Big)^{3\left(\frac{24}{8}\right)^3}\Bigg)-\Bigg(3\times\Big(\frac{16}{6}\Big)^{3\left(\frac{16}{6}\right)^3}\Bigg)+\Bigg(\frac{1}{2}\times\Big(\frac{36}{10}\Big)^{3\left(\frac{36}{10}\right)^3}\Bigg)\Bigg)$
$+\Bigg(4\times\Big(\frac{8}{4}\Big)^{3\left(\frac{8}{4}\right)^3}\Bigg)+\Bigg(4\times\Big(\frac{12}{5}\Big)^{3\left(\frac{12}{5}\right)^3}\Bigg)+\Bigg(2\times\Big(\frac{18}{7}\Big)^{3\left(\frac{18}{7}\right)^3}\Bigg)$
$-\Bigg(4\times\Big(\frac{16}{6}\Big)^{3\left(\frac{16}{6}\right)^3}\Bigg)-\Bigg(4\times\Big(\frac{24}{8}\Big)^{3\left(\frac{24}{8}\right)^3}\Bigg)-\Bigg(2\times\Big(\frac{36}{10}\Big)^{3\left(\frac{36}{10}\right)^3}\Bigg)\Bigg].$

Proof By definition, the augmented Zagreb index is derived using Theorem 1 edge partition, and we obtain the result using it

$$AZI(SR_n) = (37.926 \times 3^n) + (20.439 \times 2^n) - 155.862.$$

$$\begin{aligned}
ENT_{AZI}(SR_n) &= log\big(AZI(SR_n)\big) - \frac{log}{AZI(SR_n)}\Bigg[\sum_{st\in\Xi}\left(\frac{\Xi(s)*\Xi(t)}{\Xi(s)+\Xi(t)-2}\right)^{3\left(\frac{\Xi(s)*\Xi(t)}{\Xi(s)+\Xi(t)-2}\right)^3}\Bigg]\\
&= log\big((37.926 \times 3^n) + (20.439 \times 2^n) - 155.862\big)\\
&\quad -\frac{log}{(37.926\times 3^n)+(20.439\times 2^n)-155.862}\Bigg[4\Big(\frac{2\times4}{2+4-2}\Big)^{3\left(\frac{2\times4}{2+4-2}\right)^3} + 4\Big(\frac{3\times4}{3+4-2}\Big)^{\left(\frac{3\times4}{3+4-2}\right)}\\
&\quad +2\Big(\frac{3\times6}{3+6-2}\Big)^{\left(\frac{3\times6}{3+6-2}\right)} + \big((2+3^n)-(3+2^n)-4\big)\Big(\frac{4\times4}{4+4-2}\Big)^{\left(\frac{4\times4}{4+4-2}\right)}\\
&\quad +(2^{n+1}-4)\Big(\frac{4\times6}{4+6-2}\Big)^{\left(\frac{4\times6}{4+6-2}\right)} + (2^{n-1}-2)\Big(\frac{6\times6}{6+6-2}\Big)^{\left(\frac{6\times6}{6+6-2}\right)}\Bigg]\\
&= log\big((37.926 \times 3^n) + (20.439 \times 2^n) - 155.862\big)\\
&\quad -\frac{log}{(37.926\times 3^n)+(20.439\times 2^n)-155.862}\Bigg[3^n\left(\left(2\times\left(\frac{16}{6}\right)^{3\left(\frac{16}{6}\right)^3}\right)\right.\\
&\quad +2^n\left(2\times\left(\frac{24}{8}\right)^{3\left(\frac{24}{8}\right)^3}\right)-\left(3\times\left(\frac{16}{6}\right)^{3\left(\frac{16}{6}\right)^3}\right)+\left.\left(\frac{1}{2}\times\left(\frac{36}{10}\right)^{3\left(\frac{36}{10}\right)^3}\right)\right)\\
&\quad +\left(4\times\left(\frac{8}{4}\right)^{3\left(\frac{8}{4}\right)^3}\right)+\left(4\times\left(\frac{12}{5}\right)^{3\left(\frac{12}{5}\right)^3}\right)+\left(2\times\left(\frac{18}{7}\right)^{3\left(\frac{18}{7}\right)^3}\right)\\
&\quad -\left(4\times\left(\frac{16}{6}\right)^{3\left(\frac{16}{6}\right)^3}\right)-\left(4\times\left(\frac{24}{8}\right)^{3\left(\frac{24}{8}\right)^3}\right)-\left(2\times\left(\frac{36}{10}\right)^{3\left(\frac{36}{10}\right)^3}\right)\Bigg].
\end{aligned}$$

□

$$\begin{aligned}
6.\ ENT_{ReZG_1}(SR_n) &= log(3^n - 2^{n-1} + 2) - \frac{log}{3^n-2^{n-1}+2}\Bigg[3^n\left(2\times\left(\frac{8}{16}\right)^{\frac{8}{16}}\right)\\
&\quad +2^n\left(\left(2\times\left(\frac{10}{24}\right)^{\frac{10}{24}}\right)-\left(3\times\left(\frac{8}{16}\right)^{\frac{8}{16}}\right)+\left(\frac{1}{2}\times\left(\frac{12}{36}\right)^{\frac{12}{36}}\right)\right)\\
&\quad +\left(4\times\left(\frac{6}{8}\right)^{\frac{6}{8}}\right)+\left(4\times\left(\frac{7}{12}\right)^{\frac{7}{12}}\right)+\left(2\times\left(\frac{9}{18}\right)^{\frac{9}{18}}\right)\\
&\quad -\left(4\times\left(\frac{8}{16}\right)^{\frac{8}{16}}\right)-\left(4\times\left(\frac{10}{24}\right)^{\frac{10}{24}}\right)-\left(2\times\left(\frac{12}{36}\right)^{\frac{12}{36}}\right)\Bigg].
\end{aligned}$$

Proof Applying Theorem 1 and by definition, we attain the redefined first Zagreb index. Hence, the result follows:

$$
\begin{aligned}
ReZG_1(SR_n) &= 3^n - 2^{n-1} + 2.\\
ENT_{ReZG_1}(SR_n) &= log\big(ReZG_1(SR_n)\big)\\
&\quad - \frac{log}{ReZG_1(SR_n)}\Bigg[\sum_{st\in\Xi}\left(\frac{\Xi(s)*\Xi(t)}{\Xi(s)+\Xi(t)-2}\right)^{3\left(\frac{\Xi(s)*\Xi(t)}{\Xi(s)+\Xi(t)-2}\right)^3}\Bigg]\\
&= log(3^n - 2^{n-1} + 2) - \frac{log}{3^n-2^{n-1}+2}\Bigg[4\left(\frac{2+4}{2\times 4}\right)^{\left(\frac{2+4}{2\times 4}\right)} + 4\left(\frac{3+4}{3\times 4}\right)^{\left(\frac{3+4}{3\times 4}\right)}\\
&\quad +2\left(\frac{3+6}{3\times 6}\right)^{\left(\frac{3+6}{3\times 6}\right)} + ((2+3^n)-(3+2^n)-4)\left(\frac{4+4}{4\times 4}\right)^{\left(\frac{4+4}{4\times 4}\right)}\\
&\quad +(2^{n+1}-4)\left(\frac{4+6}{4\times 6}\right)^{\left(\frac{4+6}{4\times 6}\right)} + (2^{n-1}-2)\left(\frac{6+6}{6\times 6}\right)^{\left(\frac{6+6}{6\times 6}\right)}\Bigg]\\
&= log(3^n - 2^{n-1} + 2) - \frac{log}{3^n-2^{n-1}+2}\Bigg[3^n\left(2\times\left(\frac{8}{16}\right)^{\frac{8}{16}}\right)\\
&\quad +2^n\left(\left(2\times\left(\frac{10}{24}\right)^{\frac{10}{24}}\right)-\left(3\times\left(\frac{8}{16}\right)^{\frac{8}{16}}\right)+\left(\frac{1}{2}\times\left(\frac{12}{36}\right)^{\frac{12}{36}}\right)\right)\\
&\quad +\left(4\times\left(\frac{6}{8}\right)^{\frac{6}{8}}\right)+\left(4\times\left(\frac{7}{12}\right)^{\frac{7}{12}}\right)+\left(2\times\left(\frac{9}{18}\right)^{\frac{9}{18}}\right)\\
&\quad -\left(4\times\left(\frac{8}{16}\right)^{\frac{8}{16}}\right)-\left(4\times\left(\frac{10}{24}\right)^{\frac{10}{24}}\right)-\left(2\times\left(\frac{12}{36}\right)^{\frac{12}{36}}\right)\Bigg].
\end{aligned}
$$

□

7. $$
\begin{aligned}
ENT_{ReZG_2}(SR_n) &= log\big((4\times 3^n)+(0.3\times 2^n)-7.410\big)\\
&\quad -\frac{log}{(4\times 3^n)+(0.3\times 2^n)-7.410}\Bigg[3^n\left(2\times\left(\frac{16}{8}\right)^{\frac{16}{8}}\right)+2^n\left(\left(2\times\left(\frac{24}{10}\right)^{\frac{24}{10}}\right)\right.\\
&\quad \left.-\left(3\times\left(\frac{16}{8}\right)^{\frac{16}{8}}\right)+\left(\frac{1}{2}\times\left(\frac{36}{12}\right)^{\frac{36}{12}}\right)\right)+\left(4\times\left(\frac{8}{6}\right)^{\frac{8}{6}}\right)\\
&\quad +\left(4\times\left(\frac{12}{7}\right)^{\frac{12}{7}}\right)+\left(2\times\left(\frac{18}{9}\right)^{\frac{18}{9}}\right)-\left(4\times\left(\frac{16}{8}\right)^{\frac{16}{8}}\right)\\
&\quad -\left(4\times\left(\frac{24}{10}\right)^{\frac{24}{10}}\right)-\left(2\times\left(\frac{36}{12}\right)^{\frac{36}{12}}\right)\Bigg].
\end{aligned}
$$

Proof By definition, the redefined second Zagreb index is derived using Theorem 1 edge partition, and we obtain the result using it

$$
\begin{aligned}
ReZG_2(SR_n) &= (4\times 3^n) + (0.3\times 2^n) - 7.410.\\
ENT_{ReZG_2}(SR_n) &= log\big(ReZG_2(SR_n)\big)\\
&\quad - \frac{log}{ReZG_2(SR_n)}\left[\sum_{st\in\Xi}\left(\frac{\Xi(s)*\Xi(t)}{\Xi(s)+\Xi(t)}\right)^{\left(\frac{\Xi(s)*\Xi(t)}{\Xi(s)+\Xi(t)}\right)}\right]\\
&= log\big((4\times 3^n) + (0.3\times 2^n) - 7.410\big) - \frac{log}{(4\times3^n)+(0.3\times2^n)-7.410}\\
&\quad \times\Bigg[4\left(\tfrac{2\times4}{2+4}\right)^{\left(\frac{2\times4}{2+4}\right)} + 4\left(\tfrac{3\times4}{3+4}\right)^{\left(\frac{3\times4}{3+4}\right)} + 2\left(\tfrac{3\times6}{3+6}\right)^{\left(\frac{3\times6}{3+6}\right)} + ((2+3^n)\\
&\quad -(3+2^n)-4)\left(\tfrac{4\times4}{4+4}\right)^{\left(\frac{4\times4}{4+4}\right)} + (2^{n+1}-4)\left(\tfrac{4\times6}{4+6}\right)^{\left(\frac{4\times6}{4+6}\right)}\\
&\quad +(2^{n-1}-2)\left(\tfrac{6\times6}{6+6}\right)^{\left(\frac{6\times6}{6+6}\right)}\Bigg]\\
&= log\big((4\times 3^n) + (0.3\times 2^n) - 7.410\big) - \frac{log}{(4\times3^n)+(0.3\times2^n)-7.410}\\
&\quad \times\Bigg[3^n\left(2\times\left(\tfrac{16}{8}\right)^{\frac{16}{8}}\right) + 2^n\left(\left(2\times\left(\tfrac{24}{10}\right)^{\frac{24}{10}}\right) - \left(3\times\left(\tfrac{16}{8}\right)^{\frac{16}{8}}\right)\right.\\
&\quad \left.+\left(\tfrac{1}{2}\times\left(\tfrac{36}{12}\right)^{\frac{36}{12}}\right)\right) + \left(4\times\left(\tfrac{8}{6}\right)^{\frac{8}{6}}\right) + \left(4\times\left(\tfrac{12}{7}\right)^{\frac{12}{7}}\right)\\
&\quad +\left(2\times\left(\tfrac{18}{9}\right)^{\frac{18}{9}}\right) - \left(4\times\left(\tfrac{16}{8}\right)^{\frac{16}{8}}\right) - \left(4\times\left(\tfrac{24}{10}\right)^{\frac{24}{10}}\right)\\
&\quad -\left(2\times\left(\tfrac{36}{12}\right)^{\frac{36}{12}}\right)\Bigg].
\end{aligned}
$$

□

8. $$
\begin{aligned}
ENT_{ReZG_3}(SR_n) &= log\big((256\times 3^n) + (312\times 2^n) - 1484\big) - \frac{log}{(256\times3^n)+(312\times2^n)-1484}\\
&\quad \times\Big[3^n(2\times 128^{128}) + 2^n\left((2\times 240^{240}) - (3\times 128^{128}) + \frac{432^{432}}{2}\right)\\
&\quad +(4\times 48^{48}) + (4\times 84^{84}) + (2\times 162^{162}) - (4\times 128^{128})\\
&\quad -(4\times 240^{240}) - (2\times 432^{432})\Big].
\end{aligned}
$$

Proof Applying Theorem 1 and by definition, we attain the redefined third Zagreb index. Hence, the result follows:

$$\begin{aligned}
ReZG_3(SR_n) &= (256 \times 3^n) + (312 \times 2^n) - 1484.\\
ENT_{ReZG_3}(SR_n) &= log\big(ReZG_3(SR_n)\big) - \tfrac{log}{ReZG_3(SR_n)}\\
&\times\Big[\sum_{st\in\Xi}\big(\Xi(s)*\Xi(t)\big)\big(\Xi(s)+\Xi(t)\big)^{(\Xi(s)*\Xi(t))(\Xi(s)+\Xi(t))}\Big]\\
&= log((256\times 3^n)+(312\times 2^n)-1484) - \tfrac{log}{(256\times 3^n)+(312\times 2^n)-1484}\\
&\times\Big[4\big((2\times 4)(2+4)\big)^{(2\times 4)(2+4)} + 4\big((3\times 4)(3+4)\big)^{(3\times 4)(3+4)}\\
&+2\big((3\times 6)(3+6)\big)^{(3\times 6)(3+6)} + \big((2+3^n)-(3+2^n)-4\big)\\
&\times\big((4\times 4)(4+4)\big)^{(4\times 4)(4+4)} + (2^{n+1}-4)\big((4\times 6)\\
&\times(4+6)\big)^{(4\times 6)(4+6)} + (2^{n-1}-2)\big((6\times 6)(6+6)\big)^{(6\times 6)(6+6)}\Big]\\
&= log((256\times 3^n)+(312\times 2^n)-1484) - \tfrac{log}{(256\times 3^n)+(312\times 2^n)-1484}\\
&\times\Big[3^n(2\times 128^{128}) + 2^n\Big((2\times 240^{240}) - (3\times 128^{128})\\
&+\tfrac{432^{432}}{2}\Big) + (4\times 48^{48}) + (4\times 84^{84}) + (2\times 162^{162})\\
&-(4\times 128^{128}) - (4\times 240^{240}) - (2\times 432^{432})\Big].
\end{aligned}$$

□

Theorem 10 *Let ENT be the entropy. Then, for KS_n, the entropy of the topological indices based on the degree is calculated as follows:*

1. $ENT_{M_1}(KS_n) = log(3\times 4^{n+1}) - \frac{log}{3\times 4^{n+1}}(768\times 2^{2n})$.

Proof By definition, the first Zagreb index is derived using Theorem 2 edge partition, and we obtain the result using it

$$\begin{aligned}
M_1(KS_n) &= 3\times 4^{n+1}.\\
ENT_{M_1}(KS_n) &= log\big(M_1(KS_n)\big) - \tfrac{log}{M_1(KS_n)}\Big[\sum_{st\in\Xi}\big(\Xi(s)+\Xi(t)\big)^{\Xi(s)+\Xi(t)}\Big]\\
&= log(3\times 4^{n+1}) - \tfrac{log}{3\times 4^{n+1}}[(3\times 4^n)(2+2)^{(2+2)}]\\
&= log(3\times 4^{n+1}) - \tfrac{log}{3\times 4^{n+1}}(768\times 2^{2n}).
\end{aligned}$$

□

2. $ENT_{M_2}(KS_n) = log(3\times 4^{n+1}) - \frac{log}{3\times 4^{n+1}}(768\times 2^{2n})$.

Proof Applying Theorem 2 and by definition, we attain the second Zagreb index. Hence, the result follows:

$$\begin{aligned}
M_2(KS_n) &= 3\times 4^{n+1}.\\
ENT_{M_1}(KS_n) &= log\big(M_1(KS_n)\big) - \tfrac{log}{M_1(KS_n)}\Big[\sum_{st\in\Xi}\big(\Xi(s)*\Xi(t)\big)^{\Xi(s)*\Xi(t)}\Big]\\
&= log(3\times 4^{n+1}) - \tfrac{log}{3\times 4^{n+1}}[(3\times 4^n)(2\times 2)^{(2\times 2)}]\\
&= log(3\times 4^{n+1}) - \tfrac{log}{3\times 4^{n+1}}(768\times 2^{2n}).
\end{aligned}$$

□

3. $ENT_{HM}(KSR_n) = log(48 \times 2^{2n}) - \frac{log}{48\times 2^{2n}}(3 \times 16^{16} \times 2^{2n})$.

Proof By definition, the hyper Zagreb index is derived using Theorem 2 edge partition, and we obtain the result using it

$$\begin{aligned}
HM(KS_n) &= 48 \times 2^{2n}.\\
ENT_{HM}(KS_n) &= log\big(HM(KS_n)\big) - \frac{log}{HM(KS_n)}\Big[\sum_{st\in\Xi}\big(\Xi(s)+\Xi(t)\big)^{2\left(\Xi(s)+\Xi(t)\right)^2}\Big]\\
&= log(48 \times 2^{2n}) - \frac{log}{48\times 2^{2n}}[(3\times 4^n)(2+2)^{2(2+2)^2}]\\
&= log(48 \times 2^{2n}) - \frac{log}{48\times 2^{2n}}(3 \times 16^{16} \times 2^{2n}).
\end{aligned}$$

□

4. $ENT_F(KS_n) = log(24 \times 2^{2n}) - \frac{log}{24\times 2^{2n}}(3 \times 8^8 \times 2^{2n})$.

Proof Applying Theorem 2 and by definition, we attain the forgotten index. Hence, the result follows:

$$\begin{aligned}
F(KS_n) &= 24 \times 2^{2n}.\\
ENT_F(KS_n) &= log\big(F(KS_n)\big) - \frac{log}{F(KS_n)}\Big[\sum_{st\in\Xi}\big(\Xi(s)^2+\Xi(t)^2\big)^{\Xi(s)^2+\Xi(t)^2}\Big]\\
&= log(24 \times 2^{2n}) - \frac{log}{24\times 2^{2n}}[(3\times 4^n)(2^2+2^2)^{(2^2+2^2)}]\\
&= log(24 \times 2^{2n}) - \frac{log}{24\times 2^{2n}}(3 \times 8^8 \times 2^{2n}).
\end{aligned}$$

□

5. $ENT_{AZI}(KSR_n) = log(24 \times 2^{2n}) - \frac{log}{24\times 2^{2n}}(3 \times 8^8 \times 2^{2n})$.

Proof By definition, the augumented Zagreb index is derived using Theorem 2 edge partition, and we obtain the result using it

$$\begin{aligned}
AZI(KS_n) &= 24 \times 2^{2n}.\\
ENT_{AZI}(KS_n) &= log\big(AZI(KS_n)\big) - \frac{log}{AZI(KS_n)}\Bigg[\sum_{st\in\Xi}\left(\frac{\Xi(s)*\Xi(t)}{\Xi(s)+\Xi(t)-2}\right)^{3\left(\frac{\Xi(s)*\Xi(t)}{\Xi(s)+\Xi(t)-2}\right)^3}\Bigg]\\
&= log(24 \times 2^{2n}) - \frac{log}{24\times 2^{2n}}\Bigg[(3\times 4^n)\left(\frac{2\times 2}{2+2}\right)^{3\left(\frac{2\times 2}{2+2}\right)^3}\Bigg]\\
&= log(24 \times 2^{2n}) - \frac{log}{24\times 2^{2n}}(3 \times 8^8 \times 2^{2n}).
\end{aligned}$$

□

6. $ENT_{ReZG_1}(KS_n) = log(3 \times 4^n) - \frac{log}{3\times 4^n}(3 \times 4^n)$.

Proof Applying Theorem 2 and by definition, we attain the redefined first Zagreb index. Hence, the result follows:

$$
\begin{aligned}
ReZG_1(KS_n) &= 3 \times 4^n. \\
ENT_{ReZG_1}(KS_n) &= log\big(ReZG_1(KS_n)\big) \\
&\quad - \frac{log}{ReZG_1(KS_n)}\left[\sum_{st\in\Xi}\left(\frac{\Xi(s)*\Xi(t)}{\Xi(s)+\Xi(t)-2}\right)^{3\left(\frac{\Xi(s)*\Xi(t)}{\Xi(s)+\Xi(t)-2}\right)^3}\right] \\
&= log(3\times 4^n) - \frac{log}{3\times 4^n}\left[(3\times 4^n)\left(\frac{2+2}{2\times 2}\right)^{\left(\frac{2+2}{2\times 2}\right)}\right] \\
&= log(3\times 4^n) - \frac{log}{3\times 4^n}(3\times 4^n).
\end{aligned}
$$

□

7. $ENT_{ReZG_2}(KS_n) = log(3\times 4^n) - \frac{log}{3\times 4^n}(3\times 4^n)$.

Proof By definition, the redefined second Zagreb index is derived using Theorem 2 edge partition, and we obtain the result using it

$$
\begin{aligned}
ReZG_2(KS_n) &= 3 \times 4^n. \\
ENT_{ReZG_2}(KS_n) &= log\big(ReZG_2(KS_n)\big) - \frac{log}{ReZG_2(KS_n)}\left[\sum_{st\in\Xi}\left(\frac{\Xi(s)*\Xi(t)}{\Xi(s)+\Xi(t)}\right)^{\left(\frac{\Xi(s)*\Xi(t)}{\Xi(s)+\Xi(t)}\right)}\right] \\
&= log(3\times 4^n) - \frac{log}{3\times 4^n}\left[(3\times 4^n)\left(\frac{2\times 2}{2+2}\right)^{\left(\frac{2\times 2}{2+2}\right)}\right] \\
&= log(3\times 4^n) - \frac{log}{3\times 4^n}(3\times 4^n).
\end{aligned}
$$

□

8. $ENT_{ReZG_3}(KSR_n) = log(48\times 2^{2n}) - \frac{log}{48\times 2^{2n}}(3\times 16^{16}\times 2^{2n})$.

Proof Applying Theorem 2 and by definition, we attain the redefined third Zagreb index. Hence, the result follows:

$$
\begin{aligned}
ReZG_3(KS_n) &= 48 \times 2^{2n}. \\
ENT_{ReZG_3}(KS_n) &= log\big(ReZG_3(KS_n)\big) - \frac{log}{ReZG_3(KS_n)} \\
&\quad \times\left[\sum_{st\in\Xi}\big(\Xi(s)*\Xi(t)\big)\big(\Xi(s)+\Xi(t)\big)^{(\Xi(s)*\Xi(t))(\Xi(s)+\Xi(t))}\right] \\
&= log(48\times 2^{2n}) - \frac{log}{48\times 2^{2n}}[(3\times 4^n)\big((2\times 2)(2+2)\big)^{(2\times 2)(2+2)}] \\
&= log(48\times 2^{2n}) - \frac{log}{48\times 2^{2n}}(3\times 16^{16}\times 2^{2n}).
\end{aligned}
$$

□

Neighbourhood-Based Entropy on Topological Indices
Some neighbourhood-based topological entropy indices are considered, and discussions are provided below for SR_n and KS_n.

Theorem 11 *Let ENT represent the entropy. Then the entropy of neighbourhood-based topological indices for SR_n is calculated as follows:*

1. $$ENT_{ABC_4}(SR_n) = log\big((0.684 \times 3^n) - (0.323 \times 2^n) + 0.699\big) - \frac{log}{(0.684\times3^n)-(0.323\times2^n)+0.699}$$
$$\times\Bigg[3^n\Bigg(2\times\Big(\sqrt{\tfrac{30}{256}}\Big)^{\sqrt{\frac{30}{256}}}\Bigg) + 2^n\Bigg(\Big(\sqrt{\tfrac{34}{320}}\Big)^{\sqrt{\frac{34}{320}}}\Bigg) - \Bigg(4\times\Big(\sqrt{\tfrac{30}{256}}\Big)^{\sqrt{\frac{30}{256}}}\Bigg)$$
$$+\Bigg(\frac{1}{2}\times\Big(\sqrt{\tfrac{54}{784}}\Big)^{\sqrt{\frac{54}{784}}}\Bigg) + \Bigg(4\times\Big(\sqrt{\tfrac{20}{112}}\Big)^{\sqrt{\frac{20}{112}}}\Bigg) + \Bigg(4\times\Big(\sqrt{\tfrac{29}{238}}\Big)^{\sqrt{\frac{29}{238}}}\Bigg)$$
$$+\Bigg(2\times\Big(\sqrt{\tfrac{37}{350}}\Big)^{\sqrt{\frac{37}{350}}}\Bigg) + \Bigg(2\times\Big(\sqrt{\tfrac{31}{272}}\Big)^{\sqrt{\frac{31}{272}}}\Bigg) + \Bigg(4\times\Big(\sqrt{\tfrac{35}{340}}\Big)^{\sqrt{\frac{35}{340}}}\Bigg)$$
$$-\Bigg(8\times\Big(\sqrt{\tfrac{30}{256}}\Big)^{\sqrt{\frac{30}{256}}}\Bigg) + \Bigg(4\times\Big(\sqrt{\tfrac{40}{425}}\Big)^{\sqrt{\frac{40}{425}}}\Bigg) + \Bigg(2\times\Big(\sqrt{\tfrac{51}{700}}\Big)^{\sqrt{\frac{51}{700}}}\Bigg)$$
$$-\Bigg(4\times\Big(\sqrt{\tfrac{54}{784}}\Big)^{\sqrt{\frac{54}{784}}}\Bigg)\Bigg].$$

Proof By definition, the fourth atom bond connectivity index is derived using Theorem 5 edge partition, and we obtain the result using it

$$ABC_4(SR_n) = (0.684 \times 3^n) - (0.323 \times 2^n) + 0.699.$$
$$ENT_{ABC_4}(SR_n) = log\big(ABC_4(SR_n)\big) - \frac{log}{ABC_4(SR_n)}$$
$$\times\Bigg[\sum_{st\in\Psi}\Bigg(\sqrt{\frac{\Psi(s)+\Psi(t)-2}{\Psi(s)*\Psi(t)}}\Bigg)^{\sqrt{\frac{\Psi(s)+\Psi(t)-2}{\Psi(s)*\Psi(t)}}}\Bigg]$$
$$= log\big((0.684 \times 3^n) - (0.323 \times 2^n) + 0.699\big)$$
$$-\frac{log}{(0.684 \times 3^n) - (0.323 \times 2^n) + 0.699}\Bigg[4\Bigg(\sqrt{\frac{8+14-2}{8\times14}}\Bigg)^{\sqrt{\frac{8+14-2}{8\times14}}}$$
$$+4\Bigg(\sqrt{\frac{14+17-2}{14\times17}}\Bigg)^{\sqrt{\frac{14+17-2}{14\times17}}} + 2\Bigg(\sqrt{\frac{14+25-2}{14\times25}}\Bigg)^{\sqrt{\frac{14+25-2}{14\times25}}}$$
$$+2\Bigg(\sqrt{\frac{16+17-2}{16\times17}}\Bigg)^{\sqrt{\frac{16+17-2}{16\times17}}} + 4\Bigg(\sqrt{\frac{17+20-2}{17\times20}}\Bigg)^{\sqrt{\frac{17+20-2}{17\times20}}}$$
$$+\big((2+3^n)-(2^{n+2})-8\big)\Bigg(\sqrt{\frac{16+16-2}{16\times16}}\Bigg)^{\sqrt{\frac{16+16-2}{16\times16}}}$$
$$+(2^n-4)\Bigg(\sqrt{\frac{16+20-2}{16\times20}}\Bigg)^{\sqrt{\frac{16+20-2}{16\times20}}} + (2^{n+1}-2)$$
$$\times\Bigg(\sqrt{\frac{20+25-2}{20\times25}}\Bigg)^{\sqrt{\frac{20+25-2}{20\times25}}} + 4\Bigg(\sqrt{\frac{17+25-2}{17\times25}}\Bigg)^{\sqrt{\frac{17+20-2}{17\times25}}}$$

$$+2\left(\sqrt{\frac{25+28-2}{25\times 28}}\right)^{\sqrt{\frac{25+28-2}{25\times 28}}} + (2^{n-1}-4)$$
$$\times\left(\sqrt{\frac{28+28-2}{28\times 28}}\right)^{\sqrt{\frac{28+28-2}{28\times 28}}}\Bigg]$$
$$= log\big((0.684\times 3^n)-(0.323\times 2^n)+0.699\big)$$
$$-\frac{log}{(0.684\times 3^n)-(0.323\times 2^n)+0.699}\left[3^n\left(2\times\left(\sqrt{\frac{30}{256}}\right)^{\sqrt{\frac{30}{256}}}\right)\right.$$
$$+2^n\left(\left(\sqrt{\frac{34}{320}}\right)^{\sqrt{\frac{34}{320}}}\right)-\left(4\times\left(\sqrt{\frac{30}{256}}\right)^{\sqrt{\frac{30}{256}}}\right)$$
$$+\left(\frac{1}{2}\times\left(\sqrt{\frac{54}{784}}\right)^{\sqrt{\frac{54}{784}}}\right)+\left(4\times\left(\sqrt{\frac{20}{112}}\right)^{\sqrt{\frac{20}{112}}}\right)$$
$$+\left(4\times\left(\sqrt{\frac{29}{238}}\right)^{\sqrt{\frac{29}{238}}}\right)+\left(2\times\left(\sqrt{\frac{37}{350}}\right)^{\sqrt{\frac{37}{350}}}\right)$$
$$+\left(2\times\left(\sqrt{\frac{31}{272}}\right)^{\sqrt{\frac{31}{272}}}\right)+\left(4\times\left(\sqrt{\frac{35}{340}}\right)^{\sqrt{\frac{35}{340}}}\right)$$
$$-\left(8\times\left(\sqrt{\frac{30}{256}}\right)^{\sqrt{\frac{30}{256}}}\right)+\left(4\times\left(\sqrt{\frac{40}{425}}\right)^{\sqrt{\frac{40}{425}}}\right)$$
$$\left.+\left(2\times\left(\sqrt{\frac{51}{700}}\right)^{\sqrt{\frac{51}{700}}}\right)-\left(4\times\left(\sqrt{\frac{54}{784}}\right)^{\sqrt{\frac{54}{784}}}\right)\right].$$

□

2. $$ENT_{GA_5}(SR_n) = log\big((2\times 3^n)-(0.519\times 2^n)-0.269\big)-\frac{log}{(2\times 3^n)-(0.519\times 2^n)-0.269}$$
$$\times\left[3^n\left(2\times\left(\frac{\sqrt{256}}{16}\right)^{\frac{\sqrt{256}}{16}}\right)+\frac{2^n}{2\times 9^{\frac{4\sqrt{5}}{9}}}\left(\left(2^{\frac{8\sqrt{5}+9}{9}}\times 5^{\frac{2\sqrt{5}}{9}}\right)\right.\right.$$
$$\left.+\left(2^{\frac{4\sqrt{5}+3}{3}}\times 5^{\frac{2\sqrt{5}}{9}}\right)-\left(7\times 3^{\frac{8\sqrt{5}}{9}}\right)\right)+\left(4\times\left(\frac{\sqrt{112}}{11}\right)^{\frac{\sqrt{112}}{11}}\right)$$
$$+\left(4\times\left(\frac{2\sqrt{238}}{31}\right)^{\frac{2\sqrt{238}}{31}}\right)+\left(2\times\left(\frac{2\sqrt{350}}{39}\right)^{\frac{2\sqrt{350}}{39}}\right)$$
$$+\left(4\times\left(\frac{2\sqrt{272}}{33}\right)^{\frac{2\sqrt{272}}{33}}\right)+\left(4\times\left(\frac{2\sqrt{340}}{37}\right)^{\frac{2\sqrt{340}}{37}}\right)$$
$$-\left(8\times\left(\frac{\sqrt{256}}{16}\right)^{\frac{\sqrt{256}}{16}}\right)-\left(4\times\left(\frac{\sqrt{320}}{18}\right)^{\frac{\sqrt{320}}{18}}\right)$$
$$-\left(2\times\left(\frac{2\sqrt{500}}{45}\right)^{\frac{2\sqrt{500}}{45}}\right)+\left(4\times\left(\frac{\sqrt{425}}{12}\right)^{\frac{\sqrt{425}}{12}}\right)$$

$$+\left(2\times\left(\frac{2\sqrt{700}}{53}\right)^{\frac{2\sqrt{700}}{53}}\right)-\left(4\times\left(\frac{\sqrt{784}}{28}\right)^{\frac{\sqrt{784}}{28}}\right)\Bigg].$$

Proof Applying Theorem 5 and by definition, we attain the fifth geometric arithmetic index. Hence, the result follows:

$$GA_5(SR_n) = (2\times 3^n) - (0.519\times 2^n) - 0.269.$$

$$\begin{aligned}
ENT_{GA_5}(SR_n) =& log\big(GA_5(SR_n)\big) - \frac{log}{GA_5(SR_n)}\Bigg[\sum_{st\in\Psi}\left(\frac{2\sqrt{\Psi(s)*\Psi(t)}}{\Psi(s)+\Psi(t)}\right)^{\frac{2\sqrt{\Psi(s)*\Psi(t)}}{\Psi(s)+\Psi(t)}}\Bigg]\\
=& log\big((2\times 3^n)-(0.519\times 2^n)-0.269\big) - \frac{log}{(2\times 3^n)-(0.519\times 2^n)-0.269}\\
&\times\Bigg[\left(\frac{2\sqrt{8\times 14}}{8+14}\right)^{\frac{2\sqrt{8\times 14}}{8+14}} + 4\left(\frac{2\sqrt{14\times 17}}{14+17}\right)^{\frac{2\sqrt{14\times 17}}{14+17}} + 2\left(\frac{2\sqrt{14\times 25}}{14+25}\right)^{\frac{2\sqrt{14\times 25}}{14+25}}\\
&+2\left(\frac{2\sqrt{16\times 17}}{16+17}\right)^{\frac{2\sqrt{16\times 17}}{16+17}} + 4\left(\frac{2\sqrt{17\times 20}}{17+20}\right)^{\frac{2\sqrt{17\times 20}}{17+20}} + ((2+3^n)-(2^{n+2})-8)\\
&\times\left(\frac{2\sqrt{16\times 16}}{16+16}\right)^{\frac{2\sqrt{16\times 16}}{16+16}} + (2^n-4)\left(\frac{2\sqrt{16\times 20}}{16+20}\right)^{\frac{2\sqrt{16\times 20}}{16+20}} + (2^{n+1}-2)\\
&\times\left(\frac{2\sqrt{20\times 25}}{20+25}\right)^{\frac{2\sqrt{20\times 25}}{20+25}} + 4\left(\frac{2\sqrt{17\times 25}}{17+25}\right)^{\frac{2\sqrt{17\times 25}}{17+25}} + 2\left(\frac{2\sqrt{25\times 28}}{25+28}\right)^{\frac{2\sqrt{25\times 28}}{25+28}}\\
&+(2^{n-1}-4)\left(\frac{2\sqrt{28\times 28}}{28+28}\right)^{\frac{2\sqrt{28\times 28}}{28+28}}\Bigg]\\
=& log\big((2\times 3^n)-(0.519\times 2^n)-0.269\big) - \frac{log}{(2\times 3^n)-(0.519\times 2^n)-0.269}\\
&\times\Bigg[3^n\left(2\times\left(\frac{\sqrt{256}}{16}\right)^{\frac{\sqrt{256}}{16}}\right) + \frac{2^n}{2\times 9^{\frac{4\sqrt{5}}{9}}}\left(\left(2^{\frac{8\sqrt{5}+9}{9}}\times 5^{\frac{2\sqrt{5}}{9}}\right)+\left(2^{\frac{4\sqrt{5}+3}{3}}\times 5^{\frac{2\sqrt{5}}{9}}\right)\right.\\
&\left.-\left(7\times 3^{\frac{8\sqrt{5}}{9}}\right)\right) + \left(4\times\left(\frac{\sqrt{112}}{11}\right)^{\frac{\sqrt{112}}{11}}\right) + \left(4\times\left(\frac{2\sqrt{238}}{31}\right)^{\frac{2\sqrt{238}}{31}}\right)\\
&+\left(2\times\left(\frac{2\sqrt{350}}{39}\right)^{\frac{2\sqrt{350}}{39}}\right) + \left(4\times\left(\frac{2\sqrt{272}}{33}\right)^{\frac{2\sqrt{272}}{33}}\right) + \left(4\times\left(\frac{2\sqrt{340}}{37}\right)^{\frac{2\sqrt{340}}{37}}\right)\\
&-\left(8\times\left(\frac{\sqrt{256}}{16}\right)^{\frac{\sqrt{256}}{16}}\right) - \left(4\times\left(\frac{\sqrt{320}}{18}\right)^{\frac{\sqrt{320}}{18}}\right) - \left(2\times\left(\frac{2\sqrt{500}}{45}\right)^{\frac{2\sqrt{500}}{45}}\right)\\
&+\left(4\times\left(\frac{\sqrt{425}}{12}\right)^{\frac{\sqrt{425}}{12}}\right) + \left(2\times\left(\frac{2\sqrt{700}}{53}\right)^{\frac{2\sqrt{700}}{53}}\right) - \left(4\times\left(\frac{\sqrt{784}}{28}\right)^{\frac{\sqrt{784}}{28}}\right)\Bigg].
\end{aligned}$$

□

Theorem 12 *Let ENT be the entropy. Then, for KS_n, the entropy of the topological indices based on the neighbourhood is calculated as follows:*

1. $ENT_{ABC_4}(KS_n) = log(3\sqrt{6} \times 4^{n-1}) - \frac{log}{3\sqrt{6}\times 4^{n-1}}\left(\frac{3\times 6^{\frac{\sqrt{6}}{8}}\times 2^{2n}}{4^{\frac{\sqrt{6}}{4}}}\right).$

Proof By definition, the fourth atom bond connectivity index is derived using Theorem 6 edge partition, and we obtain the result using it

$$
\begin{aligned}
ABC_4(KS_n) &= 3\sqrt{6} \times 4^{n-1}.\\
ENT_{ABC_4}(KS_n) &= log\big(ABC_4(KS_n)\big) - \frac{log}{ABC_4(KS_n)}\Big[\sum_{st\in\Psi} \big(\Psi(s)+\Psi(t)\big)^{\Psi(s)+\Psi(t)}\Big]\\
&= log(3\sqrt{6} \times 4^{n-1}) - \frac{log}{3\sqrt{6}\times 4^{n-1}}\left[(3\times 4^n)\left(\sqrt{\frac{4+4-2}{4\times 4}}\right)^{\sqrt{\frac{4+4-2}{4\times 4}}}\right]\\
&= log(3\sqrt{6} \times 4^{n-1}) - \frac{log}{3\sqrt{6}\times 4^{n-1}}\left(\frac{3\times 6^{\frac{\sqrt{6}}{8}}\times 2^{2n}}{4^{\frac{\sqrt{6}}{4}}}\right).
\end{aligned}
$$

□

2. $ENT_{GA_5}(KS_n) = log(3 \times 4^n) - \frac{log}{3\times 4^n}(3 \times 2^{2n}).$

Proof Applying Theorem 6 and by definition, we attain the fifth geometric arithmetic index. Hence, the result follows:

$$
\begin{aligned}
GA_5(KS_n) &= 3 \times 4^n.\\
ENT_{GA_5}(KS_n) &= log\big(GA_5(KS_n)\big) - \frac{log}{GA_5(KS_n)}\left[\sum_{st\in\Psi}\left(\frac{2\sqrt{\Psi(s)*\Psi(t)}}{\Psi(s)+\Psi(t)}\right)^{\frac{2\sqrt{\Psi(s)*\Psi(t)}}{\Psi(s)+\Psi(t)}}\right]\\
&= log(3 \times 4^n) - \frac{log}{3\times 4^n}\left[(3\times 4^n)\left(\frac{2\sqrt{4\times 4}}{4+4}\right)^{\frac{2\sqrt{4\times 4}}{4+4}}\right]\\
&= log(3 \times 4^n) - \frac{log}{3\times 4^n}(3 \times 2^{2n}).
\end{aligned}
$$

□

4 Discussion on Numerical Simulation

The graph for the derivation of M-polynomial and neighbourhood M-polynomial of SR_n and KS_n has been shown in Figs. 3 and 5. Furthermore, an analogy graph illustrates the values of computed topological indices for the examined fractional structures. Figures 4 and 7 show the M-polynomial and entropy measures for degree-based topological indices. Figures 6 and 8 illustrate the M-polynomial and entropy measures for neighbourhood topological indices. The number of iterations (n) for the sequences of SR_n and KS_n and the evaluated topological values are projected in the horizontal and vertical directions. In the graphs, each topological index is shown separately. As the number of iterations progresses, readers may observe that the value of all indices increases accordingly.

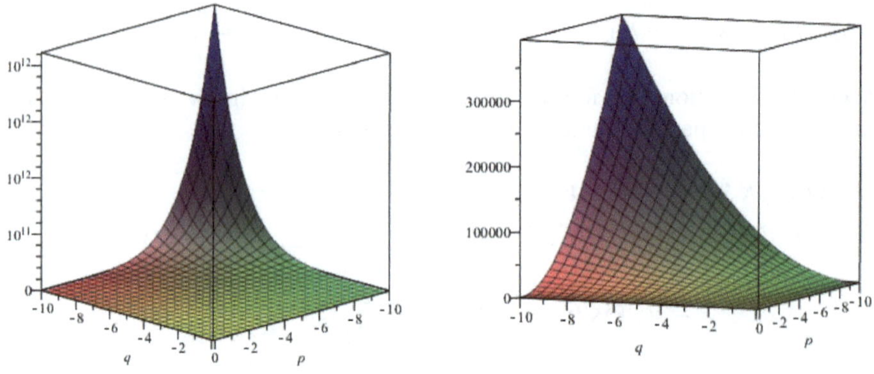

(a) SR_n (b) KS_n

Fig. 3 Degree-based M-polynomial

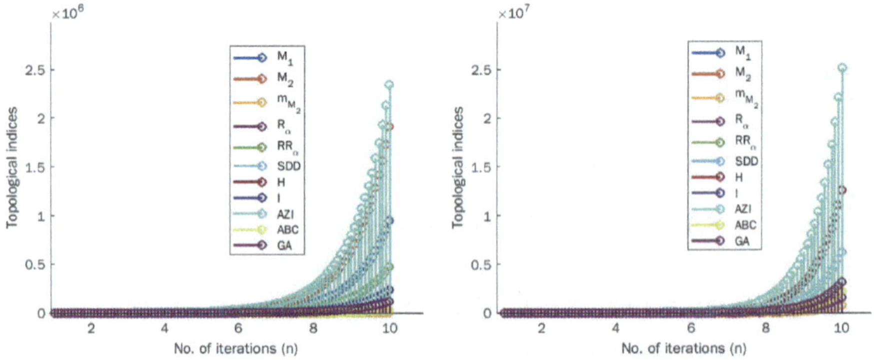

(a) SR_n (b) KS_n

Fig. 4 Degree-based topological indices

The irregularity of an object increases as the dimension of the fractal structure increases. Therefore, to study the fractal structure of an object, it is sufficient to examine the fractal dimension. In this analysis, we compared the topological indices with the fractal dimension. After some level, the number of copies in the fractal dimension is related to the topological indices. So, all computed indices of SR_n have a certain level of positive number $N \in \mathbb{N}$, such that for each $n > N$, it provides three copies. Using the same approach for KS_n, we obtain four copies after some level N.

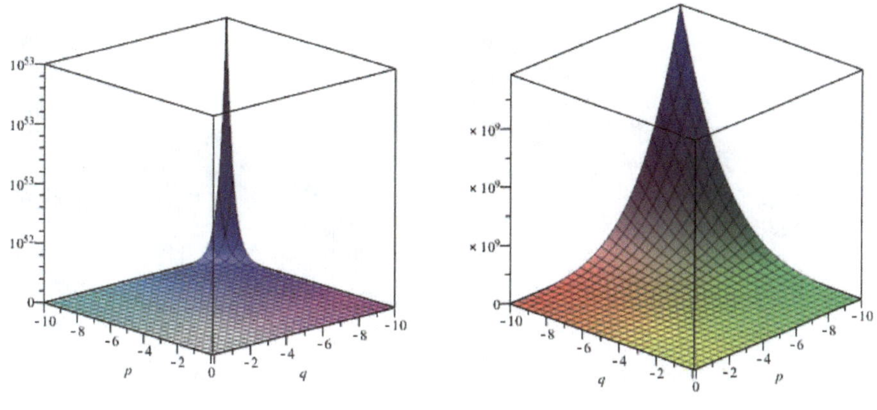

(a) SR_n (b) KS_n

Fig. 5 Neighbourhood M-polynomial

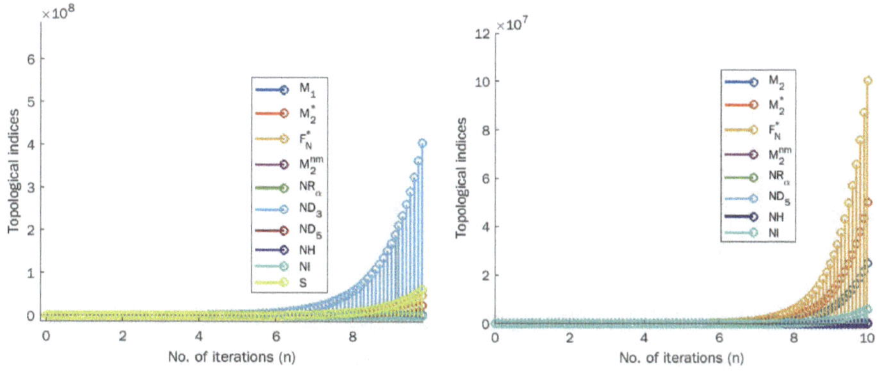

(a) SR_n (b) KS_n

Fig. 6 Neighbourhood M-polynomial topological indices

For the M-polynomial, besides all indices, the modified Zagreb index gives three copies of SR_n for the smallest possible value n. Similarly, for KS_n, all indices receive four copies for all n. For entropy measure, the redefined first Zagreb index provides the fractal dimension of SR_n. Likewise, for KS_n, the hyper Zagreb index gets the fractal dimension for the smallest value n.

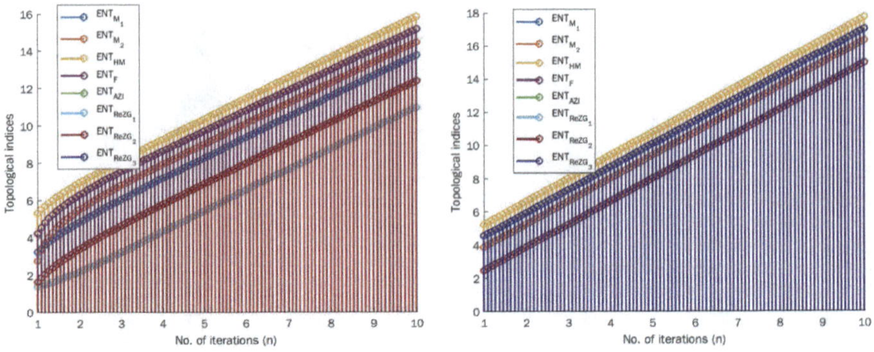

(a) SR_n (b) KS_n

Fig. 7 Entropy measure of degree-based topological indices

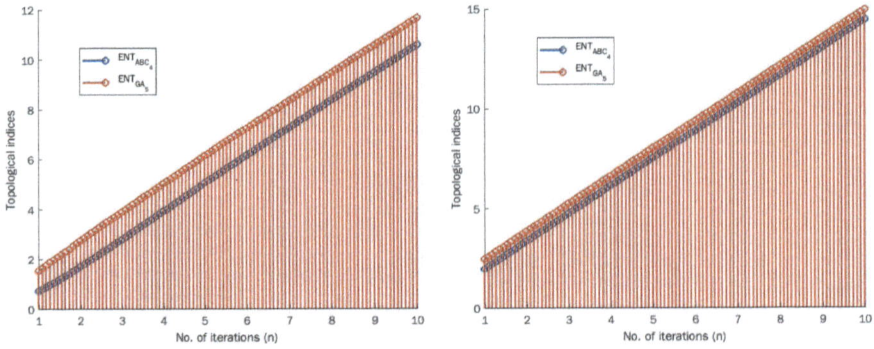

(a) SR_n (b) KS_n

Fig. 8 Entropy measure of neighbourhood topological indices

5 Conclusion

This work aims to study the fractal dimension with the aid of topological indices. To analyse that, we obtained a closely related formula of M-polynomial and entropy measure for SR_n and KS_n. Topological indices are commonly assisted in knowing the features of the physical properties of a structure, and it is a descriptor. From this perspective, topological indices can be considered a score function that maps structural properties to a real number. With this motivation, we related the indices with the fractal dimension. A graphical representation is also shown to understand the behaviour of all the indices for the fractal structure.

Conflicts of Interest The authors declare that there is no conflict of interest.
Data Availability Statement No Data is associated with the manuscript.

Acknowledgements The authors declare that there is no acknowledgement to funding bodies or any other acknowledgement.

References

1. Amić D, Bešlo D, Lucić B, Nikolić S, Trinajstić N (1998) The vertex-connectivity index revisited. J Chem Inf Comput Sci 38(5):819–822
2. Balaban T (1982) Highly discriminating distance based numerical descriptor. Chem Phys Lett 89:399–404
3. Bollobás B, Erdös P (1998) Graphs of extremal weights. Ars Combinatoria 50:225
4. Deutsch E, Klavžar S (2015) M-polynomial and degree-based topological indices. Iranian J Math Chem 6(2):93–102
5. Divya A, Manimaran A (2021) Topological indices for the iterations of Sierpiński rhombus and Koch snowflake. Eur Phys J Spec Top 230:3971–3980
6. Estrada E, Torres L, Rodríguez L, Gutman I (1998) An atom-bond connectivity index: modelling the enthalpy of formation of Alkanes. Indian J Chem 37:849–855
7. Furtula B, Graovac A, Vukičević D (2010) Augmented Zagreb index. J Math Chem 48:370–380
8. Gutman I, Das KC (2004) The first Zagreb index 30 years after. MATCH 50:83–92
9. Gutman I, Trinajstić N (1972) Graph theory and molecular orbitals, total π - electron energy of alternant hydrocarbons. Chem Phys Lett 17:535–538
10. Haoer RS (2021) Topological indices of metal-organic networks via neighborhood M-polynomial. J Discret Math Sci Cryptogr 24:369–390
11. Hutchinson JE (1981) Fractals and self similarity. Indiana Univ Math J 30(5):713–747
12. Imran M, Hafi SE, Gao W, Farahani MR (2017) On topological properties of Sierpinski networks. Chaos Solitons Fractals 98:199–204
13. Kenneth F (2003) Fractal geometry. Mathematical foundations and application. Wiley, England
14. Khalaf AJM, Hussain S, Afzal D, Afzal F, Maqbool A (2020) M-Polynomial and topological indices of book graph. J Discret Math Sci Cryptogr 23:1217–1237
15. Lokesha V, Usha A, Ranjini PS, Deepika T (2016) Harmonic index, redefined Zagreb indices of dragongraph with complete graph. Asian J Math Comput Res 9:161–166
16. Mandelbrot B (1982) The fractal geometry of nature. W.H. Freeman and Company, New York
17. Mowshowitz A, Dehmer M (2012) Entropy and the complexity of graphs revisited. Entropy 14:559–570
18. Nikolić S, Kovačević G, Miličević A, Trinajstić N (2003) The Zagreb indices 30 years after. Croat Chem Acta 76:113–124
19. Randić M (1975) Characterization of molecular branching. J Am Chem Soc 97(23):6609–6615
20. Ranjini PS, Lokesha V, Usha A (2013) Relation between phenylene and hexagonal squeeze using harmonic index. Int J Graph Theory 1:116–121
21. Rashevsky N (1955) Life, information theory, and topology. Bull Math Biol 17:229–235
22. Santo B, Hassan MK, Mukherjee S, Gowrisankar A (2020) Fractal patterns in nonlinear dynamics and applications. CRC Press, Boca Raton
23. Santo B, Easwaramoorthy D, Gowrisankar A (2021) Fractal functions, dimensions and signal analysis. Cham
24. Schroeder M (1991) Fractals, chaos, power laws: minutes from an infinite paradise. W.H. Freeman and Company, New York
25. Shannon CE (1948) A mathematical theory of communication. Bell Syst Tech J 27:379–423
26. Shirdel GH, Rezapour H, Sayadi AM (2013) The hyper Zagreb index of graph operations. Iranian J Math Chem 4:213–220
27. Vukičević D, Gasperov M (2010) Bond additive modeling 1. Adriatic indices. Croat Chem Acta 83:243–260

28. Vukičević D, Furtula B (2009) Topological index based on the ratios of geometrical and arithmetical means of end-vertex degrees of edges. J Math Chem 46:1369–1376
29. Zhong L (2012) The harmonic index for graphs. Appl Math Lett 25:561–566

Chapter 8
Stochastic Locomotive of Nanofluid(s)

Rashmi Bhardwaj and Roberto Acevedo

Abstract This study examines the convection dynamics of superparamagnetic iron oxide nanoparticles (MNPs) in blood fluid. The system under study consists of a liquid with a layer of blood and MNPs with an iron oxide core travelling through a blood artery with cholesterol deposits on the inner lining. It is heated and subjected to an external magnetic field as it passes through the blood vessel. The partial differential equations of momentum and energy conservation were utilized to create the nonlinear three-dimensional governing equations for the system under consideration. Phase portrait, time series, and Hartmann number calculations are used to examine the impact of the magnetic field on the chaotic convection of the MNPs in blood fluid. It has been found that the convection of MNPs stabilizes in a specific direction when magnetized by an external magnetic field. These points to the expanding use of magnetic resonance imaging (MRI) and magnetic drug delivery (MDD) as contrast agents in non-invasive imaging techniques.

1 Section Heading

Nanomaterials are obtained through the synthesis of nanoparticles (nanoscale particles). Among the most important nanomaterials are those based on carbon, metal, silica, and iron oxide. The most likely industrial applications are unbreakable detection, arsenic removal, drug delivery, and water treatment. The nanotechnology solutions are as follows: a nanotechnology sensor that detects unbreakable in real time, the use of iron oxide nanoparticles for arsenic removal, and nanofilters that remove heavy metals from water. It is direct to conclude that nanotechnology has industrial

R. Bhardwaj (✉)
IMA, Southend-on-Sea, UK
e-mail: rashmib@ipu.ac.in

Nonlinear Dynamics Research Lab, GGSIP University, Delhi, India

R. Acevedo
Facultad de Ingeniería y Tecnología, Universidad San Sebastian,
Bellavista 7, Santiago 8520524, Chile

G. Arulprakash et al. (eds.), *Mathematical Modelling of Complex Patterns Through Fractals and Dynamical Systems*, Studies in Infrastructure and Control,
https://doi.org/10.1007/978-981-97-2343-0_8

applications in mining based on the search for critical risks in mining processes. In the area of research, it is not within reach of the technological developments underway in developed countries. This is because investment is not focused on large-scale technological development, which represents a handicap in our development. Also, we can add that the search for sustainable strategies arises from the mitigation of emerging pollutants caused by industrial processes. For these effects, there are advances in nanomaterials focused on environmental remediation. In general terms, any material that mixed with the extracted mineral can damage the machinery or, in the worst case, get stuck and delay the entire production is referred to; on the other hand, it also has environmental applications that through nanotechnology seeks to eliminate critical contaminants, such as the removal of arsenic and the implementation of nanofilters. However, the field of nanotechnology research applied is worrying from an environmental point of view since companies are currently investing in technology that allows them to extract and process the ore with the least possible damage to the environment, but the investment area is not robustly focused on nanotechnology. The importance of seeking sustainable strategies stems from the mitigation of emerging pollutants caused by industrial processes that do not have adequate methods of protection for the environment and human beings, and the fate of pollutants is not protected, finally reaching large quantities of water (ocean, rivers, lakes, etc.).

History of Nanotechnology

Nanotechnology stems from the pioneering studies of Prof. Richard Feynman, recognized as the father of this branch of science. This author, a theoretical physicist, winner of the Nobel Prize for his contribution to quantum electrodynamics, gave a lecture (1959) in this area of knowledge at the University of Caltech (California, USA) (Fundación Argentina, s.f.). The formal beginning of nanotechnology consists primarily of the manipulation of atoms. In 1981, Heinrich Rohrer and Gerd Binning created the tunneling microscope, an instrument capable of visualizing electrically conductive surfaces at the atomic level and thus manipulating individual atoms through the current exchange. The microscope consists of a sharp metal tip, composed of an atom, connected to a current of energy; the tip must approach the surface for electron transfer to occur and the so-called tunnel effect, where the surface electrons have a certain probability of reaching the tip, as presented in Fig. 1.

The procedure described above has the following functions:

1. When the tip approaches atom by atom, an image of the surface can be captured through a sophisticated optical procedure and computer simulation and optimization tools.

2. The tip's constant proximity to the individual atom increases the energy, allowing the atom to be moved individually.

At that point of discussion, there was a theory and an instrument; from then on, nanotechnology had no limits, and advances led to the discovery of molecules. In 1985, fullerene (a molecule composed of carbon that can take geometric shapes resembling a sphere, an ellipsoid, a tube, or a ring) or more specifically, carbon 60 (C60), was discovered. Until then, only two forms of carbon were known: graphite

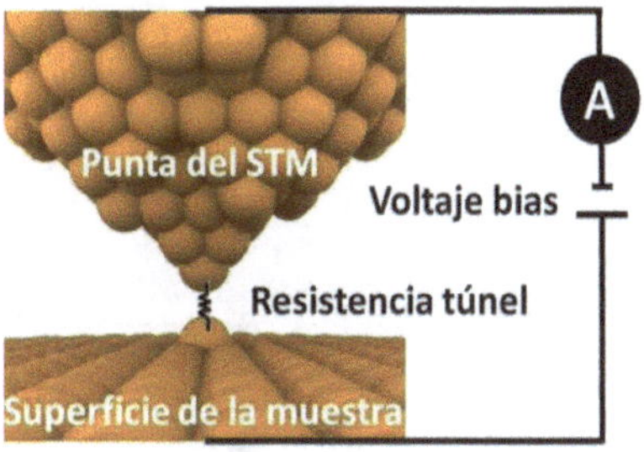

Fig. 1 Illustration of the equivalent circuit when the tip-sample system is in tunnel mode (*Note* Adapted from: "Topographic measurements on atomically flat surfaces under ambient conditions using a tunneling microscope, a didactic approach" by Delgado L, Chacón S, Sabater C, Sáenz G (2019) Uniciencia 33(1):32 (https://www.revistas.una.ac.cr/index.php/uniciencia/article/view/11103/14003). License Creative Commons Attribution-NoCommercial-SinDerivates 4.0 Unported 1984, for National University)

and diamond. Fullerene, consisting of 12 pentagons and 20 hexagons, became recognized as the third form of carbon. After the discovery of fullerene and knowing the properties of carbon in 1991, carbon nanotubes were discovered, characterized by their properties and the structure of a fullerene, able to be used in health, energy, technology, etc. In 1993, the monolayer nanotube was discovered, another form of carbon; its structure can be considered from a graphite sheet rolled on itself, depending on the degree of winding and the way the original sheet is formed, the result can lead to nanotubes of different diameter and internal geometry [1].

Previously mentioned were the most important advances in the history of nanotechnology, which came to mark the beginning of a revolution in science.

Nanotechnology

Nanotechnology was born from the branch of nanoscience, where objects are studied at the nanoscale. Nanoscience as a basic knowledge discipline has allowed the theoretical foundation and development of nanotechnology, which constitutes know-how for the production of objects, materials, instruments, structures, and systems on such a scale with a practical purpose [2] (Fig. 2).

Nanotechnology is a technology capable of manipulating materials on a nanometer scale, i.e., nanoparticles in the range of 1–100 nanometers (nm); the nanometer (nm) is equivalent to one-millionth of a meter ($1\text{ nm} = 10^{-9}\text{ m}$).

Nanoparticles

Nanostructures and nanomaterials are designed to modify properties based on nanoparticle manipulation. The influence of size on some properties is described below:

Larger Specific Area

Given the following formula, the high specific surface present in nanoparticles is defined as:

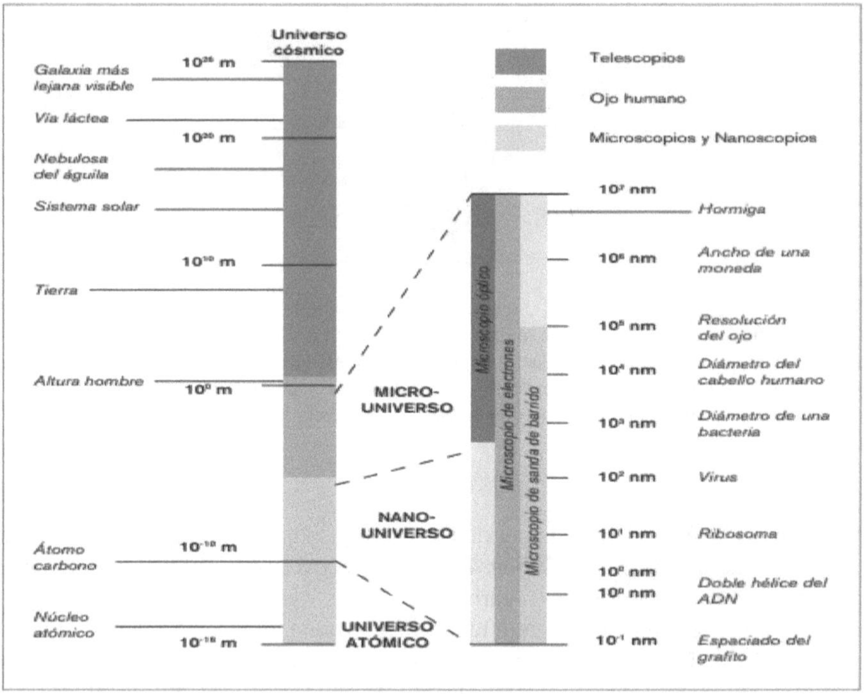

Fig. 2 Makes a comparison of metric units according to the scope of the different forms of visualization: the human eye, telescope, microscopes, and nanoscope (*Note* Adapted from "Techniques and applications of microengineering and nanoengineering" by Aguayo F, Zarzuela E, Ramón J, Córdoba A (2011) Técnica Industrial 295:27 (https://hdl.handle.net/11441/49411). Creative Commons Attribution-NonCommercial-NoDerivatives License 4.0 International 2015 for Research Deposit of the University of Seville)

$$Spherical\ surface = \frac{Superficial\ area}{Volume}.$$

Higher Hardness

According to Hall–Petch's law, hardness is inversely proportional to particle size, thanks to the grain boundary between crystals that prevents the particle from being exposed to the movements of dislocations.

Tensile Strength

Nanoparticles, being subject to stresses, have a higher tensile strength than macromolecular materials because their molecular ordering is able to withstand the deformations caused by the loads, have a high-voltage module, and are able to achieve the maximum resistance of the material. The elongation or tensile deformation of some nanoparticles can be in the range of 100–1000% before breaking, without breaking [3, 4].

Increased Chemical Reactivity
It is due to a greater surface area that facilitates interaction with other materials or substances, allowing it to cross membranes and disperse widely [5].

Nanomaterials
Nanomaterials are materials that contain particles with one or more dimensions on the nanoscale, about one nanometer to one hundred nanometers.

Iron Oxide-Based Nanomaterials
Iron oxide-based nanomaterials are nanoparticles composed primarily of iron oxides, such as magnetite or maghemite. These materials exhibit unique magnetic properties, making them highly valuable in various applications, including magnetic resonance imaging (MRI) contrast agents, targeted drug delivery, and environmental remediation. Due to their small size and large surface area, they offer enhanced reactivity and can be easily functionalized with other molecules for specific uses. Their biocompatibility and relatively low toxicity also make them suitable for biomedical applications.

Synthesis of Nanoparticles by Coprecipitation Method Fe_2O_2
Coprecipitation is the simplest and most efficient synthetic route to obtain magnetic nanoparticles [6].

The coprecipitation method is based on the synthesis of magnetite from iron salts (Fe^{3+} and Fe^{2+}) in basic and inert media [7].

$$Fe^{2+} + 2Fe^{3+} + 8OH^- \rightarrow Fe_3O_4 + 4H_2O.$$

The coprecipitation method is based on the synthesis of maghemite from iron salts (Fe^{3+} and Fe^{2+}) in the presence of oxygen [7].

$$Fe^{2+} + Fe^{3+} + O_2 + 2OH^- \rightarrow Fe_2O_3 + H_2O.$$

Particle size control, morphology, and composition are limited by particle growth kinetics [8].

Two processes are involved in coprecipitation:

a. The nucleation.
b. The growth of nuclei.

If the objective is to obtain a mono-dispersed distribution of iron oxide nanoparticles, the nucleation and growth of the colloids should be separated; in other words, nucleation should be avoided during the growth stage [9].

Nanoparticle Synthesis by Micro-Fe_2O_2 Emulsion Method
The technique consists of mixing water, oil, and a surfactant. When the oily phase is in greater quantity, and the aqueous phase is dispersed as microspheres surrounded by surfactant molecules, the method is known as reverse micelles or as water in oil (w/o). When it is the aqueous phase that is in greater measure, it is known as normal micelles or oil in water (o/w).

Synthesis of Nanoparticles Fe_2O_2 by Solvothermal Method
A common synthesis of crystalline microspheric ferrites by this technique consists of a mixture of ferric chloride ($FeCl_3$), ethylene glycol (reducing agent), sodium acetate (stabilizing agent), and polyethylene glycol (surfactant), obtaining a transparent solution that is heated inside an autoclave at 200 °C for 8–72 h [10].

Iron Oxide Nanoparticles as Catalysts
Iron oxide nanoparticles are currently used as magnetically recoverable catalysts due to their magnetic properties. With a suitable surface coating, magnetic nanoparticles can be dispersed in water, which has applications for the removal of emerging contaminants. It has been shown that by modifying and controlling the size, shape, and crystallinity of nanoparticles, precise results are obtained for a wide range of applications.

In recent years, past research has been done in medical sciences for:

- New imaging techniques for clinical diagnostics.
- New methods of drug delivery systems for treatment in complex medical cases.
- New methods for targeted drug delivery.
- New techniques for the treatment of complex medical cases.

The fundamental problems in complex medical cases are as follows:

- Specific detection of disease or problem through clinical diagnosis.
- Treatment of patients through intravenous drug delivery.

In most complex medical cases, drugs are administered in the patient's body through intravenous injection. MRI and MDD are the new advancements in the field of applied medical sciences. Nowadays, researchers are more focused on developing techniques for the detection of better images and drug delivery mechanisms, for which MNPs and SPIOs (superparamagnetic iron oxide nanoparticles) are commonly getting these days.

The effect of using an ultra-small SPIO (USIPO)-based contrast agent on the enhancement of images in cardiovascular magnetic resonance (CMR) was observed on a 50-year-old patient with myocardial infarction. In [11], it was empirically evident that USPIO-based contrast agents cannot only provide accurate detection of the infarct core but also enable the detection of the pre-infarct zone, whose precise observation through gadolinium-based contrast agents is still challenging to achieve. The various molecular MRI agents underlying this non-invasive imaging technique are reviewed for application to imaging stem cell therapy in the heart [12]. Gadolinium is paramagnetic in nature, and (https://en.wikipedia.org/wiki/Chelate) chelated gadolinium (III) has been used for more than a decade as an extracellular MR contrast agent as its detection threshold is in the micromolar range and thus superior to iodinated contrast agents, which have millimolar detection limit. Due to incredibly short intravascular half-lives, the utility of conventional extracellular chelated gadolinium is limited for most imaging applications.

Pre- and postoperative MR imaging and histochemistry of Ferumoxtran-10 (FX) with conventional gadolinium enhancements for seven brain tumor patients were compared [13]. It has been observed that FX show even a 0.15T area of enhancement for intra-operative MR as compared to gadolinium, which is more advantageous in postoperative imaging. The development of better contrast agents in CMR is fundamentally essential for a country like India, where people are more prone to coronary artery disease (CAD) [14]. Nearly half of the Indians are vegetarian but still have early-age CAD of a malignant nature due to aggressive modifications in the lifestyle from adolescence onward. Thus, detection of the disease at the right time is essential to get cured. In the CNS, SPIO MNPs have applications because they are helpful as contrast agents in MRI for tumors, neuro-inflammatory pathologies, cerebro-vasculature, or injury tracking [15–18]. For targeted drug delivery, ferrofluids with encapsulated magnetic nanoparticles may be used. MNPs carrying DNA, enzymes, and drugs present in the bloodstream can be internally dragged through a magnetic field exploiting the super-magnetic magnetization of MNPs [19–21]. The anti-inflammatory medications and anti-angiogenic chemotherapy strategies may improve disease detection, therapeutic monitoring, and treatment efficacy. To image the liver and lymphatic system in the clinical arena, selected MNPs are used [22, 23], which have evolved as a highly appealing platform for molecular MRI due to established higher safety records. First-generation MNPs like Feridex tend to form in vivo aggregates due to the thin dextran coating, which gets readily cleared from the bloodstream. Subsequently generated MNPs called MIONs are MNPs with a more extensive polymer coating that may remain mono-dispersed; its cross-linked version called CLIO is developed for imaging applications [24–26]. From the circulation by the reticuloendothelial system, MNPs intravenously injected are eventually removed; the blood half-life of MION and CLIO in humans is 24 h [27].

The impact of a magnetic field in a soaked permeable layer convection smooth movement can be delayed with observation of transitions from steady-state convection to chaos via Hopf bifurcation [28]. The development of liquid convection in a soaked permeable layer from consistent to turbulent stage with expanding Rayleigh number is studied [29] using a pseudo-spectral numerical scheme. The MNPs used in cardiac MRI have a central core of iron oxide, which measures 3–5 nm in diameter, surrounded by a dextran, starch, or polymer coat [30]. The mathematical modeling of convection of these MNPs in the blood is important as advances in several new imaging techniques like MPI, MMUS, and MPA have taken place in the past few years [31] for visualizing and determining the location of these magnetic nanoparticles in blood without knowing the dynamics of MNPs.

A mathematical model is developed to study the convection of iron oxide nanoparticles (IONPs) as MNPs flow through ferrofluid in the blood fluid layer, which is assumed to be non-Newtonian. The study shows that magnetic field channelizes and stabilizes the IONPs in the blood, and thus, SPIO nanoparticles are used in MRI and drug delivery applications.

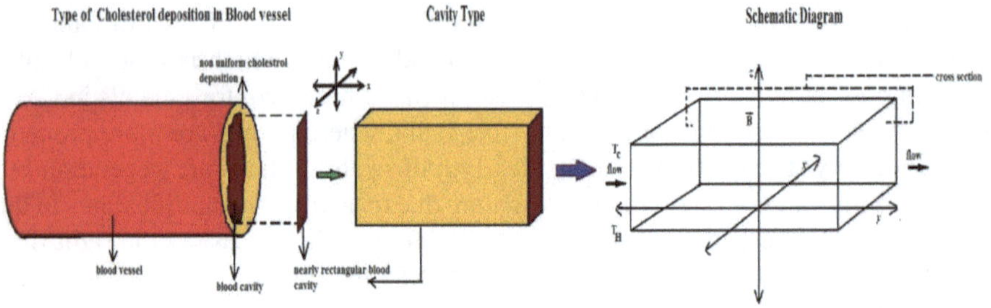

Fig. 3 Rectangular cavity for nanofluid to flow through blood

2 Mathematical Modeling

Let us consider a microscopic blood (non-Newtonian) depression through which iron oxide ferro-nanoliquid (Newtonian) streams with USPIO-MION-MNPs. The cholesterol deposition in the blood vessel is considered non-uniform for the case under approximation, and it is assumed that the nanofluid flows with blood through a rectangular cavity, as shown in Fig. 3.

The governing equations are obtained as partial differential equations from the equation of conservation of momentum and energy, which are converted to a three-dimensional nonlinear differential equation system, which is similar to Lorenz equations in dynamics. As discussed in [32], the symbols have their usual meanings:

$$\overset{\bullet}{X} = \Pr \bar{\upsilon} \left(1 + \frac{\bar{\gamma}}{(\pi^2 + k^2)}\right) [Y - X] \tag{1}$$

$$\overset{\bullet}{Y} = \left[\left(\frac{R\overline{\beta}}{\overline{\upsilon}}\right) \left(\frac{1}{\left(1 + \frac{\overline{\gamma}}{(\pi^2+k^2)}\right)}\right) X - \overline{\alpha} Y - \left(\frac{\overline{\beta}}{\overline{\gamma}}\right) \left(R - \frac{\overline{\alpha \upsilon}}{\overline{\beta}} \left(1 + \frac{\overline{\gamma}}{(\pi^2 + k^2)}\right)\right) XZ \right] \tag{2}$$

$$\overset{\bullet}{Z} = \overline{\alpha} \lambda \, (XY - Z) \,. \tag{3}$$

To study the stability analysis, Eqs. (1) to (3) can be written as

$$\overset{\bullet}{x} = c \, (y - x) \tag{4}$$

$$\overset{\bullet}{y} = jx - ay - m \, (j - a) \, xy \tag{5}$$

$$\overset{\bullet}{z} = s \, (xy - z) \,, \tag{6}$$

where

$$m = \left(1 + \frac{\overline{\gamma}}{(\pi^2 + k^2)}\right),$$

$$k = \Pr \overline{\upsilon},$$

$$l = \frac{R\overline{\beta}}{\overline{\upsilon}},$$

$$s = a\lambda,$$

$$a = \overline{\alpha},$$

$$j = \frac{l}{m},$$

$$c = km,$$

$$f = m\left[\frac{l}{m} - a\right].$$

The fixed points obtained are (0, 0, 0), $\left(\sqrt{\frac{j-a}{f}}, \sqrt{\frac{j-a}{f}}, \frac{j-a}{f}\right)$ $\left(-\sqrt{\frac{j-a}{f}}, -\sqrt{\frac{j-a}{f}}, \frac{j-a}{f}\right)$.

1. The point (0, 0, 0) is:
   - stable asymptotic if $j < a$;
   - unstable if $j > a$;
   - critical case if $j = a$.
2. The point $\left(\sqrt{\frac{j-a}{f}}, \sqrt{\frac{j-a}{f}}, \frac{j-a}{f}\right)$:

Jacobian

$$J = \begin{bmatrix} -c & c & 0 \\ a & -a & -\sqrt{f(j-a)} \\ s\sqrt{\frac{(j-a)}{f}} & s\sqrt{\frac{(j-a)}{f}} & -s \end{bmatrix}$$

and the characteristic polynomial is given as:

$$\lambda^3 + \lambda^2(s + a + c) + \lambda\,[ac + s(c + j)] + 2cs(j - a) = 0$$

$$\Rightarrow j = [4sac + sc(s + c) + ac(a + c)]\,/(cs - s^2 - as) = j_c.$$

The point $\left(\sqrt{\frac{j-a}{f}}, \sqrt{\frac{j-a}{f}}, \frac{j-a}{f}\right)$ is

- stable if $j < jc$,

- critical if j = jc,
- chaotic if j = jc.

3. The point $\left(-\sqrt{\frac{j-a}{f}}, -\sqrt{\frac{j-a}{f}}, \frac{j-a}{f}\right)$

   Jacobian

$$J = \begin{bmatrix} -c & c & 0 \\ a & -a & -\sqrt{f(j-a)} \\ s\sqrt{\frac{(j-a)}{f}} & s\sqrt{\frac{(j-a)}{f}} & -s \end{bmatrix}$$

   and the characteristic polynomial is given as:

$$\lambda^3 + \lambda^2(s + a + c) + \lambda\,[ac + s(c + j)] + 2cs(j - a) = 0$$

$$\Rightarrow j = [4sac + sc(s + c) + ac(a + c)]\,/(cs - s^2 - as) = j_c.$$

   The point $\left(-\sqrt{\frac{j-a}{f}}, -\sqrt{\frac{j-a}{f}}, \frac{j-a}{f}\right)$ is

- stable if j < jc;
- critical if j = jc;
- chaotic if j = jc.

3 Results and Discussions

The effect of a magnetic field on the dynamics of iron oxide nanofluid convection in blood is numerically studied using MATLAB for different values of Ha, keeping Pr and Ra constant. The thermophysical properties of water, blood, and Fe_3O_4 are given in Table 1, and the values for nanoparticles are given in Table 2.

Table 1 Thermophysical properties for water and blood at temperature T = 25 °C

Substance	ρ (kgm^{-3})	k (Wm^{-1}k^{-1})	C_p (Jkg^{-1}K^{-1})	$\beta \times 10^5$ (K^{-1})
H_2O	997.1	0.613	4179	21
Blood	1050.0	0.476	3780	0.146
Fe_3O_4	5180	670	9.7	0.5

Table 2 Thermophysical properties for nanoparticles

Substance	$\overline{\beta}$	$\overline{\alpha}$	$\overline{\nu}$
Fe_3O_4	0.9402	0.790396	1.140

Table 3 Observation for variation of Rayleigh number

Pr	Ha	Ra	Different stages observed in phase space
		≤24	Stable phase (inward spiral exists)
		−36	Critical phase (limit cycle exist)
		≥37	Chaotic phase (butterfly pattern exists)

Table 4 Observation for variation of Hartmann number

Pr	Ra	Ha	Different stages observed in phase space
		≤0.6	Chaotic phase (butterfly pattern exists)
		−1.0	Critical phase (limit cycle exist)
		≥1.1	Stable phase (inward spiral exists)

The state of chaos starts from Ra = 36 onward, which indicates that at higher temperatures, the cavity gets chaotic, which is in sync with the fact that if the blood vessel of a patient has a uniform deposition of a thick layer of cholesterol, then there will be a rise in the body temperature of the patient, and the patient becomes more prone to cardiac arrest as the flow becomes chaotic and due to the uniform choking no narrow spaces are available for flow as in the case of non-uniform deposition through which some regular flow is still feasible. It is observed that for patients with non-uniform cholesterol deposition, the damage is accurate, so curing through expected drug delivery is difficult. As the fluid considered in this study is iron oxide ferrofluid, it has chaotic convective flow, which reduces the flow efficiency in a channelized manner to the specific target. The transition of the phase of the convection dynamics from chaotic state at R = 37 and Ha = 0.5 to critical state at R = 37 and Ha = 0.8 and stable state at R = 37 and Ha = 1.2 is shown in Figs. 4, 5, and 6, respectively. Three possible stages of dynamics of nanofluid convection, namely, the stable stage, critical stage, and chaotic stage, are observed through numerical simulation, which is in accordance with stability analysis. The effect of magnetic and temperature variations on the nanofluid convection in a cavity is listed in Tables 3and 4, as the Hartmann and Rayleigh numbers are varied, respectively.

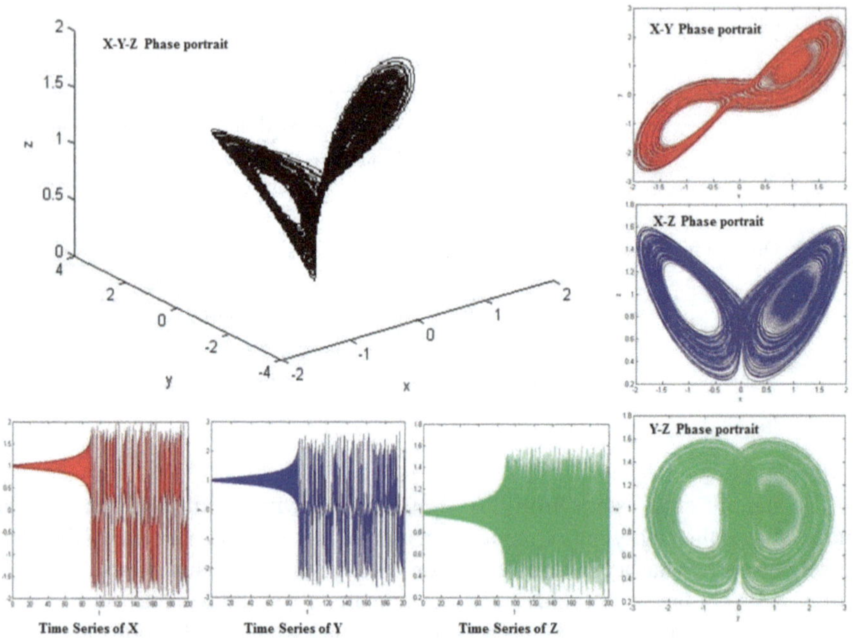

Fig. 4 Phase portraits and time series for Pr = 10, Ha = 0.5, and Ra = 37.0

The iron oxide nanofluid, being paramagnetic in nature, has random orientations of particles as per the individual spin, which is to be oriented in one direction for channelized flow to a specific target, which is useful for targeted drug delivery and MRI applications. In the worst case of chaotic convective flow, the nanoparticles randomly oriented spins should get oriented in a specific direction for stable channelized flow with the application of a magnetic field. Hartmann is directly associated with applied magnetic field intensity; thus, on increasing magnetic field, Hartmann number increases proportionally. Thus, if the Hartmann number is increased or the field is varied to higher intensity, then the chaotic phase of convective flow should transit to a stable state for channelized flow.

The main clinical application of MRI is to detect changes in vascularity, flow dynamics, and perfusion. In MRI, when an external magnetic field is applied, the nuclei of protons are aligned parallel or anti-parallel, during which the spins are processed under Larmor frequency. When nuclei are subjected to a resonance frequency in the range of radio frequency, protons absorb energy and get excited to an anti-parallel state from where, after the RF pulse disappears, the excited proton gives rise to the image by relaxing to a lower energy state through either longitudinal T1 pathway to initial state or transverse R2 pathway with dephasing of spin and so the disappearance of induced magnetization on a perpendicular plane. Based on the relaxation process, T1 and T2 contrast agents, which are usually paramagnetic complexes, iron oxide nanoparticles are commercially available [13].

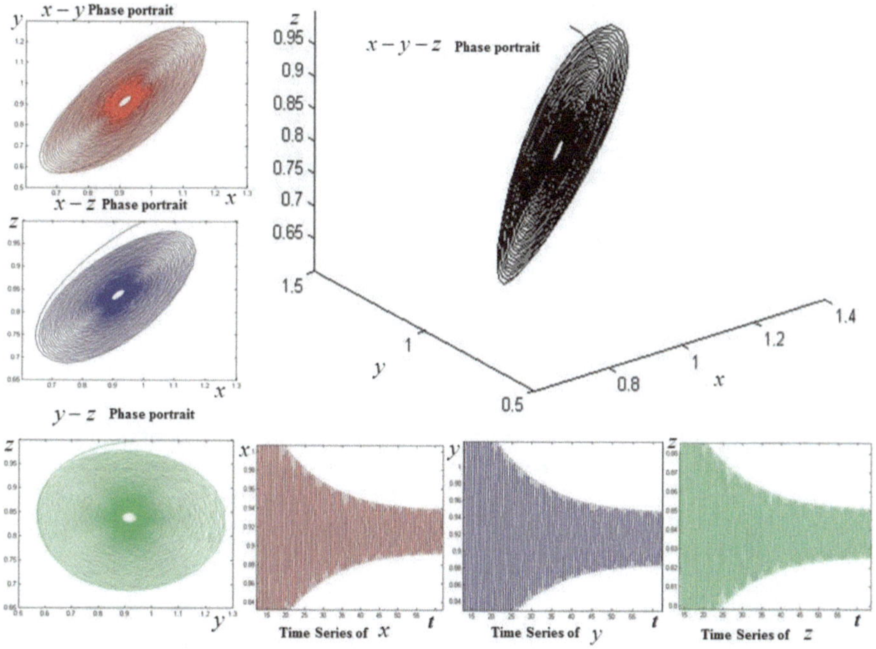

Fig. 5 Phase portraits and time series for Pr = 10, Ha = 0.8, and Ra = 37.0

The phase portrait obtained for the Fe_3O_4 convection for different values of Hartmann number variation verifies stability analysis and uses iron oxide fluids for channelized drug delivery and MRI contrast agent. When the Rayleigh number is high during a chaotic state, then by increasing the Hartmann number, stability in the convective flow of the iron oxide ferro-nanofluid is obtained. During a chaotic state, the Rayleigh number is high, and then by increasing the convective iron oxide, the ferromagnetic fluid flow becomes channelized and stable.

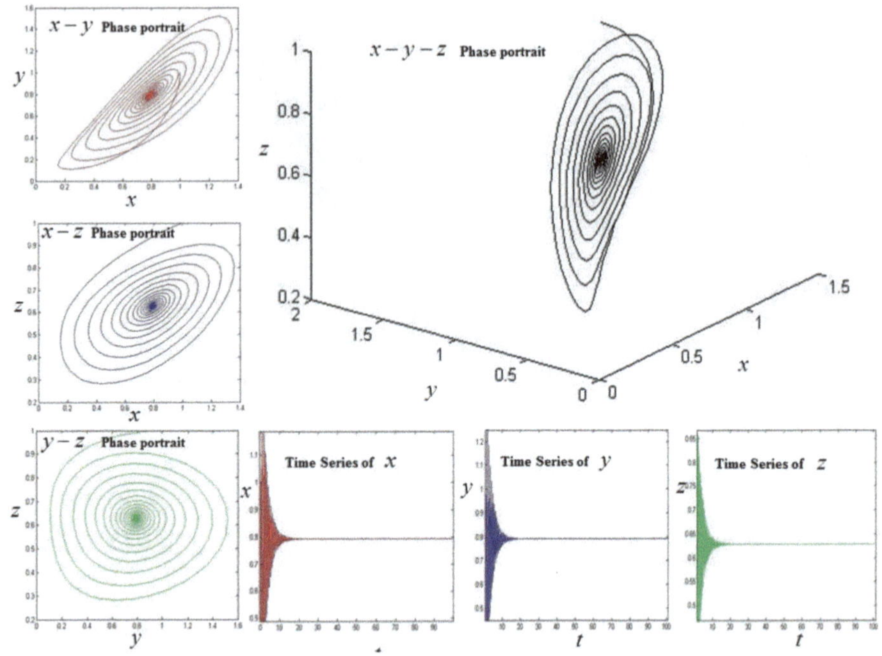

Fig. 6 Phase portraits and time series for Pr = 10, Ha = 1.2, and Ra = 36.0

4 Conclusion

The convection dynamics of ferro-nanofluid flow with magnetite (Fe_3O_4) nanoparticles through blood vessels with cholesterol deposition have been mathematically modeled for rectangular cavity systems with non-uniform depositions. The worst case is when the convective fluid flow through the cavity in cholesterol-deposited blood vessels is chaotic. Through stability analysis, two equilibrium states are determined, which are observed as stable states and chaotic states through phase portraits and time series. For fixed Prandtl and Rayleigh numbers during a chaotic state when the magnetic field through Hartmann number is varied, the transition from chaotic to stable state occurs for the paramagnetic iron oxide nanofluid convection. The observation indicates the tendency of flow in a channelized manner on applying a magnetic field to the specific target, which is helpful for magnetic target-specific drug delivery and CMR applications for enhanced MRI imaging.

References

1. Iijima S (1991) Helical microtubules of graphitic Carbon. Nature 354:56–58
2. Aguayo F, Zarzuela E, Lama J (2011) Nanotechnology and nanochemistry. University of Seville Research Repository, vol 36, pp 28–37
3. Cornejo L (2013) Application to the construction industry. https://nuevastecnologiasymateriales.com/?did=650&vp_ed_act=show_download
4. Cornejo L (2015) Some of the main nanoparticles and their properties. https://nuevastecnologiasymateriales.com/algunas-de-las-principales-nano-particulas-y-sus-propiedades/
5. Diaz F (2012) Introduction to nanomaterials. http://olimpia.cuautitlan2.unam.mx/pagina_engineering/mechanics/mat_mec/m6/Introductionalosnanomaterials.pdf
6. Li X, Xu G, Liu Y, He T (2012) Magnetic Fe3O4 nanoparticles: synthesis and application in water treatment. Nanosci & Nanotechnol-Asia 1(1):14–24
7. Escobar A, Pizzio L, Romanelli G (2018) Magnetic catalysts based on iron oxides: synthesis. Prop Appl Sci Develop 10(1):79–101
8. Wu W, Wu Z, Yu T, Jiang C, Kim W (2015) Recent progress on magnetic iron oxide nanoparticles: synthesis, surface functional strategies and biomedical applications. Sci Technol Adv Mater 16(2):1–44
9. Cortés I (2018) Ion cooling and confinement. https://bigbang.nucleares.unam.mx/~jimenez/FAMC/TrabajosFAMC2018/Cort%C3%A9s_Iliana_Fullereno.pdf
10. Deng H, Li X, Peng Q, Wang X, Chen J, Li Y (2005) Monodisperse magnetic single-crystal ferrite microspheres. Angewandte Chemie International Edition 44(18):2782–2785
11. Yilmaz S, Rösch H, Yildiz S, Klumpp US (2012) Images Cardiovasc Med 126:1932
12. Sosnovik DE, Nahrendorf M, Weissleder R (2007) New Drugs & Technol 115:2076
13. Neuwelt EA, Várallyay P, Bagó AG, Muldoon LL, Nesbit G, Nixon R (2004) Neuropathol Appl Neurobiol 30:456
14. Enas EA, Mehta J (1995) Clin Cardiol 18(3):131
15. Aaiza G, Khan L, Shafie S (2015) Nanoscale Res Lett 10:490
16. Na HB, Song IC, Hyeon T (2009) Adv Mater 21:2133
17. Mejri I, Mahmoudi A, Abbassi MA, Omri A (2014) Int J Math Comput Phys Electr Comput Eng 8(1):75
18. Weinstein JS, Varallyay CG, Dosa E, Gahramanov S, Hamilton B, Rooney WD, Muldoon LL, Neuwelt EA (2010) J Cerebr Blood Flow & Metabol 30:15
19. Garandet JP, Alboussiere T, Moreau R (1992) Int J Heat Mass Transf 35:741
20. Valvano JW, Chitsabesan B (1987) Lasers Life Sci 1:219
21. Blaney L (2007) Lehigh Rev 15:33
22. Hosseindibaeebonab M, Shafii MB, Nobakhti MH (2014) Indian J Sci Res 1(2):733
23. Harisinghani MG, Barentsz J, Hahn PF, Deserno WM, Tabatabaei S, Van de Kaa CH, de la Rosette J, Weissleder R (2003) New England J Med 348:2491
24. Lewin M, Carlesso N, Tung CH, Tang XW, Cory D, Scadden DT, Weissleder R (2000) Natl Biotechnol 18:410
25. Bekki N, Moriguchi H (2007) Phys Plasm 14, Art. no. 012306
26. Wunderbaldinger P, Josephson L, Weissleder R (2002) Acad Radiol 9(2):304
27. Wunderbaldinger P, Josephson L, Weissleder R (2002) Bioconjugated Chem 13:264
28. Idris R, Hashim I (2010) Nonlinear dynamics 62:905
29. Kimura S, Schubert G, Straus JM (1986) J Fluid Mech 166:305
30. Shen T, Weissleder R, Papisov M, Bogdanov A, Brady TJ (1993) Magn Reson Med 29:599
31. Shin T-H, Choi Y, Kim S, Cheon J (2015) Chem Soc Rev 44:4501
32. Bhardwaj R, Chawla M, Das S, Bangia A, Goncerzewicz J (2020) Math Appl 48(1):3

Chapter 9
Rössler Attractor via Fractal Functions and Its Fractal Dimension

R. Valarmathi, A. Gowrisankar, and Kishore Bingi

Abstract This chapter investigates the fractal dimension of linear fractal interpolation with various settings. Additionally, the Riemann–Liouville fractional integral of a linear fractal interpolation function with variable scaling factors is studied. Also, the fractal approximation of the Rossler attractor based on different types of parameter and how the parameters affect the fractal dimension for the time series of the Rossler attractor are investigated.

1 Introduction

The idea of a fractal interpolation function was first proposed by Barnsley [1] in 1986, which formed the basis of constructive approximation theory for non-differentiable functions. Fractal interpolation functions (FIFs) are typically produced as graphs of continuous functions using repeated function systems. Fractal interpolation is a generalization of traditional interpolation methods, as opposed to traditional methods, which produce intricate mathematical structures through a simple and recursive process and thereby give more flexibility in the approximation technique. Many researchers developed various kinds of FIFs, such as hidden variable FIFs, vector-valued FIFs, and non-affine FIFs, which are explored in [2–5]. Fractional calculus is considered a useful tool for more accurately representing the dynamic behavior of

R. Valarmathi
Department of Mathematics, School of Science and Humanities,
Vignan's Foundation for Science, Technology & Research, Guntur 522213, Andhra Pradesh, India
e-mail: valarmathi.2142@gmail.com

A. Gowrisankar (✉)
Department of Mathematics, School of Advanced Sciences, Vellore Institute of Technology,
Vellore 632 014, Tamil Nadu, India
e-mail: gowrisankargri@gmail.com

K. Bingi
Department of Electrical and Electronics Engineering Seri Iskandar,
Universiti Teknologi PETRONAS, Seri Iskandar, Malaysia
e-mail: bingi.kishore@ieee.org

G. Arulprakash et al. (eds.), *Mathematical Modelling of Complex Patterns Through Fractals and Dynamical Systems*, Studies in Infrastructure and Control,
https://doi.org/10.1007/978-981-97-2343-0_9

real-world phenomena. For example, earthquake nonlinear oscillations can be well modeled using partial derivatives. Fractional calculus has many applications in other fields of engineering and science, such as electromagnetism, viscoelasticity, fluid dynamics, electrochemistry, biological population models, optics, and signal processing. On the other hand, like the fractional calculus, the fractal interpolation functions have received considerable attention in contemporary times. Many researchers have noted the importance of using fractional calculus to study fractal functions. A literature review shows that fractional calculus, fractal functions and their fractal dimensions are significantly affected (for more details, [6–21]). The classical integral as well as Riemann–Liouville fractional integral of linear fractional interpolation function (LFIF) with constant scaling factors is proved as a linear fractional interpolation function in [22, 23]. The upper box dimension for the Riemann–Liouville fractional integral of an LFIF and the classical integral of LFIF with variable scaling factors are investigated in [24, 25]. In this chapter, the box dimension for classical integral of LFIF and Riemann–Liouville fractional integral of LFIF with variable scaling factors are examined. The variable scaling factors are chosen as a continuous function set over a closed interval of interpolation, providing a very flexible fractal interpolation function.

A chaotic attractor, also referred to as a strange attractor, is a term used to describe how chaotic systems behave. The formation of semi-stable patterns without a fixed spatial position is predicted by a strange attractor, in contrast to a normal attractor. One of these is the Rössler attractor, which consists of three nonlinear ordinary differential equations. The continuous-time dynamical system defined through these differential equations shows chaotic dynamics associated to the fractal features of the attractor. A fractal version of the chaotic dynamics system for nonlinear ion-acoustic slow and fast electrostatic structures is described using the fractal interpolation function [26]. "Theoretical and numerical analyses of a 4D non-equilibrium dynamical system under the fractal-fractional operator inside the context of the Mittag-Leffler function are investigated [27]." The fractal property of the Lorentz attractor is explored in [28]. The reconstruction of self-similar Lorenz attractors based on fractal interpolation function with function scaling parameters is studied and it is also proposed that the fractal interpolation function is very applicable to any time series/system generating chaotic attractors [29]. The Rössler attractor is studied by taking the values of the parameters in the control parameter space ($i \in [0.33, 0.557]$, $j = 2, k = 4$) and its topology characterization is investigated [30]. Based on these two studies, the present chapter is the result of our effort toward the fractal function to study the approximation of the Rössler attractor based on different types of parameters chosen in control parameter space and the fractional dimension of its time series. A brief construction of the fractal interpolation function is as follows: Let $t_0 < t_1 < \cdots < t_N$ be the partition of the closed interval $I = [t_0, t_N]$. Let a set of data points $\{(t_\iota, s_\iota) \in I \times \mathbb{R} : \iota = 0, 1, \ldots, N\}$ be given. Set $I_\iota = [t_{\iota-1}, t_\iota]$ and let $l_\iota : I \to I_\iota$ be contractive homeomorphism such that

$$l_\iota(t_0) = t_{\iota-1}, \quad l_\iota(t_N) = t_\iota. \tag{1}$$

Denote $K = I \times \mathbb{R}$, $F_\iota : K \to \mathbb{R}$ is N continuous mapping such that, for the constant $0 < r < 1$

$$F_\iota(t_0, s_0) = s_{\iota-1}, \quad F_\iota(t_N, s_N) = s_\iota, \tag{2}$$

$$|F_\iota(t, s') - F_\iota(t, s'')| \le r|s' - s''| \ \forall t \in I, s', s'' \in \mathbb{R}$$

for some $1 \le \iota \le N$. Define $w_\iota : I \times \mathbb{R} \to I_\iota \times \mathbb{R}$ by

$$w_\iota(t, s) = (l_\iota(t), F_\iota(t, s))$$

for all $t \in I$, $s \in \mathbb{R}$ and $1 \le \iota \le N$. The form of the iterated function system (IFS) is as

$$\{K; w_\iota : \iota = 1, 2, \ldots, N\}. \tag{3}$$

The IFS (3) has a unique attractor A^* which is the graph of a continuous function $f : I \to \mathbb{R}$ such that $f(t_\iota) = s_\iota$, for $0 \le \iota \le N$. The continuous function f is called FIF and satisfying the following functional equation:

$$f(t) = \alpha_\iota f(l_\iota^{-1}(t)) + q_\iota(l_\iota^{-1}(t)). \tag{4}$$

The IFS corresponding to the FIF is as follows:

$$l_\iota(t) = a_\iota t + b_\iota \text{ and } F_\iota(t, s) = \alpha_\iota s + q_\iota(t), \tag{5}$$

where $q_\iota : I \to \mathbb{R}$ is a continuous mapping that satisfies the condition imposed on F_ι. The parameters α_ι refer to the vertical scaling factor of the transformations w_ι, which must obey

$$w_\iota(t_0, s_0) = (t_{\iota-1}, s_{\iota-1}), \quad w_\iota(t_N, s_N) = (t_\iota, s_\iota).$$

Denote $|\alpha|_\infty = \max\{|\alpha_\iota| : \iota = 1, 2, \ldots, N\}$. If $q_\iota(t)$ are linear in (4), we have

$$q_\iota(t) = c_\iota t + e_\iota, \tag{6}$$

the corresponding FIF is called LFIF. The parameters a_ι, b_ι, c_ι and e_ι are determined by the end point conditions (1) and (2).

By transforming the constant scaling factor α_ι in IFS (5) into a continuous function from $[x_0, x_N]$ to $(0, 1)$ such that $\|\alpha_\iota\|_\infty < 1$, the new IFS can be obtained as follows:

$$l_\iota(t) = a_\iota t + b_\iota \text{ and } F_\iota(t, s) = \alpha_\iota(t)s + q_\iota(t). \tag{7}$$

Following that the functional equation is satisfied by the fractal interpolation function f with variable scaling factors

$$f(t) = \alpha_\iota(t) f(l_\iota^{-1}(t)) + q_\iota(l_\iota^{-1}(t)). \tag{8}$$

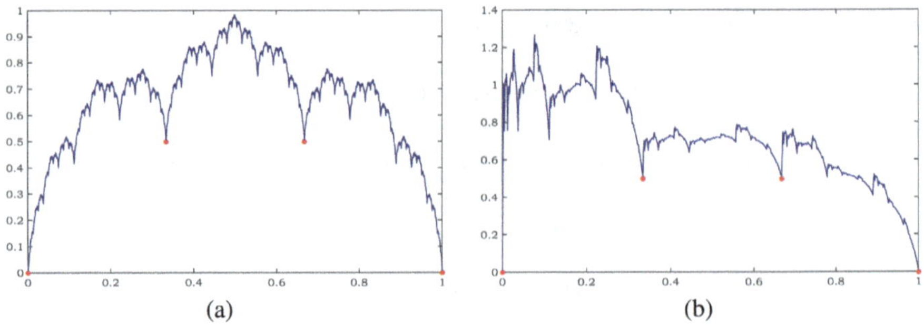

Fig. 1 LFIF: **a** constant scaling factors, **b** variable scaling factors

The linear fractal interpolation function is numerically simulated as follows.

Example 1 The fractal interpolation function f which passes through the given data point $\{(0,0),(1/3,1/2),(2/3,1/2),(1,0)\}$ with

$$l_1(t)=\frac{t}{3},\ l_2(t)=\frac{(t+1)}{3},\ l_3(t)=\frac{(t+2)}{3},$$

$$F_1(t,s)=\frac{(t+s)}{2},\ F_2(t,s)=\frac{(s+1)}{3},\ F_3(t,s)=\frac{(s-x+1)}{2}.$$

Figure 1a represents the LFIF with constant scaling factors $\alpha_\iota=0.5$ for $i=1,2,3$ and Fig. 1b depicts LFIF with variable scaling factors $\alpha_1=\frac{e^t}{4}$, $\alpha_2=0.2(e^x)$ and $\alpha_3=\frac{0.5(e^t)}{2}$.

Barnsley et al. [22] proved that the classical integral of LFIF f also LFIF determined by IFS $\{l_\iota(t),\tilde{F}_\iota(t,\tilde{s})\}_{\iota=1}^N$,

$$\tilde{f}_\iota(t,\tilde{s})=\tilde{\alpha}_\iota\tilde{s}+\tilde{q}_\iota(t), \tag{9}$$

where

$$\tilde{\alpha}_\iota=a_\iota\alpha_\iota,$$
$$\tilde{s}_\iota=\tilde{s}_0+\sum_{n=1}^{\iota}a_n[\alpha_n(\tilde{s}_N-\tilde{s}_0)+\int_{t_0}^{t_N}q_\iota],$$
$$\tilde{q}_\iota(t)=\tilde{s}_{\iota-1}-\tilde{\alpha}_\iota\tilde{s}_0+a_\iota\int_{t_0}^{t}q_\iota,$$

$$\tilde{s}_N=\tilde{s}_0+\left\{\sum_{n=1}^{N}a_n\int_{t_0}^{t_N}q_n\right\}\Big/\left\{1-\sum_{n=1}^{N}a_n\alpha_n\right\}.$$

Following is an example for classical integral of LFIF.

Integral of the LFIF passing through the data points given in Example 1 is $\tilde{f}$ with

$$\hat{F}_1(t,s) = \frac{s}{6} + \frac{t^2}{12}, \quad \hat{F}_2(t,s) = \frac{t}{6} + \frac{s}{6} + \frac{7}{36}, \quad \hat{F}_3(t,s) = \frac{t}{6} - \frac{t^2}{12} + \frac{s}{6} + \frac{17}{36}.$$

For the same vertical scaling factor as in Example 1, Fig. 2a depicts the classical integral of LFIF with constant scaling factor and Fig. 2b illustrates the classical integral of LFIF with variable scaling factors. Observe that Figs. 1a and 2a appear too irregular than those in Figs. 1b and 2b, respectively, and the shapes of the graphs are also very different. The choosing of variable scaling factors is the only reasoning for this fact. This illustrates that LFIF with variable scaling factors is more flexible than LFIF with constant scaling factors for approximating naturally occurring irregular functions.

Definition 1 ([23]) Let F be a bounded subset of $\mathbb{R}^n$ and let $\zeta_\tau(F)$ be the smallest number of sets that can cover F with a diameter of at most τ. The lower box dimension and upper box dimension of F are defined as follows:

$$\underline{\mathbb{D}}_B F = \lim_{\tau \to 0} \frac{\log \zeta_\tau(F)}{-\log \tau},$$

and

$$\overline{\mathbb{D}}_B F = \lim_{\tau \to 0} \frac{\log \zeta_\tau(F)}{-\log \tau},$$

the common value is referred to as the box dimension of F if the above two are equal. That is,

$$\mathbb{D}_B F = \lim_{\tau \to 0} \frac{\log \zeta_\tau(F)}{-\log \tau}. \tag{10}$$

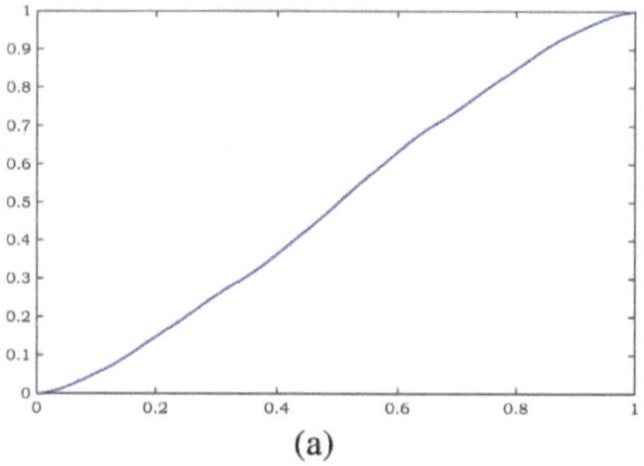

(a)

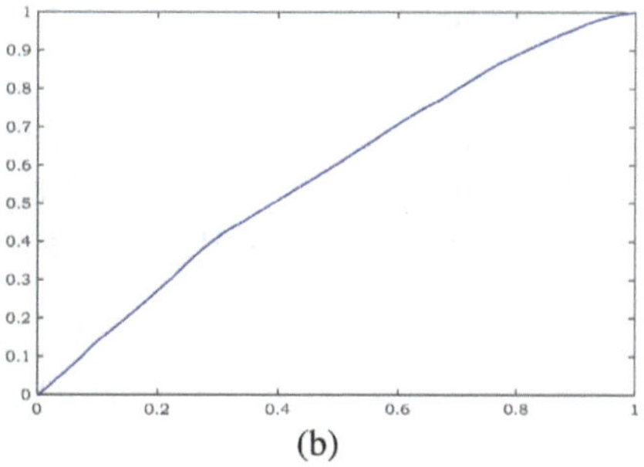

(b)

Fig. 2 Integral of LFIF: **a** constant scaling factors and **b** variable scaling factors

Let f be a continuous function and G_f be the graph of the function f, the box dimension of G_f is denoted by $\mathbb{D}_B(G_f)$.

Ruan et al. [23] proved that f is an LFIF determined by $\{l_\iota(t), F_\iota(t,s)\}_{\iota=1}^N$ where $l_\iota(t) = a_\iota t + b_\iota$ and $F_\iota(t,s) = \alpha_\iota s + q_\iota(t)$. If $\sum_{\iota=1}^N |\alpha_\iota| > 1$ and $\mathbb{D}_B(G_{q_\iota}) = 1$. Then

$$\mathbb{D}_B(G_f) = \mathcal{D}(\{a_\iota, \alpha_\iota\}) \text{ or } 1, \tag{11}$$

where G_f is graph of f and $\mathcal{D}(\{a_\iota, \alpha_\iota\})$ is the unique solution $e \in (1,2)$ of the equation $\sum_{\iota=1}^N a_\iota^{e-1}|\alpha_\iota| = 1$.

Considering that q_ι is an LFIF in all ι, Barnsley et al. [2] showed that $\mathbb{D}_B(G_f) = \mathcal{D}(\{a_\iota, \alpha_\iota\})$ iff

$$\sum_{\iota=1}^N |\alpha_\iota| > 1 \text{ and } \{t_\iota, s_\iota\}_{\iota=1}^N \text{ is not collinear} \tag{12}$$

otherwise $\mathbb{D}_B(G_f) = 1$.

If the function $\tilde{f}$ holds $\sum_{\iota=1}^N |\tilde{\alpha}_\iota| > 1$ and $\mathbb{D}_B(G_{\tilde{q}}) = 1$, then one can determine the box dimension for graph of $\tilde{f}$ is $\mathcal{D}(\{a_\iota, \tilde{\alpha}_\iota\})$ from Theorem 2 of [23], where $\mathcal{D}(\{a_\iota, \tilde{\alpha}_\iota\})$ is the unique solution $e' \in (1,2)$ of the equation $\sum_{\iota=1}^N a_\iota^{e'-1}|\tilde{\alpha}_\iota| = 1$.

Example 2 Linear fractal interpolation function f passing through the data points given in Example 1, whose graph depicted in Fig. 1a. After (12) holds, we have

$$\mathbb{D}_B(G_f) = \mathcal{D}(\{a_\iota, \alpha_\iota\}) = 1.4277$$

and $\tilde{f}$ be the integral of LFIF, if fulfills $\sum_{\iota=1}^N |\tilde{\alpha}_\iota| > 1$ and $\mathbb{D}_B(G_{\tilde{q}}) = 1$, we have

$$\mathbb{D}_B(G_{\tilde{f}}) = 1.2138.$$

2 Fractional Integral of Linear Fractal Interpolation Function

In this section, we discuss the Riemann–Liouville fractional integral of LFIF with variable scaling factors and the resulting function is proved to be an LFIF. Definition of Riemann–Liouville fractional integral for FIF f on $[t_0, t_N]$ by

$$I_{t_0}^\mu f(t) = \frac{1}{\Gamma(\mu)} \int_{t_0}^t (t-x)^{\mu-1} f(x)dx. \tag{13}$$

Theorem 1 *Let f be the LFIF with variable scaling factors determined by $\{l_\iota(t), F_\iota(t,s)\}_{\iota=1}^N$, where $F_\iota = \alpha_\iota(t)s + q_\iota(t)$. If $0 \le a_\iota^\mu \alpha_\iota(t) < 1$ with*

$\sum_{n=1}^{N} a_n^{\mu}\alpha_n(t_N) \neq 1$ *and* $\hat{s}_0 = 0$, *then the Riemann–Liouville fractional integral of* f *is also an LFIF generated by* $\{l_\iota(t), \hat{F}_{\iota,\mu}(t,\hat{s})\}_{\iota=1}^{N}$, *where* $\hat{F}_{\iota,\mu}(t,\hat{s}) = a_\iota^{\mu}\alpha_\iota(t)\hat{s} + \hat{q}_{\iota,\mu}(t)$, *for* $\iota = 1, 2, \ldots, N$,

$$\hat{q}_{\iota,\mu}(t) = \hat{s}_{\iota-1} + f_{\iota,\mu}(t) + a_\iota^{\mu}(I_{t_0}^{\mu}q_\iota)(t),$$

$$\hat{s}_\iota = \sum_{n=1}^{\iota}\left\{f_{n,\mu}(t_N) + a_n^{\mu}\alpha_n(t_N)\hat{s}_N + a_n^{\mu}(I_{t_0}^{\mu}q_n)(t_N)\right\},$$

$$\hat{s}_N = \frac{\sum_{n=1}^{N}\left\{f_{n,\mu}(t_N) + a_n^{\mu}(I_{t_0}^{\mu}q_n)(t_N)\right\}}{1 - \sum_{n=1}^{N} a_n^{\mu}\alpha_n(t_N)}.$$

Proof Given f is a linear fractal interpolation function, its Riemann–Liouville fractional integral is provided by

$$I_{t_0}^{\mu} f(l_\iota(t)) = \frac{1}{\Gamma(\mu(t))}\int_{t_0}^{l_\iota(t)} (l_\iota(t) - t)^{\mu-1} f(x)dx.$$

Applying the variable $t = l_\iota(u)$ provides

$$\begin{aligned} I_{t_0}^{\mu} f(l_\iota(t)) = &\frac{1}{\Gamma(\mu)}\int_{t_0}^{t_{\iota-1}} (t_{\iota-1} - t)^{\mu-1} f(x)dx - \frac{1}{\Gamma(\mu)}\int_{t_0}^{t_{\iota-1}} (t_{\iota-1} - t)^{\mu-1} f(x)dx \\ &+ \frac{1}{\Gamma(\mu)}\int_{t_0}^{t_{\iota-1}} (l_\iota(t) - t)^{\mu-1} f(x)dx + \frac{a_\iota}{\Gamma(\mu)}\int_{t_0}^{t} (l_\iota(t) - l_\iota(u))^{\mu-1} f(l_\iota(u))du. \end{aligned}$$

The functional Eq. (5) yields the following equation:

$$\begin{aligned} (I_{t_0}^{\mu} f)(l_\iota(t)) = \hat{s}_{\iota-1} + f_{\iota,\mu}(t) + &\frac{a_\iota^{\mu}}{\Gamma(\mu)}\int_{t_0}^{t} (t-u)^{\mu-1}\alpha_\iota(t) f(u)du \\ + &\frac{a_\iota^{\mu}}{\Gamma(\mu)}\int_{t_0}^{t} (t-u)^{\mu-1} q_\iota(u)du. \end{aligned}$$

Now applying Leibniz's rule for fractional integration, one can obtain the following:

$$\begin{aligned} (I_{t_0}^{\mu} f)(l_\iota(t)) &= \hat{s}_{\iota-1} + f_{\iota,\mu}(t) + a_\iota^{\mu}\sum_{k=0}^{\infty}\binom{\mu}{k}(I_{t_0}^{\mu+k} f(t))(D_{t_0}^{k}\alpha_\iota(t)) + a_\iota^{\mu}(I_{t_0}^{\mu}q_\iota)(t) \\ &= \hat{s}_{\iota-1} + f_{\iota,\mu}(t) + a_\iota^{\mu}(I_{t_0}^{\mu} f(t))\alpha_\iota(t) + a_\iota^{\mu}(I_{t_0}^{\mu}q_\iota)(t) \\ &= a_\iota^{\mu}\alpha_\iota(t)(I_{t_0}^{\mu} f)(t) + \hat{q}_{\iota,\mu}(t), \end{aligned}$$

where $\hat{q}_{\iota,\mu}(t) = \hat{s}_{\iota-1} + f_{\iota,\mu}(t) + a_\iota^\mu (I_{t_0}^\mu q_\iota)(t)$ and $f_{\iota,\mu}(t) = \frac{1}{\Gamma(\mu)} \int_{t_0}^{t_{\iota-1}} ((l_\iota(t) - t)^{\mu-1} - (t_{\iota-1} - t)^{\mu-1}) f(x)dx$. Hence, the Riemann–Liouville fractional integral of the LFIF with variable scaling factors $I_{t_0}^\mu f$ is also an LFIF determined by the IFS $\{l_\iota(t), \hat{F}_{\iota,\mu}(t,s)\}_{\iota=1}^N$. Consider the following equation to determine the new data points:

$$(I_{t_0}^\mu f)(l_\iota(t)) = \hat{s}_{\iota-1} + f_{\iota,\mu}(t) + a_\iota^\mu \alpha_\iota(t)(I_{t_0}^\mu f)(t) + a_\iota^\mu (I_{t_0}^\mu q_\iota)(t). \tag{14}$$

Let $x = t_N$ and $l_\iota(t_N) = t_\iota$. Then, Eq. (14) gives

$$\hat{s}_\iota - \hat{s}_{\iota-1} = f_{\iota,\mu}(t_N) + a_\iota^\mu \alpha_\iota(t_N)\hat{s}_N + a_\iota^\mu (I_{t_0}^\mu q_\iota)(t_N).$$

The system of equations $\hat{s}_\iota = \hat{s}_0 + \sum_{n=1}^{\iota}(\hat{s}_n - \hat{s}_{n-1})$ is employed to get

$$\hat{s}_\iota = \sum_{n=1}^{\iota} \left\{ f_{n,\mu}(t_N) + a_n^\mu \alpha_n(t_N)\hat{s}_N + a_n^\mu (I_{t_0}^\mu q_n)(t_N) \right\}.$$

The end point $\hat{s}_N$ is obtained by replacing $\iota = N$ in the above equation

$$\hat{s}_N = \sum_{n=1}^{N} \left\{ f_{n,\mu}(t_N) + a_n^\mu (I_{t_0}^\mu q_n)(t_N) \right\} \Big/ 1 - \sum_{n=1}^{N} a_n^\mu \alpha_n(t_N).$$

Example 3 The Riemann–Liouville fractional integral of order 0.3 of the LFIF passing through the data points given in Example 1 is $\hat{f}$ with

$$\hat{F}_1(t,s) = (0.5773)\frac{e^t}{4}s + \frac{0.5773(t)^{0.3}}{\Gamma 1.5}\left\{\frac{0.3x}{1.3}\right\},$$

$$\hat{F}_2(t,s) = 0.1785 + (0.5773)2e^t s + \frac{0.2886(t)^{0.3}}{\Gamma 1.3},$$
$$\hat{F}_3(t,s) = 1.0546 + (0.5773)\frac{0.5e^t}{2}s + \frac{0.5773(t)^{0.3}}{\Gamma 1.3}\left\{0.3 - \frac{0.3x}{1.3}\right\}.$$

The graphical illustrations of Riemann–Liouville fractional integral of order 0.3 of LFIF f are given in Fig. 3.

Remark 1 The Riemann–Liouville fractional integral of an LFIF is an LFIF with variable scaling factors and the initial condition $\hat{s}_N = 0$, according to Theorem 1. Similar to this, if the upper limit of the Riemann–Liouville fractional integral f is defined t_N as $I_{t_N}^\mu f(t) = -\frac{1}{\Gamma(\mu)}\int_t^{t_N}(t-x)^{\mu-1} f(x)dx$, then function $I_{t_N}^\mu f(t)$ is again an LFIF generated by the IFS $\{l_\iota(t), \hat{F}_{\iota,\mu}(t,\hat{s})\}_{\iota=1}^N$, where $\hat{F}_\iota(t,\hat{s}) = a_\iota^\mu \alpha_\iota \hat{s} + \hat{q}_{\iota,\mu}(t)$,

$$\hat{q}_{\iota,\mu}(t) = \hat{s}_\iota - f_{\iota,\mu}(t) + a_\iota^\mu (I_{t_N}^\mu q_\iota)(t),$$

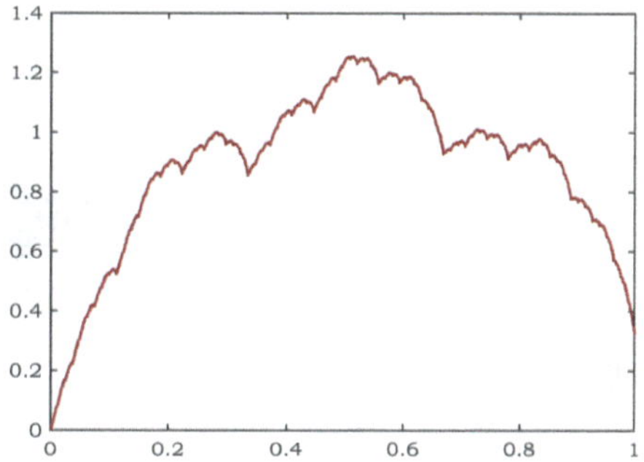

(a)

Fig. 3 Riemann–Liouville fractional integral of order 0.3 of LFIF f

$$\hat{s}_{i-1} = -\sum_{n=i}^{N} \left\{ f_{n,\mu}(t_0) - a_n^{\mu} d_n \hat{s}_0 - a_n^{\mu} (I_{t_N}^{\mu} q_n)(t_0) \right\},$$

$$\hat{s}_0 = -\frac{\sum_{n=1}^{N} \left\{ f_{n,\mu}(t_N) - a_n^{\mu} (I_{t_0}^{\mu} q_n)(t_0) \right\}}{1 - \sum_{n=1}^{N} a_n^{\mu} d_n}.$$

If we choose $\alpha(t) = \alpha$, the result of Theorem 1 yields the result given by Huo-Jun Ruan et al. [23].

3 Fractal Approximation of the Rössler Attractor

This section explores the Rössler attractor in perspective of fractal function and its fractal dimension. The Rössler system, first investigated by Otto Rössler in the 1976s, is a system of three nonlinear ordinary differential equations whose attractor is known as the Rössler attractor. These differential equations form a continuous-time dynamical system with chaotic dynamics corresponding with the attractor's fractal features [31]. Unlike Lorenz's, the Rössler equations are simpler because they contain only one nonlinear term. The Rössler system is defined as follows:

$$\left.\begin{array}{l} \frac{dx(\xi)}{d\xi} = -y(\xi) - z(\xi), \\ \frac{dy(\xi)}{d\xi} = x(\xi) + iy(\xi), \\ \frac{dz(\xi)}{d\xi} = j + z(\xi)(x(\xi) - k). \end{array}\right\} \tag{15}$$

Here $x(\xi)$, $y(\xi)$, $z(\xi)$ are dynamical variables defining the phase space; ξ stands for time; and the parameters of the original classical Rössler system are $i = 0.2$, $j = 0.2$ and $k = 5.7$, the graph of which is shown in Fig. 4. From graph of the Rössler attractor, it can be seen that fractal, i.e., self-similarity, repeating pattern, and scale invariant are visualized.

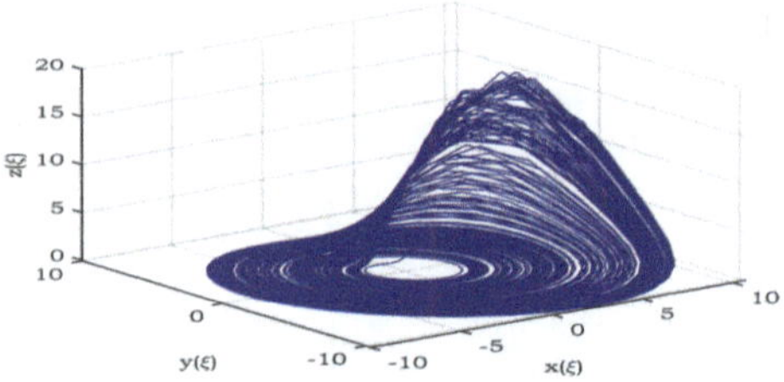

Fig. 4 Classical Rössler attractor phase space trajectory $i = 0.2$, $j = 0.2$ and $k = 5.7$

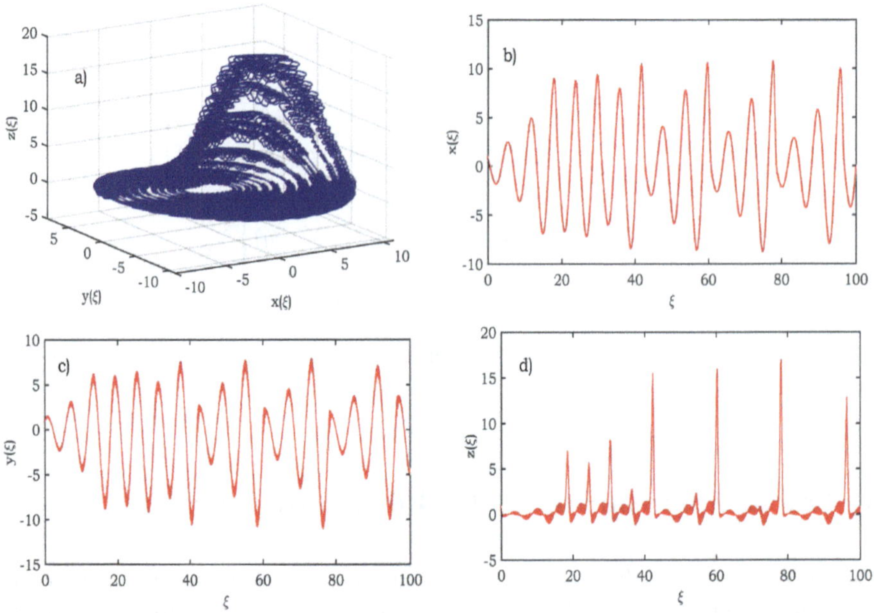

Fig. 5 Fractal phase and time portraits of Rössler system **a** $i = 0.2$, $j = 0.2$ and $k = 5.7$, **b** x versus time ξ, **c** y versus time ξ, **d** z versus time ξ

The fractal interpolation function is very applicable to any chaotic attractor [29], and since the Rössler attractor is one of the chaotic attractors, we can use the fractal function to approximate it. No previous studies have dealt with the fractal perspective of Rössler attractor, and thus the present study focuses on the fractal perspective of Rössler attractor as well as fractal dimension for its time series by choosing different types of parameter. A fractional analog of the time series of the Rössler attractor

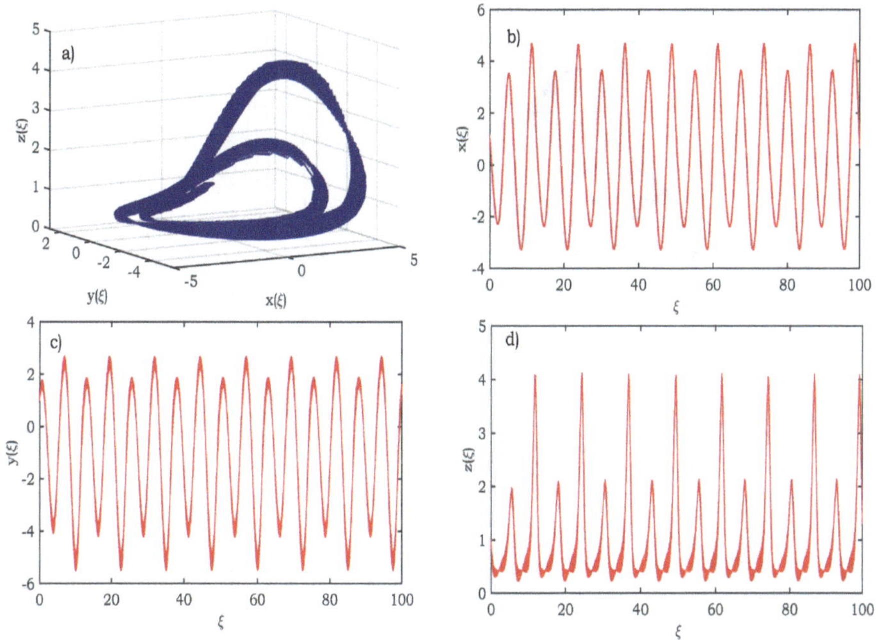

Fig. 6 Fractal phase and time portraits of Rössler system **a** $i = 0.33$, $j = 2$ and $k = 4$, **b** x versus time ξ, **c** y versus time ξ, **d** z versus time ξ

can be obtained using the fractional function (5). For $i = 0.2$, $j = 0.2$ and $k = 5.7$, fractional approximation of the Rössler attractor is depicted in Fig. 5a–d. By using the parameter values in the control parameter space ($i \in [0.33, 0.557]$, $j = 2$, $k = 4$), the Rössler attractor is studied detail in [31–33]. It is also discussed that the number of unstable periodic orbits of Rössler attractor increases as the value of parameter i increases. Similarly, by increasing the value of parameter i and keeping the value of other two parameters j and k fixed at 2 and 4, its Rössler attractor fractal approximation is investigated using Eq. (5) and its graph is illustrated in Figs. 6a–d, 7a–d, and 8a–d.

The fractal dimension of the time series is studied in the following. We already investigated the fractal dimension for the graph of the LFIF in Sect. 1, the same way fractal dimension for the graph of the time series is determined by using Eq. 12. We examined the fractal dimension of the time series with parameter i set to 0.3, 0.4, and 0.5 and the other two parameters j and k fixed at 2 and 3. The obtained fractal dimension is shown in Table 1. As the value of parameter i increases, Table 1 demonstrates that the fractal dimension of the time series decreases, and as a result, Figs. 6, 7, and 8 clearly illustrates that the irregularity of the time series decreases significantly.

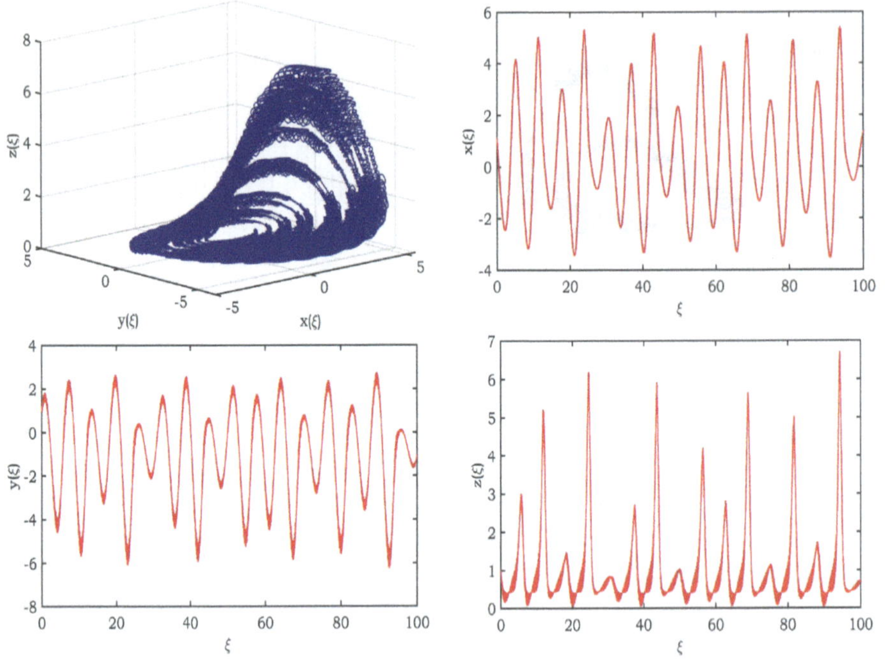

Fig. 7 Fractal phase and time portraits of Rössler system **a** $i = 0.4$, $j = 2$ and $k = 4$, **b** x versus time ξ, **c** y versus time ξ, **d** z versus time ξ

Table 1 Fractal dimension for fractal function of $x(\xi)$, $y(\xi)$ and $z(\xi)$ for ($i \in [0.33, 0.5]$, $j = 2$ and $k = 2$)

Parameter i	Fractal dimension ($\mathbb{D}_B$)		
	$(\xi, x(\xi))$	$(\xi, y(\xi))$	$(\xi, z(\xi))$
0.33	1.8089	1.7611	1.7301
0.4	1.7455	1.7026	1.6483
0.5	1.6866	1.6332	1.5868

4 Conclusion

In this chapter, the box dimension for the classical integral of LFIF and its Riemann–Liouville fractional integral is studied by choosing the vertical scaling factors as a continuous function, and the resulting function is proved to be LFIF. In addition, the fractal approximation of the Rössler attractor is described with various parameters. Also, fractal dimension of time series graphs is explored by choosing different types of parameters.

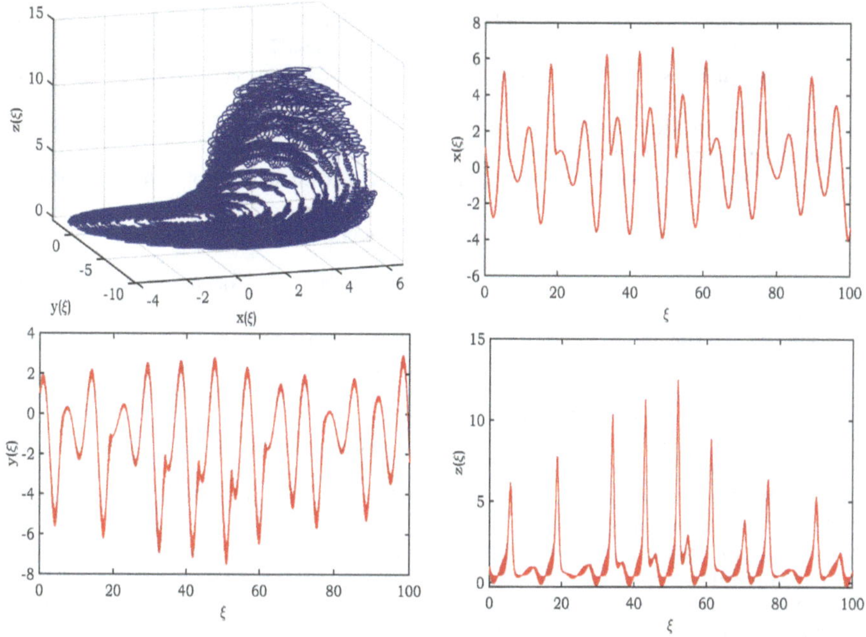

Fig. 8 Fractal phase and time portraits of Rössler system **a** $i = 0.5$, $j = 2$ and $k = 4$, **b** x versus time ξ, **c** y versus time ξ, **d** z versus time ξ

References

1. Barnsley MF (1986) Fractal functions and interpolation. Constr Approx 2(01):303–329
2. Barnsley MF, Elton J, Hardin D, Massopust P (1989) Hidden variable fractal interpolation functions. SIAM J Math Anal 20(05):1218–1242
3. Chandra S, Abbas S, Verma S (2022) Bernstein super fractal interpolation function for countable data systems. Numer Algorithms :1–25
4. Priyanka TMC, Gowrisankar A (2021) Riemann-Liouville fractional integral of non-affine fractal interpolation function and its fractional operator. Eur Phys J Spec Top 230(21):3789–3805
5. Valarmathi R, Gowrisankar A (2023) On the variable order fractional calculus characterization for the Hidden variable fractal interpolation function. Fractal Fract 7(01):34
6. Peng WL, Yao K, Zhang X, Yao J (2019) Box dimension of Weyl-Marchaud fractional derivative of linear fractal interpolation functions. Fractals 27(04):1950058
7. Liang YS, Su WY (2016) Fractal dimensions of fractional integral of continuous functions. Acta Math Sin Engl Ser 32(12):1494–1508
8. Wu XE, Liang YS (2017) Relationship between fractal dimensions and fractional calculus. Nonlinear Sci Lett A 8(01):77–89
9. Xiao W (2017) Relationship of upper box dimension between continuous fractal functions and their Riemann-Liouville fractional integral. Fractals 29(08):2150264
10. Liang YS, Wang HX (2021) Upper Box dimension of Riemann-Liouville fractional integral of fractal functions. Fractals 29(01):2150015
11. Lei J, Liu K, Dai Y (2020) Box dimensions of the Riemann-Liouville fractional integral of Hölder continuous multivariate functions. Fractals 28(06):2050113

12. Priyanka TMC, Gowrisankar A (2021) Analysis on Weyl-Marchaud fractional derivative for types of fractal interpolation function with fractal Dimension. Fractals 29(07):2150215
13. Golmankhaneh AK, Welch K, Serpa C, Jørgensen PE (2023) Non-standard analysis for fractal calculus. J Anal :1–22
14. Gowrisankar A, Khalili Golmankhaneh A, Serpa C (2021) Fractal calculus on fractal interpolation functions. Fractal Fract 5(04):157
15. Akhtar MN, Prasad MGP, Navascués MA (2016) Box dimensions of α-fractal functions. Fractals 24(03):1650037
16. Prithvi BV, Katiyar SK (2022) Interpolative operators: fractal to multivalued fractal. Chaos Solitons Fractals 164:112449
17. Golmankhaneh AK, Welch K, Tunç C, Gasimov YS (2023) Classical mechanics on fractal curves. Eur Phys J Spec Top :1–9
18. Sahu DR, Chakraborty A, Dubey RP (2010) K-iterated function system. Fractals 18(01):139–144
19. Serpa C, Buescu J (2010) Explicitly defined fractal interpolation functions with variable parameters. Chaos Solitons Fractals 75:76–83
20. Easwaramoorthy D, Uthayakumar R (2011) Analysis on fractals in fuzzy metric spaces. Fractals 19(03):379–386
21. Selmi B, Mabrouk AB (2022) On the mixed multifractal formalism for vector-valued measures. Proyecciones (Antofagasta, On line). 41(05):1015–1032
22. Barnsley MF, Harrington AN (1989) The calculus of fractal interpolation functions. J Approx Theory 57(01):14–34
23. Ruan HJ, Su WY, Yao K (2009) Box dimension and fractional integral of linear fractal interpolation functions. J Approx Theory 161(01):187–197
24. Liang YS, Wang HX (2021) Upper box dimension of Riemann-Liouville fractional integral of fractal functions. Fractals 29(01):2150015
25. Agathiyan, Gowrisankar A, Priyanka TMC (2022) Construction of new fractal interpolation functions through integration method. Results Math 77(03):1–20
26. Prasad PK, Gowrisankar A, Saha A, Banerjee S (2020) Dynamical properties and fractal patterns of nonlinear waves in solar wind plasma. Phys Scr 95(06):065603
27. Haidong Q, ur Rahman M, Al Hazmi SE, Yassen MF, Salahshour S, Salimi M, Ahmadian A (2023) Analysis of non-equilibrium 4D dynamical system with fractal fractional Mittag Leffler kernel. Eng Sci Technol Int J 37:101319
28. Viswanath D (2004) The fractal property of the Lorenz attractor. Phys D Nonlinear Phenom 190(1–2):115–128
29. Fataf NAA, Gowrisankar A, Banerjee V (2020) In search of self-similar chaotic attractors based on fractal function with variable scaling approximately. Phys Scr 95(07):075206
30. Letellier C, Dutertre P, Maheu B (1995) Unstable periodic orbits and templates of the Rössler system: toward a systematic topological characterization. Chaos Interdiscip J Nonlinear Sci 5(01):271–282
31. Mindlin GM, Gilmore R (1992) Topological analysis and synthesis of chaotic time series. Phys D Nonlinear Phenom 58(1–4):229–242
32. Grebogi C, Ott E, Yorke JA (1987) Chaos, strange attractors, and fractal basin boundaries in nonlinear dynamics. Science 238(4827):632–638
33. Mindlin GB, Hou XJ, Solari HG, Gilmore R, Tufillaro NB (1990) Classification of strange attractors by integers. Phys Rev Lett 64(20):2350

Chapter 10
Dynamical Behaviour of Ocean Waves Described by the Geophysical-Burgers' Equation

Bikram Mondal, Yogesh Chettri, Alireza Abdikian, and Asit Saha

Abstract Phase Plane analysis is applied to study nonlinear waves described by the Geophysical-Burgers' equation. A travelling wave transformation is considered to convert the geophysical-Burgers' equation to a dynamical system. All equilibrium points of the corresponding dynamical system are obtained and analysed based on the corresponding eigenvalues. Phase portrait for the dynamical system is plotted. Solutions of kink and anti-kink waves corresponding to heteroclinic orbits and periodic waves corresponding to periodic orbits are obtained. Effect of various parameters on these wave solutions is shown.

1 Introduction

Tsunamis are large waves mainly caused by underwater earthquakes or volcanic activity. These waves are set in motion by the shaking of the Earth during an earthquake. Ocean wave does not gain much amplitude in deep sea. But, when the wave moves towards the land area, the sea turns shallow and the amplitude of the wave starts to increase. Furthermore, the velocity of these waves is determined by the depth of the ocean rather than its proximity to the source location. In the depths of the ocean, these waves can outpace even the swiftest airplanes, showcasing remarkable velocity. However, as they approach shallower waters, their speed tends to decrease. Tsunami waves are more devastating than tidal waves due to their greater height, due to which they penetrate much further inland on the mainland.

B. Mondal
Department of Mathematics, Biswa Bangla Biswavidyalaya, Shivpur, West Bengal, India

Y. Chettri · A. Saha (✉)
Department of Mathematics, Sikkim Manipal Institute of Technology, Sikkim Manipal University, Majitar, Rangpo 737136, East Sikkim, India
e-mail: asit_saha123@rediffmail.com

A. Abdikian
Department of Physics, Malayer University, 65719-95863 Malayer, Iran

G. Arulprakash et al. (eds.), *Mathematical Modelling of Complex Patterns Through Fractals and Dynamical Systems*, Studies in Infrastructure and Control,
https://doi.org/10.1007/978-981-97-2343-0_10

In 1974, Burgers [1] introduced Burgers' equation and discussed its application to fluid dynamics. While not directly focussed on ocean waves, it laid the foundation for the study of the geophysical-Burgers' equation. In 1895, Korteweg and de Vries [2, 3] discovered a nonlinear evolution equation called the Korteweg–de Vries (KdV) equation. This equation has huge real-life applications in numerous branches of natural science and engineering. The KdV equation [4, 5] when coupled with the Coriolis effect parameter gives the geophysical-KdV equation that can describe ocean waves [6]. Recent studies have highlighted the role of the Coriolis force [7] in influencing nonlinear tsunami waves. Whitham [8], in 1974, provided a comprehensive treatment of linear and nonlinear waves, including discussions on Burgers' equation and its relevance to wave dynamics.

The Korteweg–de Vries equation along with its modified forms plays an important role in the study of solitary wave [9, 10]. Wijngaarden [11] and Gao [12] studied the propagation of liquids composed of turbulence and gas bubbles. In 1983, Mei [13] discussed Burgers' equation in the context of wave propagation and transformation. In 2016, Bruhl et al. [14] investigated long nonlinear cosine wave in shallow water. Johnson [15], in 2019, reported the existence of nonlinear waves on an elastic tube.

Kalisch [19] focussed on numerical methods for solving shallow-water flow equations, including Burgers' equation, which is frequently used for modelling ocean waves. In 2021, Saha et al. [20] with the help of phase plots and time series plots investigated modified geophysical Korteweg–de Vries equation (mGKdV). The mGKdV equation was transformed to a dynamical system using travelling wave transformation. Applying phase plot analysis, presence of the solitary, periodic and superperiodic tsunami waves were shown. They showed the effect of Coriolis parameter ω_0 and velocity of travelling wave on the tsunami waves. In 2019, Turgut et al. [21] investigated ocean wave via geophysical-KdV equation. But they did not study the ocean flow due to the presence of dissipation. So, in this paper, a nonlinear evolution equation of the geophysical-Burgers' equation which describes ocean waves is considered. Phase plane analysis is applied to geophysical-Burgers' equation to obtain kink, anti-kink and periodic wave features of the ocean waves. Effect of various parameters on these wave features has been illustrated.

2 Nonlinear Evolution Equation

For a given field $u(x, t)$ and diffusion coefficient ν, the general form of Burgers' equation in one space dimension is given by [1]:

$$\frac{\partial u}{\partial t} + u\frac{\partial u}{\partial x} = \nu\frac{\partial^2 u}{\partial t^2}.$$

The KdV equation is a nonlinear, dispersive partial differential equation for a function ϕ of two dimensionless real variables, x and t which denotes space and time respectively. The general form of the KdV equation is given by [9, 10]:

$$\frac{\partial \phi}{\partial t} - 6\phi \frac{\partial \phi}{\partial x} + \frac{\partial^3 \phi}{\partial x^3} = 0$$

The general form of geophysical-KdV equation is given by [21]

$$\frac{\partial u}{\partial t} - \omega_0 \frac{\partial u}{\partial x} + \frac{3}{2} u \frac{\partial u}{\partial x} + \frac{1}{6} \frac{\partial^3 u}{\partial x^3} = 0,$$

where u denotes the free surface advancement and ω_0 represent Coriolis effect parameter.

The general form of geophysical-Burgers' equation which describes ocean waves is given by

$$\frac{\partial \phi}{\partial t} - \omega_0 \frac{\partial \phi}{\partial x} + \frac{3}{2} \phi \frac{\partial \phi}{\partial x} + \eta \frac{\partial^2 \phi}{\partial x^2} = 0, \tag{1}$$

where η is the dissipative coefficient and ω_0 is the Coriolis parameter. This equation describes nonlinear ocean waves.

3 Phase Plane Analysis

Phase plane analysis is a graphical as well as analytical method for studying dynamical systems. Phase plane analysis provides motion trajectories, qualitative features of the trajectories and information regarding the stability of the equilibrium points.

Linear System

Let us consider a two-dimensional linear system as

$$\begin{cases} \dot{x} = ax + by, \\ \dot{y} = cx + dy, \end{cases} \tag{2}$$

where $\dot{x} = \frac{dx}{dt}$, $\dot{y} = \frac{dy}{dt}$, and a, b, c, d are real parameters.

The equilibrium point of the system (2) is $(-\frac{b}{a}, \frac{cb}{ad})$.

Equation (2) can be written as a matrix form:

$$\dot{X} = AX, \tag{3}$$

where

$$A = \begin{bmatrix} a & b \\ c & d \end{bmatrix}, \tag{4}$$

and $$X = \begin{bmatrix} x \\ y \end{bmatrix}. \tag{5}$$

Then the characteristic equation of A is written as

$$det(A - \lambda I) = 0,$$

where I is the identity matrix of order two.

$$\text{or,} \quad \begin{vmatrix} a-\lambda & b \\ c & d-\lambda \end{vmatrix} = 0,$$

$$\text{or,} \quad (a-\lambda)(d-\lambda) - bc = 0,$$

$$\text{or,} \quad ad + \lambda^2 - a\lambda - d\lambda - bc = 0,$$

$$\text{or,} \quad \lambda^2 - \lambda(a+d) + ad - bc = 0,$$

$$\text{or,} \quad \lambda^2 - \tau\lambda + \Delta = 0, \tag{6}$$

where $\tau = trace(A) = a + d$ and $\Delta = det(A) = ad - bc$.
Then, we have the eigenvalues of A as

$$\lambda_{1,2} = \frac{\tau \pm \sqrt{\tau^2 - 4\Delta}}{2}. \tag{7}$$

Classification of Equilibrium Points

To classify the equilibrium points of the system (2), we consider the obtained eigenvalues as $\lambda_1 = \frac{\tau+\sqrt{\tau^2-4\Delta}}{2}$ and $\lambda_2 = \frac{\tau-\sqrt{\tau^2-4\Delta}}{2}$.

1. **CASE 1**: $\Delta < 0$
   If $\Delta < 0$, it implies that the eigenvalues are real with opposite signs, hence the equilibrium point is a saddle point.
2. **CASE 2**: $\Delta > 0$
   a. If $\tau^2 - 4\Delta > 0$, then the equilibrium points are nodes (stable nodes for $\tau < 0$ and unstable nodes for $\tau > 0$).
   b. If $\tau^2 - 4\Delta < 0$, then we have a pair of complex eigenvalues:
      $\lambda_1 = \alpha + i\beta$ and $\lambda_2 = \alpha - i\beta$,
      where, $\alpha = \frac{\tau}{2}$ and β is the imaginary part.
      i. If $\alpha = 0$ (i.e.$\tau = 0$), then $\lambda_1 = i\beta$; and $\lambda_2 = -i\beta$. The eigenvalues are purely imaginary, so the corresponding equilibrium point is a centre.
      ii. If $\alpha > 0$ (i.e.$\tau > 0$), then we get an unstable spiral at that equilibrium point.
      iii. If $\alpha < 0$ (i.e.$\tau < 0$), then we get an stable spiral at that equilibrium point.

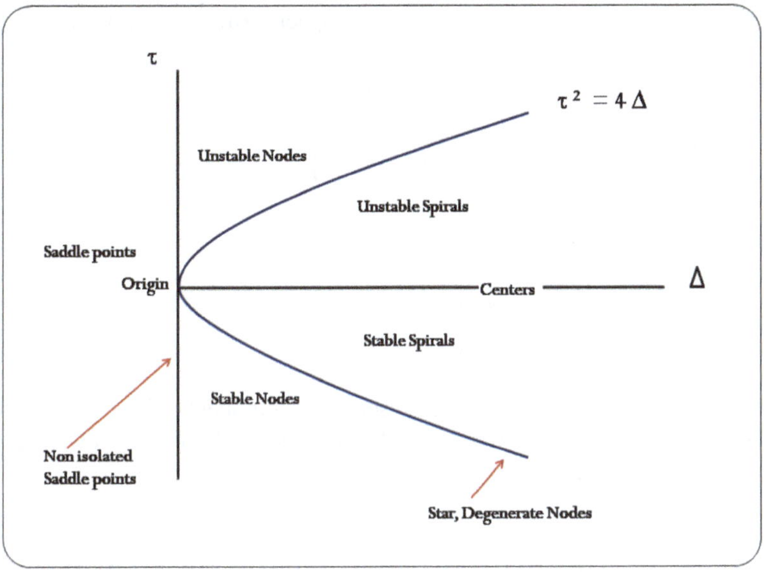

Fig. 1 Stability of equilibrium points for the system (2)

3. **CASE 3**: $\Delta = 0$
   If $\Delta = 0$, it implies that at least one of the eigenvalues is zero. Then the origin ceases to be an isolated equilibrium point. Instead, there exists either an entire line or a plane of equilibrium points when A = 0.

Stability of equilibrium points is shown in Fig. 1.

Figure 1 shows that spirals, nodes and saddle points are the primary equilibrium points found in open areas of the (Δ, τ) plane. Centres, degenerate nodes, star and non-isolated equilibrium points are marginal instances that occur along curves in the (Δ, τ) plane. Among these marginal cases, centres hold the greatest significance, frequently appearing in frictionless mechanical systems characterized by energy conservation.

Nonlinear System

On the phase plane, the general form of a vector field is

$$\begin{cases} \dot{x} = f_1(x, y), \\ \dot{y} = f_2(x, y), \end{cases} \tag{8}$$

where f_1 and f_2 are given functions.

This system can be represented in more compact form using vector notation as

$$\dot{X} = F(X), \tag{9}$$

where $X = (x, y)$ and $F(X) = \Big(f_1(X), f_2(X) \Big)$.

Suppose that (x^*, y^*) is a fixed point, that is,

$$\begin{cases} f_1(x^*, y^*) = 0, \\ f_2(x^*, y^*) = 0. \end{cases} \tag{10}$$

Now let $u = x - x^*$ and $v = y - y^*$ represent the components of a small disturbance from the fixed point. To observe whether the disturbance decays or grows, we derive differential equations for u and v.

For u, we have

$$\dot{u} = \dot{x}, \tag{11}$$

$$\text{or,} \quad \dot{u} = f_1(x^* + u, y^* + v). \tag{12}$$

Using Taylor's series expansion, we get

$$\dot{u} = f_1(x^*, y^*) + u\frac{\partial f_1}{\partial x} + v\frac{\partial f_1}{\partial y} + O(u^2, v^2, uv), \tag{13}$$

$$\text{or,} \quad \dot{u} = u\frac{\partial f_1}{\partial x} + v\frac{\partial f_1}{\partial y} + O(u^2, v^2, uv). \tag{14}$$

Here $O(u^2, v^2, uv)$ represents the quadratic terms in u and v. These quadratic terms are very small as u and v are small, hence we can neglect them.

Similarly for v, we get

$$\dot{v} = u\frac{\partial f_2}{\partial x} + v\frac{\partial f_2}{\partial y} + O(u^2, v^2, uv). \tag{15}$$

Hence, we get

$$\begin{bmatrix} \dot{u} \\ \dot{v} \end{bmatrix} = \begin{bmatrix} \frac{\partial f_1}{\partial x} & \frac{\partial f_1}{\partial y} \\ \frac{\partial f_2}{\partial x} & \frac{\partial f_2}{\partial y} \end{bmatrix} \begin{bmatrix} u \\ v \end{bmatrix}. \tag{16}$$

The matrix $J = \begin{bmatrix} \frac{\partial f_1}{\partial x} & \frac{\partial f_1}{\partial y} \\ \frac{\partial f_2}{\partial x} & \frac{\partial f_2}{\partial y} \end{bmatrix}$ denotes the Jacobian matrix at the equilibrium point (x^*, y^*). The system (16) is a linear system and it can be analysed as the previous analysis of linear system.

4 Phase Plane Analysis of the Geophysical-Burgers' Equation

Now, we apply phase plane analysis to the geophysical-Burgers' equation to study different nonlinear wave features. Let us consider a travelling wave transformation

$$\xi = x - vt, \tag{17}$$

where v is the speed of the travelling wave.

Then, we have

$$\frac{\partial \phi}{\partial t} = \frac{\partial \phi}{\partial \xi}\frac{\partial \xi}{\partial t} = -v\frac{d\phi}{d\xi}, \tag{18}$$

$$\frac{\partial \phi}{\partial x} = \frac{\partial \phi}{\partial \xi}\frac{\partial \xi}{\partial x} = \frac{d\phi}{d\xi}, \tag{19}$$

$$\text{and} \qquad \frac{\partial^2 \phi}{\partial x^2} = \frac{d^2\phi}{d\xi^2}. \tag{20}$$

Now using Eqs. (18)–(20) in Eq. (1), we get

$$-(v+\omega_0)\frac{d\phi}{d\xi} + \frac{3}{2}\phi\frac{d\phi}{d\xi} + \eta\frac{d^2\phi}{d\xi^2} = 0. \tag{21}$$

Integrating equation (21) with respect to ξ, we get

$$-(v+\omega_0)\phi + \frac{3}{2}\frac{\phi^2}{2} + \eta\frac{d\phi}{d\xi} = c_1,$$

where c_1 is an integrating constant.

Using boundary conditions $\phi \to 0$, $\frac{d\phi}{d\xi} \to 0$ as $\xi \to \pm\infty$, we get $c_1 = 0$.

Thus, we get

$$-(v+\omega_0)\phi + \frac{3}{2}\frac{\phi^2}{2} + \eta\frac{d\phi}{d\xi} = 0,$$

$$\text{or,} \qquad \frac{d\phi}{d\xi} = -(\frac{3}{4\eta})\phi^2 + (\frac{v+\omega_0}{\eta})\phi. \tag{22}$$

Now using Eq. (22) in Eq. (21), we get

$$-(v+\omega_0)(-(\frac{3}{4\eta})\phi^2+(\frac{v+\omega_0}{\eta})\phi)+\frac{3}{2}\phi(-(\frac{3}{4\eta})\phi^2+(\frac{v+\omega_0}{\eta})\phi)+\eta\frac{d^2\phi}{d\xi^2}=0$$

$$\text{or,}\quad \eta^2\frac{d^2\phi}{d\xi^2}=(-(v+\omega_0)+\frac{3}{2}\phi)(\frac{3}{4}\phi^2-(v+\omega_0)\phi)$$

$$\text{or,}\quad \frac{d^2\phi}{d\xi^2}=\frac{9}{8\eta^2}\phi^3-\frac{9}{4\eta^2}(v+\omega_0)\phi^2+\frac{(v+\omega_0)^2}{\eta^2}\phi. \tag{23}$$

Equation (23) can be written as

$$\begin{cases}\frac{d\phi}{d\xi}=z,\\ \frac{dz}{d\xi}=\frac{9}{8\eta^2}\phi^3-\frac{9}{4\eta^2}(v+\omega_0)\phi^2+\frac{(v+\omega_0)^2}{\eta^2}\phi.\end{cases} \tag{24}$$

For the equilibrium points of the system (24), we have

$$\frac{d\phi}{d\xi}=0 \text{ and } \frac{dz}{d\xi}=0.$$

$$\text{Now}\quad \frac{d\phi}{d\xi}=0 \implies z=0$$

$$\text{and}\quad \frac{dz}{d\xi}=0 \implies \frac{9}{8\eta^2}\phi^3-\frac{9}{4\eta^2}(v+\omega_0)\phi^2+\frac{(v+\omega_0)^2}{\eta^2}\phi=0$$

$$\implies \phi=0 \text{ or, } 9\phi^2-18(v+\omega_0)\phi+8(v+\omega_0)^2=0$$

$$\text{or, } \phi=\frac{18(v+\omega_0)\pm\sqrt{324(v+\omega_0)^2-288(v+\omega_0)^2}}{18}$$

$$\text{or, } \phi=\frac{18(v+\omega_0)\pm\sqrt{36(v+\omega_0)^2}}{18}$$

$$\text{or, } \phi=\frac{18(v+\omega_0)\pm 6(v+\omega_0)}{18}$$

$$\text{or, } \phi=\frac{4(v+\omega_0)}{3} \text{ and } \phi=\frac{2(v+\omega_0)}{3}.$$

So, the equilibrium points of the system (24) are $E_0(0,0)$, $E_1(\frac{4(v+\omega_0)}{3},0)$ and $E_2(\frac{2(v+\omega_0)}{3},0)$.

Now Jacobian matrix of the system (24) is given by

$$J = \begin{bmatrix} \frac{\partial \dot{\phi}}{\partial \phi} & \frac{\partial \dot{\phi}}{\partial z} \\ \frac{\partial \dot{z}}{\partial \phi} & \frac{\partial \dot{z}}{\partial z} \end{bmatrix},$$

where $\dot{\phi} = \frac{d\phi}{d\xi}$ and $\dot{z} = \frac{dz}{d\xi}$.

$$\text{or,} \qquad J = \begin{bmatrix} 0 & 1 \\ \frac{27}{8\eta^2}\phi^2 - \frac{9(\nu+\omega_0)}{2\eta^2} + \frac{(\nu+\omega_0)^2}{\eta^2} & 0 \end{bmatrix}.$$

For equilibrium point $E_0(0, 0)$,

$$J_{(0,0)} = \begin{bmatrix} 0 & 1 \\ \frac{(\nu+\omega_0)^2}{\eta^2} & 0 \end{bmatrix},$$

$$\Delta_{(0,0)} = \begin{vmatrix} 0 & 1 \\ \frac{(\nu+\omega_0)^2}{\eta^2} & 0 \end{vmatrix} = -(\frac{(\nu+\omega_0)}{\eta})^2$$

$$\text{and} \qquad \tau_{(0,0)} = 0.$$

In this case, the eigenvalues are

$$\lambda = \frac{\tau^2 \pm \sqrt{\tau^{-}4\Delta}}{2}$$

$$\text{or,} \quad \lambda = \pm\frac{(\nu+\omega_0)}{\eta}.$$

Thus, eigenvalues corresponding to $E_0(0, 0)$ are $\lambda_1 = \frac{(\nu+\omega_0)}{\eta}$ and $\lambda_2 = -\frac{(\nu+\omega_0)}{\eta}$. Here, $\frac{(\nu+\omega_0)}{\eta}$ and $-\frac{(\nu+\omega_0)}{\eta}$ are two real eigenvalues with opposite signs. So, $E_0(0, 0)$ is a saddle point.

Now for equilibrium point $E_1(\frac{4(\nu+\omega_0)}{3}, 0)$, we have

$$J_{(\frac{4(\nu+\omega_0)}{3},0)} = \begin{bmatrix} 0 & 1 \\ (\frac{27}{8\eta^2})(\frac{4(\nu+\omega_0)}{3})^2 - \frac{9.4}{2\eta^2}(\frac{(\nu+\omega_0)^2}{3}) + \frac{(\nu+\omega_0)^2}{\eta^2} & 0 \end{bmatrix},$$

$$\Delta_{(\frac{4(\nu+\omega_0)}{3},0)} = \begin{vmatrix} 0 & 1 \\ \frac{6(\nu+\omega_0)^2}{\eta^2} - \frac{6(\nu+\omega_0)^2}{\eta^2} + \frac{(\nu+\omega_0)^2}{\eta^2} & 0 \end{vmatrix} = -\left(\frac{(\nu+\omega_0)}{\eta}\right)^2$$

$$\text{and} \qquad \tau_{(\frac{4(\nu+\omega_0)}{3},0)} = 0.$$

Here, eigenvalues are $\lambda = \frac{\tau^2 \pm \sqrt{\tau - 4\Delta}}{2} \implies \lambda = \pm\frac{(v+\omega_0)}{\eta}$ So, eigenvalues corresponding to $E_1(\frac{4(v+\omega_0)}{3}, 0)$ are $\lambda_1 = \frac{(v+\omega_0)}{\eta}$ and $\lambda_2 = -\frac{(v+\omega_0)}{\eta}$.

Since $\frac{(v+\omega_0)}{\eta}$ and $-\frac{(v+\omega_0)}{\eta}$ are two real eigenvalues with opposite signs, Therefore $E_1(\frac{4(v+\omega_0)}{3}, 0)$ is a saddle point.

Now for equilibrium point $E_2(\frac{2(v+\omega_0)}{3}, 0)$, we have

$$J_{(\frac{2(v+\omega_0)}{3},0)} = \begin{bmatrix} 0 & 1 \\ (\frac{27}{8\eta^2})(\frac{4(v+\omega_0)^2}{9}) - \frac{9.2}{2\eta^2}(\frac{(v+\omega_0)^2}{3}) + \frac{(v+\omega_0)^2}{\eta^2} & 0 \end{bmatrix},$$

$$\Delta_{(\frac{2(v+\omega_0)}{3},0)} = \begin{vmatrix} 0 & 1 \\ (\frac{3(v+\omega_0)^2}{2\eta^2}) - \frac{3(v+\omega_0)^2}{\eta^2} + \frac{(v+\omega_0)^2}{\eta^2} & 0 \end{vmatrix} = -\frac{1}{2}\Big(\frac{(v+\omega_0)}{\eta}\Big)^2$$

$$\text{and} \qquad \tau_{(\frac{2(v+\omega_0)}{3},0)} = 0.$$

Therefore, eigenvalues are $\lambda = \frac{\tau^2 \pm \sqrt{\tau - 4\Delta}}{2}$

$$\text{or, } \lambda = \pm\frac{(v+\omega_0)}{\sqrt{2}\eta} i.$$

Thus eigenvalues corresponding to $E_2(\frac{2(v+\omega_0)}{3}, 0)$ are $\lambda_1 = \frac{(v+\omega_0)}{\sqrt{2}\eta} i$ and $\lambda_2 = -\frac{(v+\omega_0)}{\sqrt{2}\eta} i$.

Since $\frac{(v+\omega_0)}{\sqrt{2}\eta} i$ and $-\frac{(v+\omega_0)}{\sqrt{2}\eta} i$ are two complex eigenvalues with real part equal to zero. Therefore, $E_2(\frac{2(v+\omega_0)}{3}, 0)$ is a centre.

The phase portrait of the system (24) is shown in the following figure.

In the phase portrait shown in Fig. 2, E_0 and E_1 are saddle points and E_2 is a centre. Phase portrait shows a family of periodic orbits and heteroclinic orbits (one passing through E_0 towards E_1 and other passing through E_1 towards E_0). Similar phase portraits can be obtained for other values of A and v. As those phase portraits carry same features, we have ignored them. Heteroclinic orbits correspond to Kink and anti-kink waves and periodic orbits correspond to periodic wave solution.

Nonlinear Wave Features

In this section, we show nonlinear wave features in the considered system.

Periodic Wave Solution

Corresponding to the periodic orbit obtained in the phase portrait for the system (24), the periodic wave feature of the geophysical-Burgers' Eq. (1) can be found.

The graph for periodic wave features and effects of various parameters are shown in Figs. 3, 4 and 5.

From Fig. 3, we can observe that the amplitude of the wave rises with increasing value of v but the width decreases with as the value of v increases. From Fig. 4,

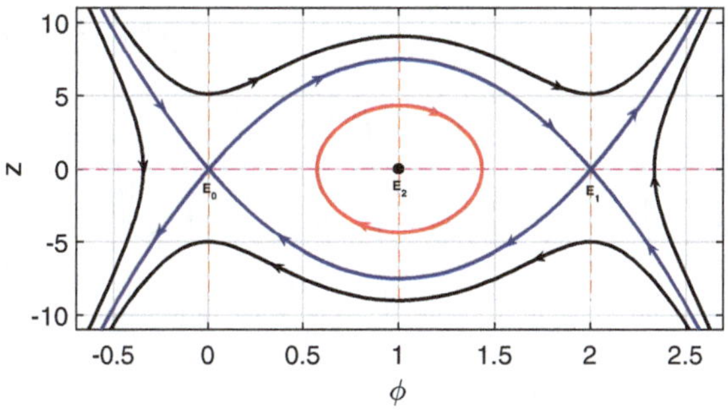

Fig. 2 Phase portrait of the dynamical system (24) for $\eta = 0.1$, $v = 1$ and $\omega_0 = 0.5$

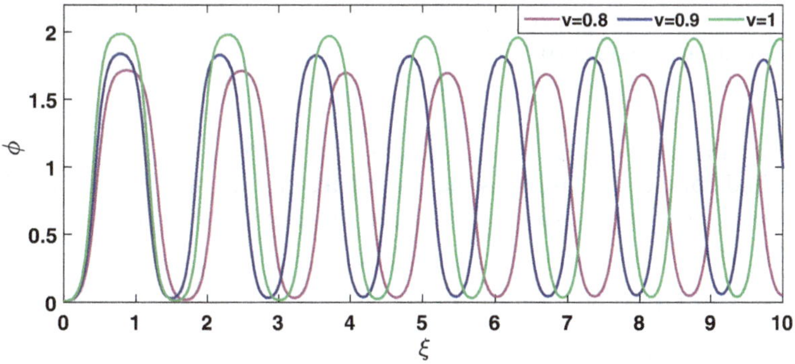

Fig. 3 Effect of v on the periodic wave features of the dynamical system (24) for $\eta = 0.1$ and $\omega_0 = 0.5$ with initial condition $(0.01, 0)$

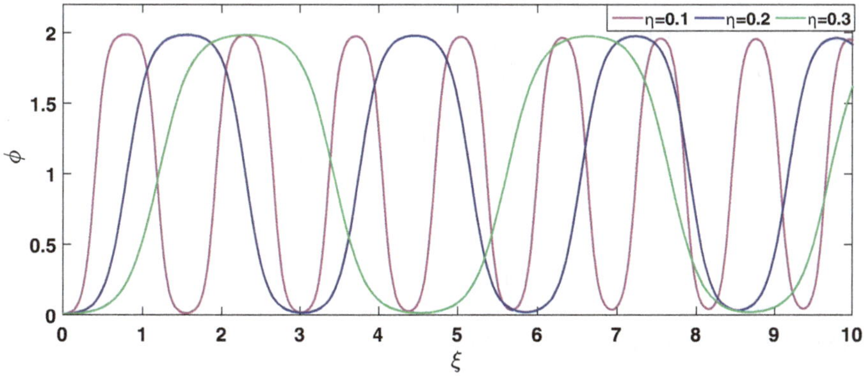

Fig. 4 Effect of η on the periodic wave features of the dynamical system (24) for $v = 1$ and $\omega_0 = 0.5$ with initial condition $(0.01, 0)$

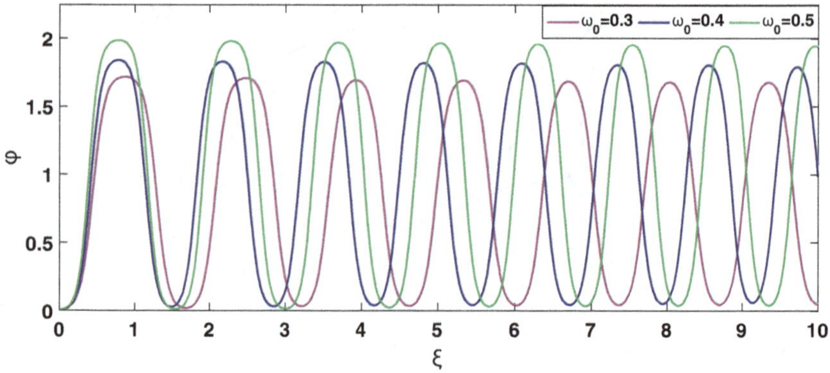

Fig. 5 Effect of ω_0 on the periodic wave features of the dynamical system (24) for $v = 1$ and $\eta = 0.1$ with initial condition $(0.01, 0)$

we observe that as the value of η increases, the width of the wave increases, but the amplitude of the wave remains the same. Likewise, from Fig. 5, we see that the amplitude of wave increases and the width decreases with the increasing value of the parameter ω_0.

Kink and Anti-kink Wave Solutions

Now, Hamiltonian function for the dynamical system (24) is given by

$$H(\phi, z) = \frac{z^2}{2} - \frac{9}{8\eta^2}\frac{\phi^4}{4} + \frac{9(v+\omega_0)}{4\eta^2}\frac{\phi^3}{3} - \frac{(v+\omega_0)^2}{\eta^2}\frac{\phi^2}{2}. \tag{25}$$

Now at $(0, 0)$, $H(0, 0) = 0 = h$ (say).

So, $H(\phi, z) = h$ for $(0, 0)$, we have

$$\frac{z^2}{2} - \frac{9}{\eta^2}\frac{\phi^4}{32} + \frac{3(v+\omega_0)}{\eta^2}\frac{\phi^3}{4} - \frac{(v+\omega_0)^2}{\eta^2}\frac{\phi^2}{2} = 0$$

$$\text{or,} \qquad z = \pm\sqrt{\frac{9}{\eta^2}\frac{\phi^4}{16} - \frac{3(v+\omega_0)}{2\eta^2}\phi^3 + \frac{(v+\omega_0)^2}{\eta^2}\phi^2}. \tag{26}$$

Using Eq. (26) in the first equation of system (24), we get

$$\frac{d\phi}{d\xi} = \pm\sqrt{\frac{9}{\eta^2}\frac{\phi^4}{16} - \frac{3(v+\omega_0)}{2\eta^2}\phi^3 + \frac{(v+\omega_0)^2}{\eta^2}\phi^2}.$$

$$\text{or,} \qquad \frac{d\phi}{\sqrt{\frac{9}{16}\phi^4 - \frac{3(v+\omega_0)}{2}\phi^3 + (v+\omega_0)^2\phi^2}} = \pm\frac{1}{\eta}d\xi$$

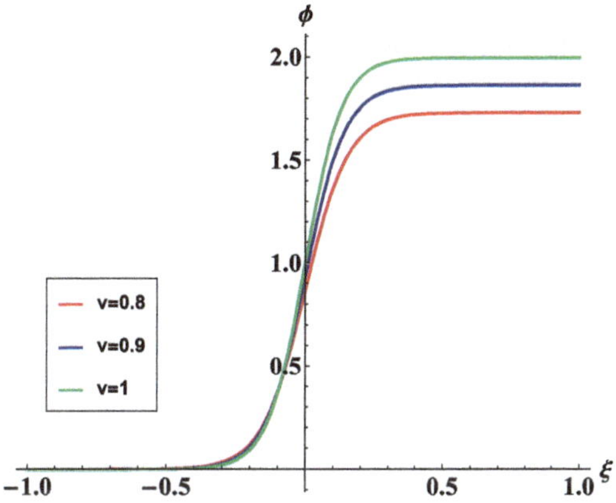

Fig. 6 Effect of v on the kink wave features of Eq. (1) for $\eta = 0.1$ and $\omega_0 = 0.5$

$$\text{or,} \qquad \frac{d\phi}{\phi\sqrt{\frac{9}{16}\phi^2 - \frac{3(v+\omega_0}{2})\phi + (v+\omega_0)^2}} = \pm\frac{1}{\eta}d\xi$$

$$\text{or,} \qquad \frac{d\phi}{\phi\sqrt{(\frac{3}{4}\phi - (v+\omega_0))^2}} = \pm\frac{1}{\eta}d\xi$$

$$\text{or,} \qquad \frac{d\phi}{(\frac{3}{4}\phi^2 - (v+\omega_0)\phi)} = \pm\frac{1}{\eta}d\xi$$

$$\text{or,} \qquad \frac{d\phi}{(\phi^2 - \frac{4(v+\omega_0)\phi}{3})} = \pm\frac{3}{4\eta}d\xi$$

$$\text{or,} \qquad \frac{d\phi}{(\phi - \frac{2(v+\omega_0)}{3})^2 - (\frac{2(v+\omega_0)}{3})^2} = \pm\frac{3}{4\eta}d\xi. \tag{27}$$

Integrating Eq. (27) with respect to ξ, we get

$$\frac{1}{(\frac{2(v+\omega_0)}{3})}tanh^{-1}\left(\frac{\phi - \frac{2(v+\omega_0)}{3}}{\frac{2(v+\omega_0)}{3}}\right) = \pm\frac{3}{4\eta}\xi$$

$$\text{or,} \qquad \phi = \frac{2(v+\omega_0)}{3}\left[1 \pm tanh\left(\frac{(v+\omega_0)}{2\eta}\xi\right)\right]. \tag{28}$$

Equation (28) is the obtained solution for the geophysical-Burgers' equation. The graph for kink and anti-kink waves features and effects of various parameters are shown in Figs. 6, 7, 8, 9, 10 and 11.

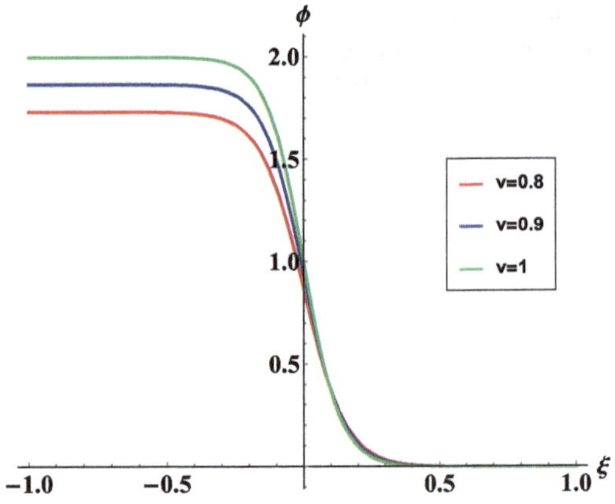

Fig. 7 Effect of v on the anti-kink wave features of Eq. (1) for $\eta = 0.1$ and $\omega_0 = 0.5$

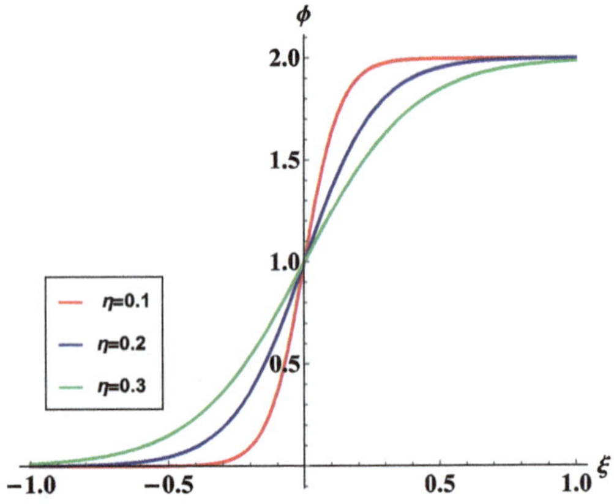

Fig. 8 Effect of η on the kink wave features of Eq. (1) for $v = 1$ and $\omega_0 = 0.5$

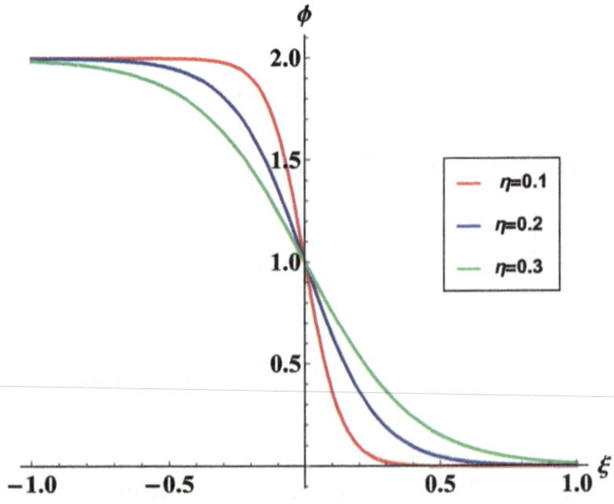

Fig. 9 Effect of η on the anti-kink wave features of Eq. (1) for $v = 1$ and $\omega_0 = 0.5$

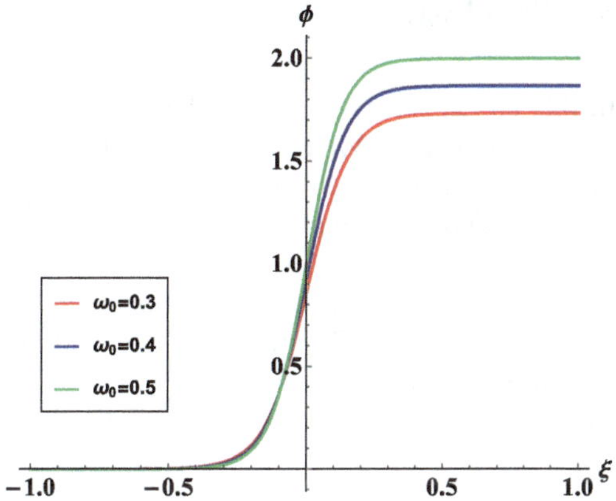

Fig. 10 Effect of ω_0 on the kink wave features of Eq. (1) for $v = 1$ and $\eta = 0.1$

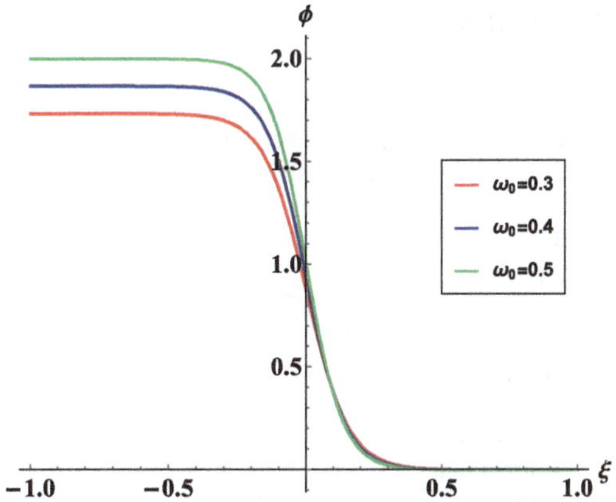

Fig. 11 Effect of ω_0 on the anti-kink wave features of Eq. (1) for $v = 1$ and $\eta = 0.1$

From Figs. 6 and 7, we can see that the amplitude of the kink and anti-kink waves increases with the increasing values of the parameters v but it has no effect on the smoothness of the kink and anti-kink waves.

From Figs. 8 and 9, we see that the smoothness of the kink and anti-kink waves increases as the value of the parameters η increases, while their amplitude remains the same.

From Figs. 10 and 11, we can see that the amplitude of the kink and anti-kink waves increases with the increasing values of the parameters ω_0, while the smoothness of these waves remains the same.

5 Conclusion

In this paper, we have studied the ocean waves in the framework of geophysical-Burgers' equation. We have applied phase plane analysis to study nonlinear waves described by the geophysical-Burgers' equation. A travelling wave transformation was considered to transform the geophysical-Burgers' equation to a dynamical system. All equilibrium points of the corresponding dynamical system were obtained and analysed based on the corresponding eigenvalues. Phase portrait for the dynamical system was plotted. Solutions of Kink and anti-kink waves corresponding to heteroclinic orbits and periodic waves corresponding to periodic orbits were obtained. Effect of various parameters on these wave solutions was also shown.

References

1. Burgers JM (1974) The nonlinear diffusion equation. J Fluid Mech 54(4):665–671
2. Korteweg DJ, de Vries G (1895) XLI. On the change of form of long waves advancing in a rectangular canal, and on a new type of long stationary waves. https://doi.org/10.1080/14786449508620739
3. Wazwaz A-M (2017) A two-mode modified KdV equation with multiple soliton solutions. Appl Math Lett 70:1–6. https://doi.org/10.1016/j.aml.2017.02.015
4. van Wijmgaarden L (1972) On the motion of gas bubbles in a perfect fluid. Ann Rev Fluid Mech 4:369–373. https://doi.org/10.1007/BF00037735
5. Stuhlmeier R (2009) KdV theory and the Chilean tsunami of 1960. Discret Contin Dyn Syst Ser B 12:623–632. https://doi.org/10.3934/dcdsb.2009.12.623
6. Karunakar P, Chakraverty S (2019) Effect of Coriolis constant on Geophysical Korteweg-de Vries equation. J Ocean Eng Sci 4(2):113–121. https://doi.org/10.1016/j.joes.2019.02.002
7. Kirby JT, Shi F, Tehranirad B, Harris JC, Grilli ST (2013) Dispersive tsunami waves in the ocean: model equations and sensitivity to dispersion and Coriolis effects. Ocean Model 62:39–55. https://doi.org/10.1016/j.ocemod.2012.11.009
8. Whitham GB (1974) Linear and nonlinear waves
9. Korteweg DJ, deVries G (1895) On the change of form of long waves advancing in a rectangular canal, and on a new type of long stationary waves. Philos Mag Ser 39:422–443
10. Wazwaz A-M (2017) A two-mode modified KdV equation with multiple soliton solutions. Appl Math Lett 70:1–6
11. van Wijngaarden L (1972) On the motion of gas bubbles in a perfect fluid. Annu Rev Fluid Mech 34:343–349
12. Gao G (1985) A theory of interaction between dissipation and dispersion of turbulence. Sci Sin Ser A 28:616–627
13. Mei CC (1983) The applied dynamics of ocean surface waves. World Scientific Publishing Company
14. Brühl M, Oumeraci H (2016) Analysis of long-period cosine-wave dispersion in very shallow water using nonlinear Fourier transform based on KdV equation. Appl Ocean Res 61:81–91
15. Johnson RS (2019) A nonlinear equation incorporating damping and dispersion. J Fluid Mech 42:49–60
16. Lakshmanan M, Rajasekar S (2003) Nonlinear dynamics. Springer, Heidelberg
17. Saha A (2012) Bifurcation of traveling wave solutions for the generalized KP-MEW equations. Commun Nonlinear Sci Numer Simul 17:3539. https://doi.org/10.1016/j.cnsns.2012.01.005

18. Saha A (2017) Bifurcation, periodic and chaotic motions of the modified equal width burgers (MEW-Burgers) equation with external periodic perturbation. Nonlinear Dyn 87:2193–2201. https://doi.org/10.1007/s11071-016-3183-5
19. Kalisch H (2016) Numerical methods for shallow-water flow. CRC Press
20. Saha A, Harshvardhan A , Tyagi M, Jha A (2021) Proceedings of the sixth international conference on mathematics and computing. Springer, pp 65–73
21. Ak T, Saha A, Dhawan S, Kara AH (2020) Investigation of Coriolis effect on oceanic flows and its bifurcation via geophysical Korteweg–de Vries equation. Numer Methods Part Differ Equ :1–20. https://doi.org/10.1002/num.22469
22. Nakoulima O, Kharif C, Pelinovsky E (2020) Dynamics of unsteady Cnoidal waves in shallow water. Nonlinear Process Geophys 27(1):47–62
23. Saha A, Pradhan B, Banerjee S (2020) Bifurcation analysis of quantum ion-acoustic kink, anti-kink and periodic waves of the Burgers equation in a dense quantum plasma. Eur Phys J Plus 135(2):216
24. Benton ER, Platzman GW (1972) A table of solutions of one-dimensional Burgers equation. Quart Appl Math 30:195–212